HEIDELBERG SCIENCE LIBRARY | Volume 5/6

DAVID C. MILLER

The New Cosmos

Albrecht Unsöld

Translated by William H. McCrea

With 137 Figures

SPRINGER-VERLAG NEW YORK INC. | 1969

First published in 1967

Heidelberger Taschenbücher
Band 16/17
„Der neue Kosmos“

Library of Congress Catalog Card Number 68-8301

Printed in Western Germany

Title No. 3915

Dedicated to M. G. J. Minnaert

From the Author's Preface

This book may serve to present the modern view of the universe to a large number of readers whose over-full professional commitments leave them no time for the study of larger monographs... Such a book should not be too compendious. Accordingly, the author has been at pains to allow the leading ideas of the various domains of astronomical investigation to stand out plainly in their scientific and historical settings; the introductory chapters of the three parts of the book..., in the framework of historical surveys, should assist the general review. With that in mind, the title was chosen following Alexander von Humboldt's well known book "*Kosmos, Entwurf einer physischen Weltbeschreibung*" (1827—1859). On the other hand, particular results — which admittedly first lend colour to the picture — are often simply stated without attempting any thorough justification.

The reader seeking further information will find guidance in the Bibliography. This makes no pretentions to completeness or historical balance. References in the text or in captions for the figures, by quoting authors and years, make it possible for the reader to trace the relevant publications through the standard abstracting journals.

I wish to thank my colleagues V. Weidemann, E. Richter and B. Baschek for their critical reading of the book and for much helpful counsel, and H. Holweger for his tireless collaboration with the proofs. Similarly, my thanks are due to Miss Antje Wagner for the careful preparation of the typescript.

April 1966 Albrecht Unsöld

Kiel, Institute for Theoretical
Physics and University Observatory

Translator's Foreword

Many graduates in mathematics and physics turn to research in optical or radio astronomy, in astrophysics, in space science or in cosmology. Professor UNSÖLD has provided the concise but comprehensive introduction to modern astronomy that all such students need at the outset of their work. Scientists in other fields who follow current advances in astronomy will find in it a compact work of reference to provide the background for their reading. Professor UNSÖLD has had in mind the widest possible circle of readers who for any reason want to know what modern astronomy is about and how it works. It is a privilege to help to extend this circle of readers to include those who prefer to read the book in English.

I have sought to put the work into serviceable English while losing as little as possible of the force and economy of Professor UNSÖLD's own masterly writing. It has not been my concern to create any illusion that the work was originally written in English.

I thank Professor UNSÖLD for reading my translation and for all the helpful comments which have, indeed, convinced me that he could himself have written his book in English better than anyone else. He has also supplied corrections of a few minor errors in the German text. I am grateful to Dr. JOHN HAZLEHURST for preparing all the diagrams and their legends for the English edition. I thank Miss SHIRLEY ANSELL for taking infinite trouble with the typescript and the proofs.

Winter 1968 WILLIAM H. MCCREA

Falmer, Sussex
Astronomy Centre
of the University of Sussex

Contents

Part I

Classical Astronomy

Part II

Sun and Stars

Astrophysics of Individual Stars

Part I
Classical Astronomy

1. Stars and Men: Observing and Thinking

Historical Introduction to Classical Astronomy

Through all the ages the stars have run their courses uninfluenced by man. So the starry heavens have ever symbolized the "Other"—Nature, Deity—the antithesis of the "Self" with its world of inner consciousness, desires and activities. The history of astronomy forms one of the most stirring chapters in the history of the human spirit. All along the emergence of new modes of thought has interlocked with the discovery of new phenomena, often made with the use of new-fashioned instruments.

Here we cannot recount the great contributions of the peoples of the ancient orient, the Sumerians, Babylonians, Assyrians and Egyptians. Also we must forbear to describe what was in its own way the highly developed astronomy of the peoples of the far east, the Chinese, Japanese and Indians.

The concept of the cosmos and of its investigation in our sense of the word goes back to the Greeks. Discarding all notions of magic and aided by an immensely serviceable language, they set about constructing forms of thought that made it possible step by step to "understand" the cosmical manifestations.

How daring were the ideas of the Presocratics! Six centuries before Christ, *Thales of Miletus* was already clear about the Earth being round, he knew that the Moon is illumined by the Sun, and he had predicted the solar eclipse of the year 584 B. C. Is it not just as important, however, that he sought to refer the entire universe to a single basic principle, namely "water"?

The little that we know about *Pythagoras* (mid-6th century B. C.) and his school has an astonishingly modern ring. Here already is talk of the sphericity of the Earth, Moon and Sun, of the rotation of the Earth and of the revolution of at least the two inferior planets Mercury and Venus round the Sun.

Since after the dissolution of the Greek states science had found a new home in Alexandria, there the quantitative investigation of the heavens based upon systematic measurements made rapid advances. Rather than dwell here upon the numerical results, it is of particular interest to note how the great Greek astronomers above all dared to apply *geo*metrical laws to the cosmos. *Aristarchus of Samos,* who lived in the first half of the third century B. C. sought to compare quantatively the Sun-Earth and Moon-Earth distances as well as the diameters of these three heavenly bodies. His starting point was that at the Moon's first and third quarters the Sun-Moon-Earth triangle is right-angled at the Moon. Following this first measurement in outer space, Aristarchus was the first to teach the heliocentric world-system. He appreciated its portentous consequence that the distances of the fixed stars must be stupendously greater than the distance of the Sun from the Earth. How far he was ahead of his time is best shown by the fact that the succeeding generation proceeded to forget this great discovery of his. Soon after these notable achievements of *Aristarchus, Eratosthenes* carried out between Alexandria and Syene the first measurement of a degree of arc on the Earth's surface. He compared the difference in latitude between the two places along a well-used caravan route, and he inferred thus early fairly exact values of the circumference and diameter of the Earth. The greatest observer of antiquity was *Hipparchus* (about 150 B. C.) whose star-catalogue was scarcely surpassed in accuracy even in the 16th century. Even though his equipment naturally did not suffice for him significantly to improve the determination of the fundamental parameters of the planetary system, he nevertheless succeeded in making the important discovery of precession, that is, of the advance of the equinoxes and so of the difference between the tropical and sidereal year.

Within the compass of Greek astronomy, the theory of planetary motion, which we must now discuss, had to remain a problem in geometry and kinematics. Gradual improvement of observations on the one hand, and the development of new mathematical approaches on the other hand, provided the material from which *Philolaus, Eudoxus, Heracleides, Apollonius* and others strove to construct a representation of planetary motions by means of ever more complicated interlacings of circular motions. It was much later that the astronomy and planetary theory of antiquity first attained its definitive form through *Claudius Ptolemy* who about 150 A. D. in Alexandria wrote the thirteen books of his Handbook of Astronomy (Mathematics) *Μαθηματικῆς Συντάξεως βιβλία ιγ*. The "Syntax" later earned the nickname *μεγίστη* (greatest), from which the arabic title *Almagest* ultimately emerged. Although the contents of the Almagest rested extensively upon the observations and investigations of *Hipparchus, Ptolemy* did nevertheless contribute something new, especially to the theory of planetary motion. Here we need give only a brief sketch of *Ptolemy's* geocentric world system: The Earth rests at the centre of the cosmos. The motions of the Moon and the Sun around the sky may be fairly well represented simply by circular paths. Ptolemy described the motions of the planets using the theory of epicycles. A planet moves round a circle, the so-called epicycle, the immaterial centre of which moves around the Earth in a second circle, the deferent. Here we need not go into the elaboration of the system using further circles, some of them eccentric, and so on. In Chapter 6 we shall consider the connexion between this system and the heliocentric world system of *Copernicus,* and the way they differ. In its intellectual attitude, the Almagest shows clearly the influence of Aristotelian philosophy, or rather of *Aristotelianism.* Its scheme of thought, which had developed from the tools of vital investigations into what eventually became the dogmas of a rigid doctrine, contributed not a little to the astonishing historical durability of the Ptolemaic system.

We cannot here trace in detail the way in which, after the decline of the Academy at Alexandria, first the nestorian Christians in Syria and then the Arabs in Baghdad took over and extended the work of Ptolemy.

Translations and commentaries upon the Almagest formed the essential sources for the first western textbook of astronomy, the *Tractatus de Sphaera* of *Johannes de Sacrobosco,* an Englishman by birth who taught in the University of Paris until his death in 1256. The *Sphaera* was time and again re-issued and commented upon; it was the text for academic instruction even to the time of *Galileo.*

Suddenly an altogether new spirit in thought and life showed itself in the fifteenth century, first in Italy and soon afterwards in the north. Today for the first time we are beginning to appreciate the penetrating meditations of Cardinal *Nicolaus Cusanus* (1401—1464). It is of the greatest interest to see how with *Cusanus* ideas about the infinity of the universe and about the quantitative investigation of nature sprang from religious and theological reflections. Towards the end of the century (1492) came the discovery of America by *Christopher Columbus,* who expressed the new feeling about the world in his classic remark, "il mondo e poco". A few years later *Nicholas Copernicus* (1473—1543) initiated the heliocentric world-system.

Part of the intellectual background of the new thinking came from the fact that after the sack of Constantinople by the Turks in 1453 many learned works of antiquity were made available in the west by Byzantine scholars. Certain very fragmentary traditions about the heliocentric systems of the ancients had obviously made a deep impression upon Copernicus. Furthermore, we notice a trend away from the rigid doctrine of *Aristotle* towards the much more vital mode of thought of the Pythagoreans and the Platonists. The "platonic" concept is that the advance of knowledge consists in a progressive adjustment of our inner world of ideas and thought-forms to the ever more fully investigated external world of phenomena. It has been shared by all the important investigators in modern times from *Cusanus* through *Kepler* to *Niels Bohr.* Finally, with the rise of industry, the question was no longer, "What does Aristotle say about so and so?" but, "How does one do so and so?"

About 1510 *Copernicus* sent to several notable astronomers in the form of a letter a communication, first re-discovered in 1877, *Nicolai Copernici de Hypothesibus Motuum Caelestium a se Constitutis Commentariolus,* which already contained most of the results of his masterpiece, *De Revolutionibus Orbium Coelestium Libri VI,* first published in Nuremberg in 1543, the year of *Copernicus's* death.

Throughout his life *Copernicus* held to the idea of the perfection of circular motion that was inescapable throughout ancient and mediaeval times and he never contemplated any other motions.

Following the traditions of the Pythagoreans and Platonists, *Johannes Kepler* (1571—1630) was the first to suceed in attaining a more general standpoint in "mathematico-physical aesthetics". Starting from the observations of *Tycho Brahe* (1546—1601), which far excelled all earlier ones in accuracy, he discovered his three laws of planetary motion (see Chapter 3). *Kepler* found the first two as a result of an incredibly burdensome trigonometric computation, using *Tycho's* observations of Mars, in his *Astronomia nova, seu physica coelestis tradita commentariis de motibus stellae Martis ex observationibus G. V. Tychonis Brahe* (Prague, 1609). The third of *Kepler's* laws was announced in the *Harmonices mundi libri V* (1619). *Kep-*

ler's astronomical telescope, his Rudolphine Tables (1627), and much else, can be only barely mentioned here.

About the same time in Italy, *Galileo Galilei* (1564—1642) directed the telescope, which he had constructed in 1609, on the heavens and discovered in rapid succession: the maria, the craters and other mountain-formations on the Moon, the many stars in the Pleiades and the Hyades, the four moons of Jupiter and their free revolution round the planet, the first indication of Saturn's rings, sunspots. *Galileo's Sidereus nuncius* (1610), in which he described his discoveries with the telescope, the *Dialogo delli due massimi sistemi del mondo, Tolemaico e Copernicano* (1632), and the book he produced after his condemnation by the Inquisition, the *Discorsi e dimostrazioni matematiche intorno a due nuove scienze attenenti alla meccanica ed ai movimenti locali* (1638), with its beginning of theoretical mechanics, are all masterpieces not only scientifically but also artistically in the way they are presented. The observations with the telescope, the observations of the supernovae of 1572 by *Tycho Brahe* and of 1604 by *Kepler* and *Galileo,* and finally the appearance of several comets, all promoted what was perhaps the most important perception of the times. Contrary to the Aristotelian view, it was that there is no basic difference between celestial and terrestrial matter, and that he same physical laws hold good in the realm of astronomy as in the realm of terrestrial physics. The Greeks had perceived this already so far as geometry was concerned. Recalling the case of *Copernicus* makes the difficulty of the concept evident. But it was this that gave wings to the enormous upsurge of natural science at the beginning of the seventeenth century. Also *W. Gilbert's* investigations of magnetism and electricity, *Otto von Guericke's* experiments with the air-pump and the electrical machine and much else sprang from the transformation of the astronomical outlook.

We cannot here pay tribute to the many observers and theorists who built up the new astronomy among whom such notabilities as *Hevelius, Huygens, Halley* were outstanding.

A whole new epoch of natural philosophy began with *Isaac Newton* (1642—1727). His masterpiece, *Philosophiae naturalis principia mathematica* (1687), with the help of his newly created infinitesmal calculus (fluxions), placed theoretical mechanics for the first time upon a sure foundation. In combination with his law of gravitation, it accounted for *Kepler's* laws, and at a single stroke it founded the whole of terrestrial and celestial mechanics.

In the domain of optics, *Newton* invented the reflecting telescope and discussed the interference-phenomena of "Newton's rings". Almost by the way, *Newton* developed the basic ideas for many branches of theoretical physics.

We may compare with him only the "princeps mathematicorum" *Karl Friedrich Gauss* (1777—1855) to whom astronomy owes the theory of determination of orbits, important contributions to celestial mechanics and advanced geodesy, as well as the method of least squares. Never has any other mathematician combined sure judgment in opening up new fields of investigation with such preeminent skill in solving special problems.

Again this is not the place to commemorate the great workers in celestial mechanics from *Euler* through *Lagrange* and *Laplace* to *Henri Poincaré*. Also we shall be able to mention the great observers like *F. W. Herschel* and *J. F. W. Herschel, F. W. Bessel, F. G. W. Struve* and O. *W. Struve* only in connexion with their discoveries. As the conclusion of this review let a single historic date be recorded — that of the measurement of the first trigonometric stellar parallaxes and so of the distances of stars by *F. W. Bessel* (61 Cygni) and *F. G. W. Struve* (Vega) in the year 1838. This conspicuous achievement of the technique of astronomical measurement forms basically the starting point for the modern advance into cosmic space.

(We shall preface Part II with some historical remarks on astrophysics and Part III with some on galactic research as well as cosmogony and cosmology.)

2. Celestial Sphere: Astronomical Coordinates: Geographic Latitude and Longitude

Man's imagination has from ancient times made star-pictures out of easily recognizable groupings of stars (Fig. 2.1). In the northern sky we easily recognize the Great Bear (Plough). We find the Polestar if we produce the line joining the two brightest stars of the Great Bear by about five times its length. The Polestar is the brightest star in the Little Bear; if we extend the line past it to about the same distance on the other side, we come to the "W" of Cassiopeia. With the help of a celestial globe or a star-map, other constellations are easily found. In 1603, J. *Bayer* in his *Uranometria nova* denoted the stars in each constellation in a generally decreasing order of brightness as α, β, γ ...

Nowadays these Greek letters are supplemented by the numbering introduced by the first Astronomer Royal, *J. Flamsteed*, in his *Historia Coelestis Britannica* (1725). The Latin names of the constellations are usually abbreviated to three letters.

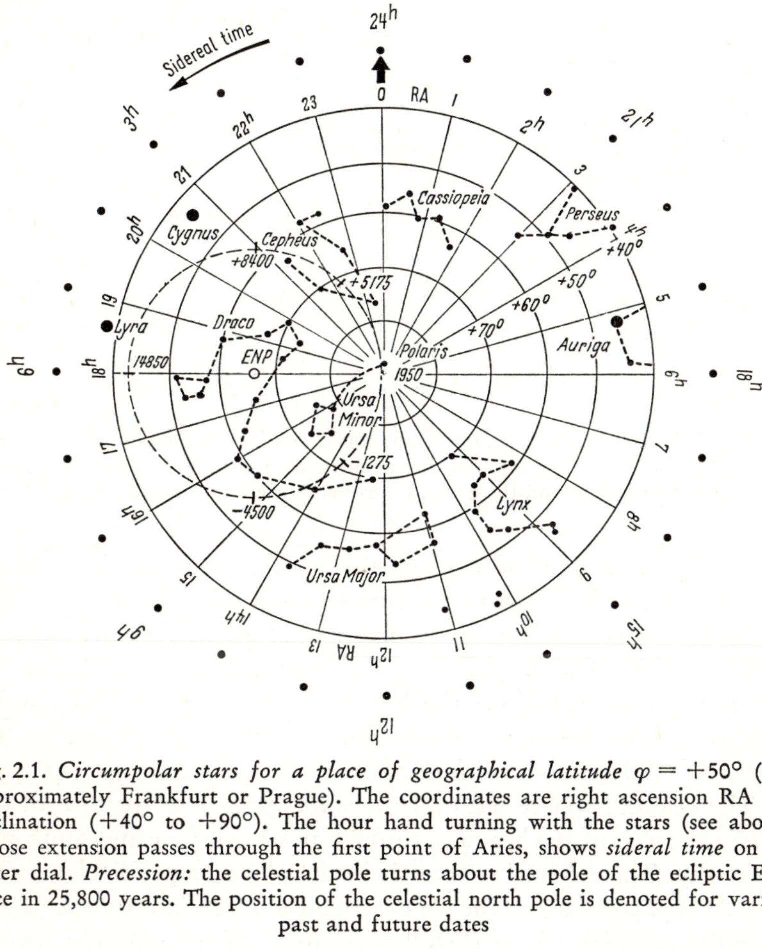

Fig. 2.1. *Circumpolar stars for a place of geographical latitude* $\varphi = +50°$ (e.g. approximately Frankfurt or Prague). The coordinates are right ascension RA and declination (+40° to +90°). The hour hand turning with the stars (see above), whose extension passes through the first point of Aries, shows *sideral time* on the outer dial. *Precession:* the celestial pole turns about the pole of the ecliptic ENP once in 25,800 years. The position of the celestial north pole is denoted for various past and future dates

As an example, the second brightest star in the Great Bear (Ursa Major) is known as β U Ma or 48 U Ma (read as 48 Ursae Majoris).

The *celestial sphere* is, mathematically speaking, the infinitely distant sphere upon which we see the stars to be projected. On this sphere we distinguish (Fig. 2.2):

a) The *horizon* with the directions north, west, south, east.

b) Vertically above us the *zenith* and below us the *nadir*.

c) The *meridian* through the celestial pole, the zenith, the south-point, the nadir and the north-point.

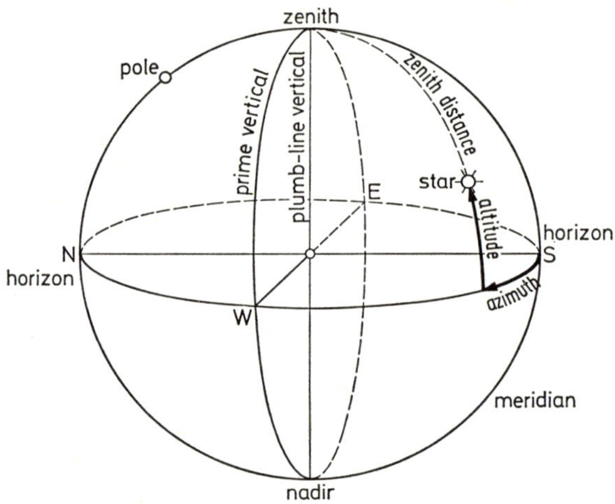

Fig. 2.2. *Celestial sphere. Horizon* with north, south, east and west points. The (celestial) *meridian* passes through the north point, the (celestial) pole, the zenith, the south point and the nadir. Coordinates are altitude and azimuth

d) The *prime vertical* through the zenith, the west-point and the east-point, at right angles to the horizon and to the meridian.

In the coordinate system so determined we decribe the instantaneous position of a star by specifying two angles (Fig. 2.2): a) The *azimuth* is reckoned along the horizon in the sense SWNE, and is measured either from the south-point or the north-point. b) The *altitude* = 90 degrees − *zenith distance.*

The celestial sphere with all the stars appears to rotate each day around the *axis* of the heavens, through the north-pole and south-pole of the sky. The *celestial equator* is in the plane perpendicular to this axis. At a particular instant we describe the *place* (position) *of a star* upon the celestial sphere (Fig. 2.3), thought of as infinitely remote, by the *declination* δ, reckoned positive from the equator towards the north-pole and negative towards the south-pole, and the *hour angle t*, reckoned from the meridian in the sense of the daily motion, that is, westward from the meridian.

In the course of a day, a star traverses a *parallel-circle* on the sphere; it reaches its greatest altitude on the meridian at upper *culmination* and its least altitude at *lower culmination.*

On the celestial equator, we now mark the *first point of Aries* ♈, which we shall explain in the next section as being the position of the Sun at the time of the spring-equinox (March 21). Its hour angle gives

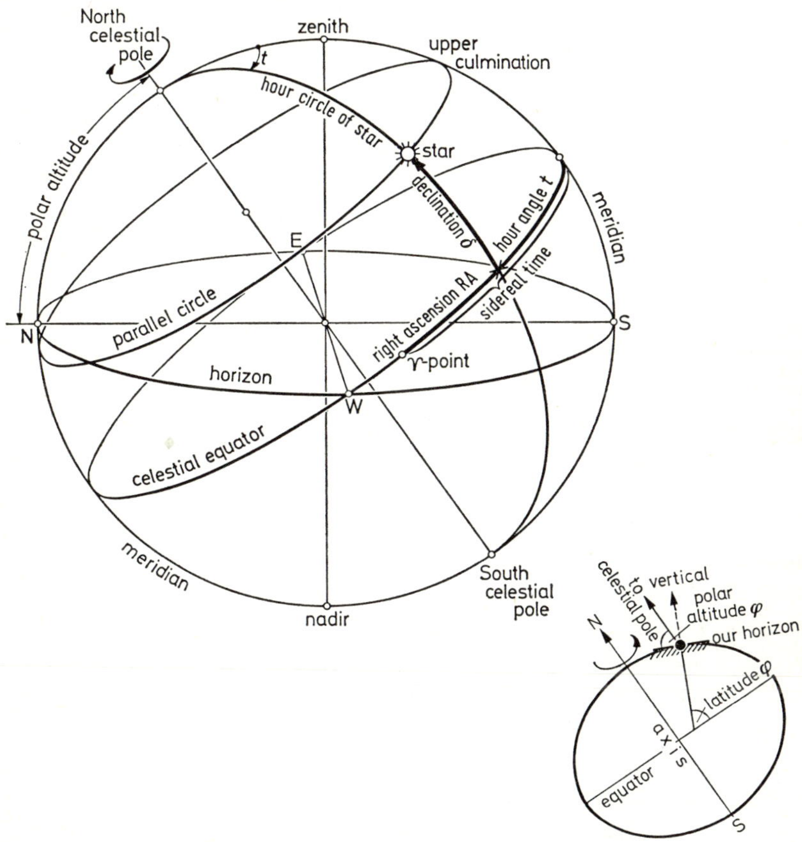

Fig. 2.3. *Celestial coordinates.* Right ascension RA and declination δ. Hour angle t=sidereal time minus right ascension RA. Lower right: the earth (flattening exaggerated). Polar altitude = geographical latitude φ

sidereal time τ. In Fig. 2.1, if we think of the arrow (which indicates the point ♈ on the celestial equator) as moving round with the stars, it will indicate sidereal time on the "clock-dial" shown round the outside of the diagram.

Finally, we can define the position of a star on the celestial sphere independently of the time of day. We call the arc of the equator from ♈ to the hour-circle of the star its *right ascension* RA. It is reckoned in hours, minutes and seconds. Thus 24^h corresponds to $360°$ so that

$$1^h = 15° \qquad 1° = \tfrac{1^h}{15} = 4^m$$
$$1^m = 15' \qquad 1' = 4^s$$
$$1^s = 15''$$

We see at once from Fig. 2.3 that

$$\text{hour angle } t = \text{sidereal time } \tau \text{ minus right ascension RA.} \tag{2.1}$$

We have already introduced our second coordinate of the star, the *declination* δ.

If the astronomer wants to point his telescope at some particular star, planet or other celestial object, he extracts its right ascension RA and declination δ from a star-catalogue, he reads the sidereal time τ on his sidereal clock, and then he sets on the graduated circles of his instrument the hour angle t in h, m, s as derived from (2.1) and the declination δ in degrees, positive to the north, negative to the south. Specially accurately determined positions of the so-called *fundamental stars,* particularly for time-determination (see below), together with positions of the Sun, Moon, planets, and so on, are to be found in the astronomical almanacs or *ephemerides,* the most important of which is "The Astronomical Ephemeris".

The Copernican system interpreted the apparent rotation of the celestial sphere by asserting that the Earth rotates about its axis once in 24 hours. The *horizon* is in the tangent plane of the Earth, or more precisely that of a water-level, at our position. The *zenith* corresponds to the direction of the plumb-line perpendicular to this plane, which is that of local gravity (including the effect of centrifugal force arising from the Earth's rotation). The pole-height (= altitude of the pole above the horizon is seen from Fig. 2.3 to be equal to the *geographical latitude* φ (= angle between the plumb-line and the plane of the equator). The pole-height is easily measured as the mean value of the altitude of the Pole-star, or any circumpolar star, at superior and inferior culmination.

Geographical longitude l corresponds to the hour angle. If the hour angle t of one and the same star is observed simultaneously at Greenwich (prime meridian, $l_G = 0°$) and, say, Kiel, then the difference is the geographical longitude of Kiel l_K. While the determination of geographical *latitude* requires only the measurement of angles, the measurement of *longitude* demands precise time-transfer. In former days, time as indicated by the motion of the Moon or of Jupiter's satellites was employed. The invention of the mariners' chronometer by John Harrison (about 1760—65), was a big advance, as was later the transmission of time-signals first by telegraph and then by radio.

We remark further: At a place of (north) latitude φ a star of declination δ reaches altitude $h_{\max} = \delta + 90° - \varphi$ at *superior culmination,* and reaches $h_{\min} = \delta - (90° - \varphi)$ at *inferior culmination. Circumpolar stars* with $\delta > 90° - \varphi$ are always above the horizon; stars with $\delta < -(90° - \varphi)$ are never above the horizon.

In measuring star-altitudes h the refractive effect of the Earth's atmosphere has to be taken into account. The apparent raising of a star (apparent altitude — true altitude) is called the *refraction.* For average conditions of pressure and temperature in the atmosphere the refraction is as shown:

Star-altitude h	0	5	10	20	40	60	90 degrees
Refraction Δh	34′50″	9′45″	5′16″	2′37″	1′09″	33″	0″

The refraction decreases a little with increasing atmospheric temperature and with decreasing atmospheric pressure, for example in a region covered by a depression or on mountains.

3. Motion of the Earth:

Seasons and the Zodiac: Day, Year and Calendar

We now consider the orbital motion, or *revolution,* of the Earth around the Sun, in the sense of *Copernicus,* and also the rotation of the Earth around its axis as well as the motion of the axis itself. Here we adopt the observer's standpoint. We shall develop Newton's theory of the motions of the Earth and the planets in accordance with his theories of mechanics and gravitation in Chapter 6.

The apparent annual motion of the Sun across the heavens *Copernicus* ascribed to the description by the Earth of an almost circular path round the Sun. The plane of this path meets the celestical sphere in a great circle called the *ecliptic* (Fig. 3.1). This cuts the celestial equator at an angle 23°27′, the *obliquity of the ecliptic.* Thus in its annual motion round the Sun, the axis of the Earth maintains a fixed direction in space, making an angle $90° - 23°27' = 66°33'$ with the plane of the Earth's orbit.

Here is a short summary of the circumstances of the onset of the *seasons* for the northern hemisphere of the Earth:

Beginning of seasons			Coordinates of the Sun		Zodiacal constellation entered by the Sun
			RA	δ	
Spring	March 21	Vernal equinox	0^h	$0°$	Aries (Ram) ♈
Summer	June 22	Summer solstice	6^h	$23°27'$	Cancer (Crab) ♋
Autumn	September 23	Autumnal equinox	12^h	$0°$	Libra (Scales) ♎
Winter	December 22	Winter solstice	18^h	$-23°27'$	Capricorn (Sea-goat) ♑

At the equinoxes the day- and night-segments of the Sun's diurnal motion are each 12 hours. The Sun reaches its greatest altitude at a place north of the tropics of latitude φ at noon on the summer

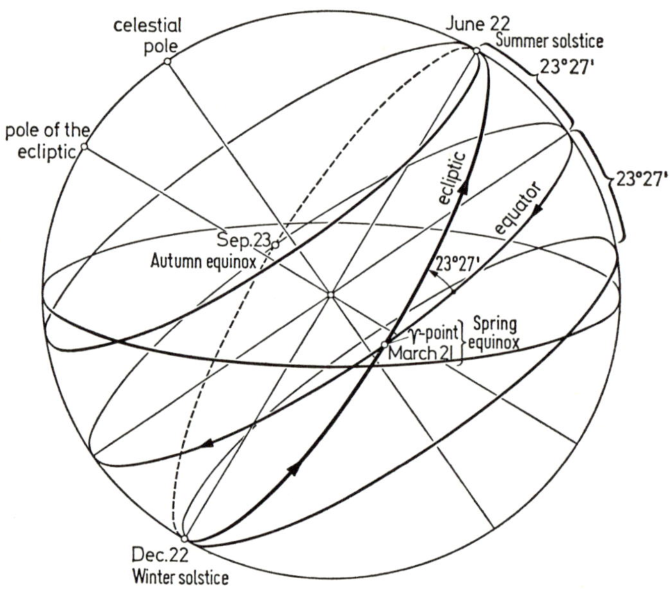

Fig. 3.1. Apparent annual motion of the Sun among the stars. Ecliptic. Seasons

solstice on June 22 with $h = 90° - \varphi + 23°27'$, and on December 22 it reaches its lowest noon-altitude with $h = 90° - \varphi - 23°27'$. The Sun can reach the zenith in geographical latitudes up to $\varphi = 23°27'$, the tropic of Cancer. On the other hand, north of the arctic circle $\varphi \geqq 90° - 23°27' = 66°33'$ around the time of the winter solstice the Sun remains below the horizon; around the time of the summer solstice, the "midnight sun" behaves as a circumpolar star.

In the southern hemisphere summer occurs when winter occurs in the north, the tropic of Capricorn corresponds to the tropic of Cancer, and so on.

The *Zodiac* is an 18°-wide belt of the sky, centred on the ecliptic. From ancient times, mankind has divided it up into twelve equal parts denoted by the *Signs of the Zodiac* (Fig. 3.2).

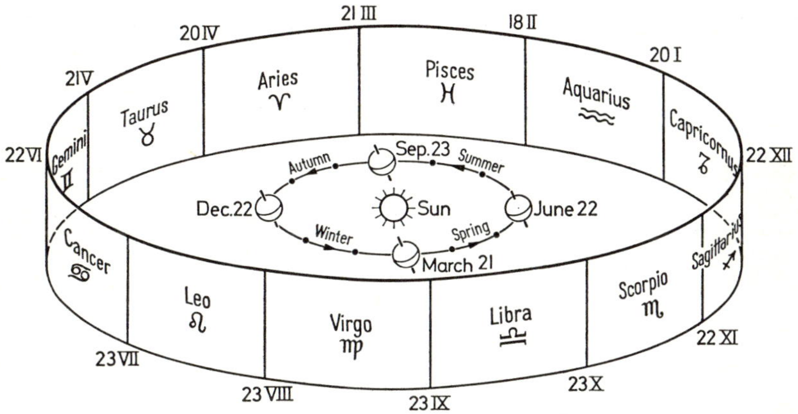

Fig. 3.2. *Path of the Earth around the Sun.* Seasons. Zodiac ("zone of animals") and its signs. The Earth is at perihelion (closest to the Sun) on January 2 and at aphelion (furthest from the Sun) on July 2

For describing the motion of the Earth and the planets it is convenient to employ a coordinate system determined by the ecliptic and its poles. Ecliptical (or astronomical) *longitude* is measured from the first point of Aries along the ecliptic, analogously to right ascension, in the sense of the annual motion of the Sun. Ecliptical (or astronomical) *latitude* is measured analogously to declination, at right angles to the ecliptic. Astronomical latitude and longitude must not be confused with geographic latitude and longitude!

Astronomers of antiquity knew of the non-uniformity of the apparent annual motion of the Sun, and *J. Kepler* (1571—1630) recognized this as a consequence of the first two of his laws of planetary motion, to which we return more fully in Chapter 5.

Kepler's first law. Each planet moves in an ellipse with the Sun at one focus.

Kepler's second law. The radius vector from the Sun to any one planet describes equal areas in equal times.

Kepler's third law. For any two planets the squares of the periods are proportional to the cubes of the semi-major axes of their orbits.

The geometrical characteristics of the orbit of the Earth or of any other planet round the Sun are shown in Fig. 3.3. The semi axis we denote by a; the distance from the centre to a focus we denote by ae and we call the fraction e the *eccentricity* of the orbit. At *perihelion* the distance of the Earth from the Sun is $r_{\min} = a(1-e)$; at aphelion, the distance is $r_{\max} = a(1+e)$. The daily motion of the Sun in the heavens, (the angle turned through in one day by the radius vector to the Earth) at perihelion and aphelion, according to Kepler's second law, are in the ratio $(r_{\max}/r_{\min})^2 = \left(\frac{1+e}{1-e}\right)^2$; the corresponding values of apparent diameter of the Sun's disk are in the ratio $\frac{1+e}{1-e}$. Measurements of both of these ratios agree in giving for the eccentricity of the Earth's orbit $e = 0.01674$. At the present time, the Earth goes through perihelion about January 2. The approximate coincidence of this date with the beginning of the calendar year is fortuitous.

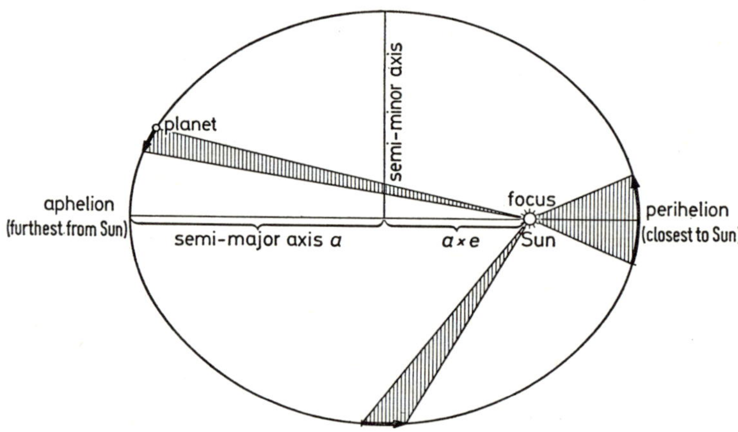

Fig. 3.3. *Elliptical orbit of a planet.* Semi-major axis a; eccentricity e; distance from centre to focus (Sun) ae. (The eccentricity of planetary orbits is much smaller than shown here)

The motion of the Earth about its mass-centre proceeds quite independently of its orbital motion. The *rotation of the Earth* about its axis, taking this for the moment to be fixed in space, mirrors the daily passage of the stars around the sky, which by definition takes one sidereal day or 24^{h} sidereal time. In recent years, people have compared with the utmost precision sidereal time determined astronomically by the passage of the stars across the meridian with time given by *quartz clocks,* which is determined by electrically excited and recorded vibrations of a quartz block or ring, or that given by *atomic clocks* where the atomic or molecular frequency of vibration serves as the time-standard. There are found to be systematic differences which show that the period of rotation of the Earth, as measured by "physical" time-standards, is not constant but shows partly irregular and partly seasonal fluctuations.

The motions of the planets and of the Sun (or, alternatively, of the Earth) and particularly of the Moon also show small common *deviations* from the ephemerides calculated in accordance with Newtonian mechanics and gravitation. This has, in fact, to do with a secular (progressive) increase in the length of the day, which may be ascribed to the frictional effect of tidal currents (Chapter 6) in narrow straits. A further part of the deviations exhibits no such evident regularity. The comparison of the deviations for different heavenly bodies compels us to ascribe them to deviations of "astronomical" measures of time from "physical" time concerned in Newton's laws.

By reason of these experiences, astronomers resolved in 1960 to abandon the definition of a measure of time based upon the rotation of the Earth, and to base all astronomical ephemerides on a time reckoning founded upon the fundamental laws of physics, so-called *ephemeris time.* The small correction ephemeris time *minus* universal time (see below) can be determined only in retrospect if great accuracy is required; however, for most purposes predictions may be made with sufficient accuracy with the use of suitable extrapolation.

Hipparchus discovered that the point of the vernal equinox is not fixed in the sky but it advances by some $50''$ a year. This means that, since ancient times, the point has moved out of the constellation Aries into the constellation Pisces. This precession of the equinoxes depends on the fact that the celestial pole moves round the sky with a period of 25800 years on a circle of radius $23°27'$ having its centre at the pole of the ecliptic, this being fixed among the stars. In other words, the Earth's axis describes a cone of semi-angle $23°27'$ about the axis of the Earth's orbit once in 25800 years.

Since the precession displaces the coordinate-frame of right ascension RA and declination δ relative to the stars, star-positions and star-catalogues must be expressed with reference to the equinox to which the stated values of RA and δ refer. Since the star-places change on account of the proper motions (Chapter 23), the *epoch* of the observations is also reduced to the same point of time. Table 3.1 gives the correction for right ascension RA (for different values of RA and δ) and for declination (for different values of RA) consequent upon ten years' precession.

Table 3.1. *Precession for ten-year interval*

a) ΔRA in minutes of time (+increase, −decrease)

Hours RA for northern objects		6	7 5	8 4	9 3	10 2	11 1
Declination | δ | in degrees		m	m	m	m	m	m
	80°	+1.77	+1.73	+1.60	+1.40	+1.14	+0.84
	70°	1.12	1.10	1.04	0.94	0.82	0.67
	60°	0.898	0.885	0.846	0.785	0.705	0.612
	50°	0.778	0.768	0.742	0.700	0.645	0.581
	40°	0.699	0.693	0.674	0.644	0.606	0.560
	30°	0.641	0.636	0.624	0.603	0.576	0.546
	20°	0.593	0.590	0.582	0.570	0.553	0.533
	10°	0.552	0.550	0.546	0.540	0.532	0.522
	0°	+0.512	+0.512	+0.512	+0.512	+0.512	+0.512
Hours RA for southern objects		18	19 17	20 16	21 15	22 14	23 13

12 0	13 23	14 22	15 21	16 20	17 19	18	
m	m	m	m	m	m	m	
+0.51	+0.19	−0.12	−0.38	−0.58	−0.70	−0.75	80°
0.51	0.35	+0.21	+0.08	−0.02	−0.08	−0.10	70°
0.512	0.412	+0.319	+0.240	+0.178	+0.140	+0.126	60°
0.512	0.444	+0.380	+0.324	+0.282	+0.256	+0.247	50°
0.512	0.464	+0.419	+0.380	+0.350	+0.332	+0.335	40°
0.512	0.479	+0.448	+0.421	+0.401	+0.388	+0.384	30°
0.512	0.491	+0.472	+0.455	+0.442	+0.434	+0.431	20°
0.512	0.502	+0.492	+0.484	+0.478	+0.476	+0.473	10°
+0.512	+0.512	+0.512	+0.512	+0.512	+0.512	+0.512	0°
0 12	1 11	2 10	3 9	4 8	5 7	6	

b) $\Delta\delta$ in minutes of arc (+increase, which implies a decrease of $|\delta|$ in the southern sky)

Hours RA	0 24	1 23	2 22	3 21	4 20	5 19
$\Delta\delta$ minutes of arc	+3′.34	+3′.23	+2′.89	+2′.36	+1′.67	+0′.86

6 18	7 17	8 16	9 15	10 14	11 13	12
0′.0	−0′.86	−1′.67	−2′.36	−2′.89	−3′.23	−3′.34

Precession with the 25800 year-period is superimposed upon a somewhat similar motion of smaller amplitude having a 19-year period; this is called *nutation*. Finally, the rotation-axis of the Earth fluctuates by amounts of the order ±0″.2 relative to the body of the Earth, which can be analysed into an irregular part and a periodic part for which the so-called Chandler period of 433 days is recognized. The corresponding *polar wandering* is kept under observation at a series of stations. We return in Chapter 6 to the explanation of the various motions of the Earth's axis.

We first give further consideration to the problem of *time-keeping*. Our daily life is determined by the position of the Sun. So there was introduced

$$\text{true solar time} = \text{hour angle of the Sun.}$$

This is the time that a simple sundial shows, 12^h corresponding to the upper transit of the Sun. On account of the non-uniform orbital motion of the Earth (Kepler's second law) and of the obliquity of the ecliptic, true solar time does not proceed uniformly. Consequently we have recourse to *mean solar time*. We imagine a *mean sun* that traverses the equator uniformly in the same time that the true Sun takes for its annual passage round the ecliptic. The hour angle of this imagined mean sun defines mean solar time. The difference

$$\text{true solar time} - \text{mean solar time} = \text{equation of time}$$

is composed of two terms depending upon the eccentricity of the Earth's orbit and on the obliquity of the ecliptic. Its extreme values are

	February 12	May 14	July 26	November 4
Equation of time	-14^m20^s	3^m45^s	-6^m23^s	16^m23^s.

Mean solar time is different for each meridian. For the purpose of ordinary intercourse, people have therefore agreed that throughout certain zones they will employ the time appropriate to one selected meridian. In Germany and central Europe they use Central European Time (Mittel-europäische Zeit MEZ) which is local mean solar time on the meridian 15° E. In western Europe they use the time of the Greenwich Prime Meridian.

For scientific purposes, for instance for astronomical and geophysical measurements at stations that are often distributed all over the globe, scientists everywhere use

Universal time (UT) = mean solar time of the Greenwich meridian.

We reckon in 24 hours starting with 0^h at midnight. For example 12^h UT is 13^h MEZ.

Using a small variable correction which is published in the *Astronomical Ephemeris* (e. g. for 1965 it is $+35^s$) we pass from UT to the already described

Ephemeris time (ET).

For astronomical observation we need also the relation between mean *solar time* and *sidereal time*. Relative to the equinox, the "mean sun" moves in a year (= 365 days) through 360° (= 24^h) from west to east. The mean solar day is therefore $24^h/365$ or 3^m56^s longer than the sidereal day. In a month the sidereal clock gains about 2^h on the "ordinary" UT or MEZ clock. To amplify this, we quote the sidereal time for 0^h local time on certain dates. As is known, this is equal to the hour angle of the point ♈ and equal to the right ascension RA of stars passing the meridian at midnight (at the longitude where the observations are made).

0^h local time (midnight)	January 1	April 1	July 1	October 1
RA on meridian and sidereal time	6^h42^m	12^h37^m	18^h35^m	0^h38^m

The unit of time for longer intervals is the *year*. We define

sidereal year = 365.25636 mean solar days

which is the time between two successive passages of the Sun through the same point in the sky. It is also the true period of revolution of the Earth ("sidereal" from Latin *sidus*, a star).

A *tropical year* = 365.24220 mean solar days is the time between two successive passages of the Sun through the vernal equinox. Since this advances westwards by 50″.3 in a year, the tropical year is correspondingly shorter than the sidereal year. The seasons and the calendar depend upon the tropical year. As unit of ephemeris time the ephemeris second has been chosen; this is the 31 556 925.9747 part of the tropical year 1900 ("tropic" from Greek *τρόπειν*, to turn).

Since for practical reasons every year should consist of an integral number of days, in ordinary life we use

$$\text{civil year} = 365.2425 = 365 + \tfrac{1}{4} - \tfrac{3}{400}\ \text{mean solar days}$$

corresponding to the intercalary prescription of the *Gregorian calendar* introduced in 1582 by Pope *Gregory XIII*. After 3 years of 365 days follow a leap-year (date divisible by 4) of 366 days except for the years of the centuries *not* divisible by 400. Here we cannot discuss the older *Julian calendar* of *Julius Caesar* BC 45, or other interesting questions of *chronology*. Modern proposals for calendar reform seek to ensure that the beginning of the year and the first days of the month shall always fall on the same days of the week. Also they would fix the present movable feasts, particularly Easter (the first Sunday after the first full moon after the spring equinox) and Whitsunday (50 days after Easter).

For ease of chronological reckoning over long periods of time especially for such applications as to observations and ephemerides of variable stars we seek to avoid the irregularities in the length of years and months. Following the proposal of *J. Scaliger* (1582) we therefore simply count the so-called *Julian days* consecutively. The Julian day begins at 12^h UT (mean Greenwich noon). The beginning of the Julian day 0 is put at 12^h UT on January 1 of the year 4713 BC. On 1965 January 1 at 12^h UT began Julian day 2 438 762.

4. Moon: Lunar and Solar Eclipses

The Moon appears to us as a disk with average diameter 31′ and so with just about the same apparent size as the Sun. Its distance from the Earth can be got by triangulation from two observatories sufficiently far apart (say along the same meridian). Astronomers call the angle subtended at the Moon by the Earth's *equatorial radius the equatorial horizontal parallax of the Moon.* Its value in the mean is 3422″.6. Since the Earth's radius is known to be 6378 km, we derive for the mean distance of the Moon from the centre of the Earth

$$60.3 \text{ Earth radii} = 384\,400 \text{ km}$$

and hence for the radius of the Moon

$$0.272 \text{ Earth radius} = 1738 \text{ km}.$$

We shall concern ourselves with the physical structure of the Earth and the Moon in Chapter 7. Here we consider the path and motion of the Moon from the observational standpoint.

The Moon circles the Earth, in the same sense as the Earth goes round the Sun, in a *sidereal month* = 27.32 days. That is to say, after this time the Moon returns to the same place among the stars.

The origin of the *phases of the Moon* is illustrated in Fig. 4.2. Their period, the *synodic month* = 29.53 days (1 to 3 in Fig. 4.1), after which the Moon returns to the same position relative to the Sun, is longer than the sidereal month (1 to 2 in Fig. 4.1). Each day the Moon moves eastwards $360°/29.53 = 12^\circ\!.2$, relative to the *Sun*, and $360°/27.32 = 13^\circ\!.2$ relative to the *stars*.

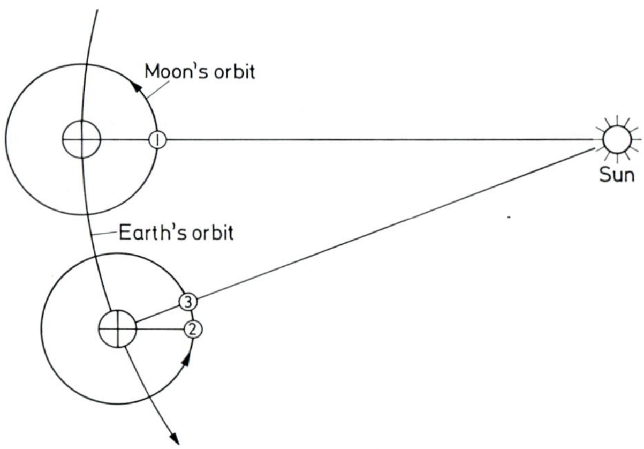

Fig. 4.1. *The synodic month (1—3) is longer than the sidereal month* (1—2), since the Earth has meanwhile moved further along its orbit

The difference between the sidereal and synodic daily motion of the Moon is equal to the daily motion of the Sun and so is $360°/365 \approx 1^\circ\!.0$. We can express this

$$\frac{1}{\text{sideral month}} - \frac{1}{\text{sideral year}} = \frac{1}{\text{synodic month}}.$$

Actually the path of the Moon round the Earth is an ellipse of eccentricity $e = 0.055$. The point in the path where the Moon approaches nearest to the Earth is called *perigee* (analogous to perihelion

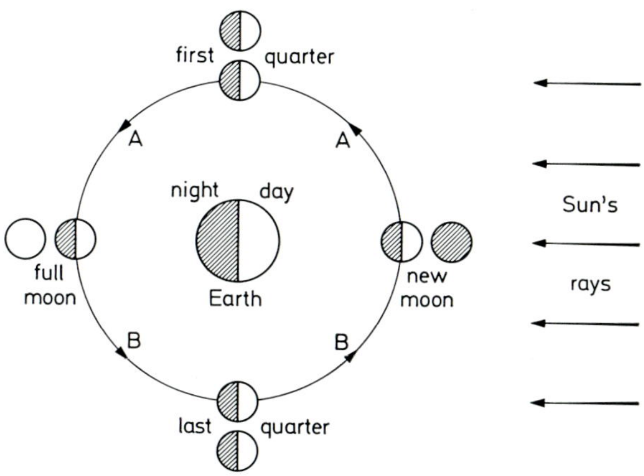

Fig. 4.2. Phases of the Moon. The Sun is on the right. The outer pictures show the phases as seen from the Earth. Waxing moon A; waning moon B

in the motion of the Earth); where it is furthest from the Earth is called *apogee*. The plane of the Moon's orbit has an inclination $i \approx 5°$ to the plane of the Earth's orbit (ecliptic). The Moon rises "above" the ecliptic from south to north at the *ascending node*, and passes "below" the ecliptic (in the sense of dwellers in the northern hemisphere) at the *descending* node.

In consequence of the perturbations produced by the attraction of the Sun and the planets the Moon's orbit further performs the following motions:

1. The perigree moves round the Earth, in the plane of the Moon's orbit, with "direct" motion, i. e. in the sense of the motion of the Earth, with a period of 8.85 years.

2. The line of nodes, in which the planes of the orbits of the Moon and the Earth intersect, has "retrograde" motion in the ecliptic, i. e. in the opposite sense to the motion of the Earth, with a period of 18.61 years, the so-called *nutation-period*.

This regression of the nodes causes among other things the "nodding" of the Earth up to a maximum of 9″, the *nutation* of the Earth's axis already mentioned.

The mean time between successive passages of the Moon through the same node is called the *draconitic* month = 27.2122 days. It is important for the computation of eclipses (see below).

If we could view the paths of the Moon and the Earth and the Sun from a space vehicle, we should confirm that, in agreement with a simple calculation, the path of the Moon is always concave towards the Sun (Fig. 4.3).

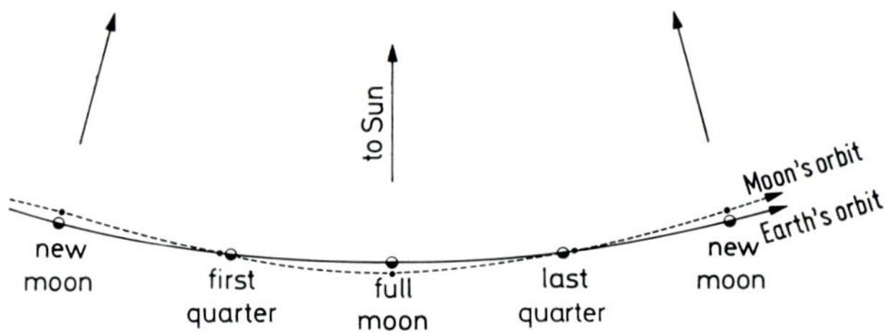

Fig. 4.3. The orbits of the Earth and Moon around the Sun

Here we shall now consider the *rotation of the Moon* and its motion relative to its mass-centre. This can be measured very accurately from observations of a sharply defined crater, or the like, on the Moon's disk.

The fact that, by and large, the Moon always presents the same face to us depends upon its rotation period being equal to its period of revolution, i. e. to the sidereal month. The equalization of the two periods has clearly been brought about by the tidal interaction of the Moon and the Earth (Chapter 6).

Precise observation shows nevertheless that the face of the Moon wobbles somewhat. The so-called *geometrical librations* of the Moon arise from the following causes:

a) The equator of the Moon and the plane of the Moon's orbit are inclined at an angle of about 6°.7: the consequent *libration in latitude* is thus about ±6°.7.

b) In agreement with the law of inertia, the Moon's rotation is uniform. In accordance with Kepler's second law, taking account of the eccentricity of the orbit, the Moon's revolution is not uniform. Hence arises the *libration in longitude* of about ±7°.6.

c) The equatorial radius of the Earth subtends an angle of 57′ at the centre of the Moon, this being the horizontal parallax of the Moon. The daily rotation of the Earth therefore produces a daily libration.

In addition there is the considerably smaller physical libration, which stems from the fact that the Moon's diameter directed towards the Earth is somewhat larger than diameters in directions at right angles to this. The momental ellipsoid of the Moon performs small oscillations in the gravitational field, mainly that produced by the Earth.

Altogether these librations result in our being able to see from the Earth 59 per cent of the Moon's surface.

Having studied the motions of the Sun, Earth and Moon we turn our attention to the impressive spectacles of lunar and solar eclipses.

A lunar eclipse occurs when the full moon enters the shadow cast by the Earth. As in the case of shadows cast by terrestrial objects, we distinguish between the full shadow, the *umbra,* and the surrounding region of part-shadow, the *penumbra.* If the Moon is completely immersed in the region of the full shadow of the Earth, we speak of a *total lunar eclipse;* if only part of the Moon is in shadow, we have a partial *lunar* eclipse. In consequence of known geometrical relationships, a lunar eclipse can last for at most $3^h\,40^m$ and totality at most $1^h\,40^m$. Since the light from the Sun is scattered by the Earth's atmosphere, the outer edge of the penumbra on the Moon is quite blurred, and the umbra is noticeably un-sharp. Since the blue light experiences stronger absorption than the red light, the penumbra, and to a less extent also the umbra, appears to be pervaded by reddish copper-coloured light. Exact photometric observations of lunar eclipses can give information about the high layers of the Earth's atmosphere.

If the Moon, at the time of new Moon, passes in front of the Sun, then a solar eclipse takes place (Fig. 4.4). This can be *partial* or *total.* If the apparent diameter of the Moon is smaller than that of the Sun, then a central coverage results only in an *annular* eclipse. In a partial eclipse of the Sun, the observer is in the penumbra of the Moon, while during totality he is in the umbra. In the case of an annular eclipse, the vertex of the umbra is between the observer and the Moon.

In regard to the astrophysical study of the outermost layers of the Sun and of interplanetary material in its neighbourhood, total solar eclipses are of special significance because the bright light from the solar disk is completely cut off *outside* the Earth's atmosphere. Since the zone of totality on the Earth's surface is relatively narrow, the atmosphere is nevertheless slightly illuminated by scattered light coming in from the sides, even during totality. Relative to the Sun, the Moon regresses corresponding to the duration of the synodic month, on the average 0″.51 per second on the sky; this corresponds to a stretch of 370 km on the Sun's surface. Eclipse observations with

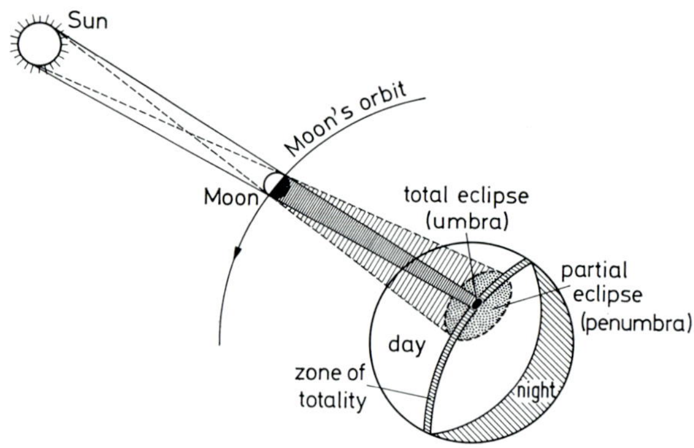

Fig. 4.4. Eclipse of the Sun (schematic). The Moon crosses the Sun's disk from west to east. Total and partial eclipse of the Sun are observed in the regions traversed by the umbra and penumbra, respectively

good time resolution therefore give an angular resolution which exceeds that of available telescopes (see below).

Occultations of stars by the Moon, which like solar eclipses have to be predicted specially for any given place of observation, are very sharply defined in time, since the Moon possesses no atmosphere. They are important for checking the Moon's path, the determination of fluctuations in the Earth's rotation, and checking ephermeris time. Since the Moon regresses in the mean by 0″.55 per second relative to the stars, by photometric observations of occultations of the stars using precise time resolution, it is even possible in favourable cases to measure the angular diameter of the infinitesimal disks of the stars concerned. Even more important are occultations by the Moon of astronomical objects for *radio astronomy,* whereby significantly better angular resolution can be attained than by any other technique.

It was known to the cultures of the ancient East that solar and lunar eclipses (in the following, we say simply "eclipses" for short) repeat themselves with a period of $18^y\ 11^d.33$, the so-called Saros cycle. This cycle depends upon the fact that an eclipse can occur only if the Sun *and* Moon are fairly near to a node of the lunar orbit. The time that the Sun takes from a lunar node to return to the same node is somewhat shorter than one tropical year, on account of the regression of the lunar nodes, being equal to 346.62 days. This time is called an eclipse year. As one easily verifies, the Saros period comprises a whole number of synodic months *and* a whole number of eclipse years, in fact

223 Synodic months = 6585.32 days
and 19 eclipse years = 6585.78 days

besides which we have

239 anomalistic months = 6585.54 days
(anomalistic month from perigee to perigee = 27.555 days).

After $18^{y}\,11^{d}\!.33$ the eclipse pattern in fact repeats itself with great accuracy. As one can show from the orbits of the Earth and the Moon, taking account of the diameters of these bodies, in a single year at most three lunar eclipses and five solar eclipses can occur. At any one place a lunar eclipse, which can indeed be seen over an entire hemisphere of the Earth, can be observed fairly frequently, while a total solar eclipse is an extremely rare occurrence.

5. Planetary System

The planets that have been known since ancient times (with their time-honoured symbols) are Mercury ☿, Venus ♀, Mars ♂, Jupiter ♃, and Saturn ♄. Mankind has continually sought to transform the apparent capriciousness of their paths into a gradual unfolding of regularities.

In Chapter 1 we have briefly mentioned the concern of the ancients for the significance of planetary motions. Here we adopt the standpoint of the heliocentric world-view, as *Nicholas Copernicus* presented it in 1543. However, we shall drop the restriction to circular motions that *Copernicus* had retained as a relic of Aristotelian concepts, and adopt the elliptic orbits of *Johannes Kepler* and his three laws of planetary motion (1609 and 1619). Thereby we arrive at the threshold of modern mathematico-physical thought that took significant shape at the hands of *Galileo Galilei* (1564—1642) and that in *Isaac Newton's Principia* (1687) was epitomized in the statement of classical mechanics and gravitational theory. We shall concern ourselves with the physical structure of the planets in Chapter 7.

The occurrence of *direct* (west-east) motion with an interval of backwards or *retrograde* (east-west) motion is clarified in Fig. 5.1 by the example of planet Mars.

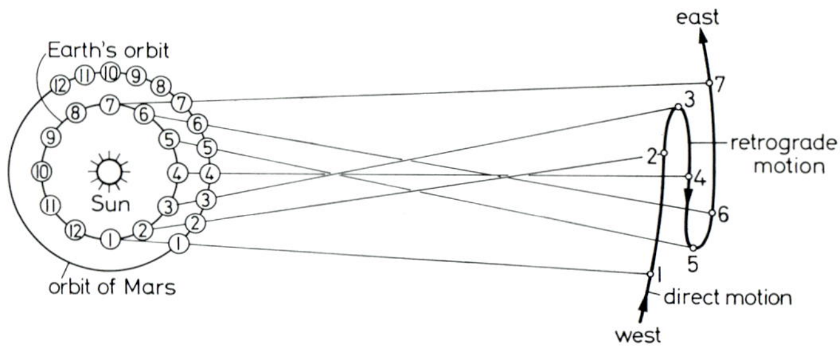

Fig. 5.1. *Direct (west-east) and retrograde (east-west) motion of the planet Mars.* The positions of the Earth and Mars in their orbits are numbered from month to month. At 4 Mars is in opposition to the Sun; it is overtaken by the Earth at this point and is therefore in retrograde motion. This is the time when it is closest to the Earth and most favourably placed for observation.
The orbit of Mars is inclined at $1\overset{\circ}{.}9'$ to that of the Earth (i. e. to the ecliptic)

Fig. 5.2 shows the path of an *inner planet*, e. g. Venus round the Sun, as seen from our standpoint on the more slowly revolving Earth. The planet is nearest to us at *inferior conjunction:* then it moves away from the Sun in the sky and reaches its *greatest westerly elongation* of 48° as a *morning star*. At superior conjunction Venus is at its furthest distance from the Earth and is again close to the Sun in the sky. It then moves further away and reaches its *greatest easterly elongation* of 48° as an *evening star*. The ratio of the orbital radii of Venus and the Earth is determined by the maximum elongation of $\pm 48^\circ$, and in the case of Mercury $\pm 28^\circ$. The phases of Venus and the corresponding changes in its apparent diameter (from $9\overset{''}{.}9$ to $64\overset{''}{.}5$) were discovered by *Galileo* with his telescope. They verify that the Sun is at the centre of the (true) path of Venus. As we can see from Fig. 5.2, Venus attains its greatest apparent brightness near its greatest elongations. At inferior conjunction, Venus (and Mercury) may pass in front of the Sun's disk. Such a transit of Venus was formerly of interest as a means of measuring the Sun's distance or solar parallax (see below).

An outer planet, Mars ♂ for example, is nearest to us at *opposition* (Fig. 5.3); it culminates there at midnight by true local time, has its greatest apparent diameter and is most favourably placed for observation. When in *conjunction* it is near the Sun in the sky. Unlike the cases of the Moon and the inner planets, the phases of an outer planet do not range the whole way from "full" to "new". We term *phase-*

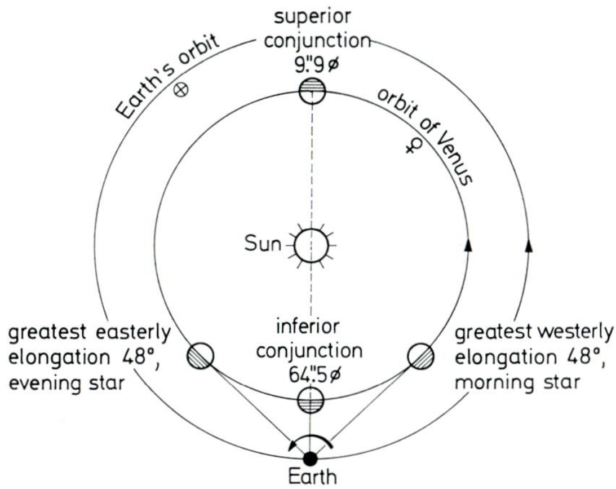

Fig. 5.2. *Orbit and phases of Venus, an inner planet.* The elongation of Venus on the sky cannot exceed $\pm 48°$ ($\pm 28°$ for Mercury). The phases resemble those of the Moon. Greatest brightness occurs near maximum elongation

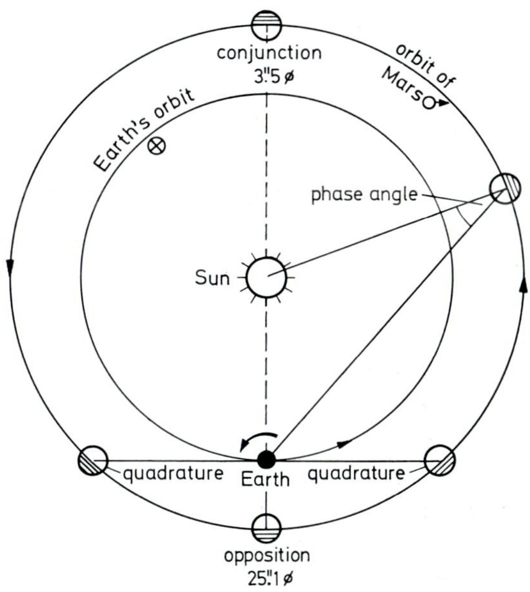

Fig. 5.3. *Orbit and phases of Mars, an outer planet.* Greatest brightness and greatest angular diameter (25″.1) occur at opposition

angle φ the angle subtended by the Sun and the Earth at the planet. The fraction $\varphi/180°$ of the planet's hemisphere that faces the Earth is in darkness. As one easily verifies, the phase angle of an outer planet is greatest in *quadrature* i. e. when the planet and the Sun are seen in directions at 90° to each other. The greatest phase angle of Mars is 47° (Fig. 5.3) and of Jupiter only 12°.

We give the name *sidereal period* to the true time of revolution of a planet round the Sun. The *synodic period* is the time of revolution relative to the Sun, that is the time between two successive corresponding conjuctions, etc. As in the case of the Moon, we have for a planet, by subtraction of angular speeds,

$$\frac{1}{\text{synodic period}} = \left| \frac{1}{\text{sidereal period}} - \frac{1}{\text{sidereal period of the Earth}} \right| \tag{5.1}$$

In the case of Mars, for example, from the direct observation of the synodic period of 780 days and knowing the sidereal year to be 365 days, the sidereal period of the planet is found to be 687 days.

Kepler first determined the true shape of the orbit of Mars by combining pairs of observations of Mars that were separated in time by the sidereal period of Mars, and in which therefore Mars must be at the same position in its orbit. From such pairs of observations of Mars separated by 687 days, he could "triangulate" the positions of Mars and so determine its true path. Two fortunate circumstances helped Kepler in due course to derive his first two laws of planetary motion: the fact that Appollonius of Perga had investigated the conic sections mathematically, and the fact that of the planets known at that time Mars happened to have the greatest eccentricity $e = 0.093$. He found his third law ten years later, proceeding from his unshakable conviction that the motions of the planets must somehow manifest the "harmony" of the universe.

The complete description of the path of a planet or a comet (see below) round the Sun requires the orbital elements shown in Fig. 5.4.

1. Semi-major axis a. We express it either in terms of the semi-major axis of the Earth = 1 astronomical unit (a. u.) or else in kilometers.
2. Eccentricity e [perihelion distance $a(1-e)$; aphelion distance $a(1+e)$].
3. Inclination of the plane of the orbit to the ecliptic i.
4. Longitude of the ascending node ☊ (angle from the equinoctial point ♈ to the ascending node).
5. Displacement ω of perihelion from the node (angle between the ascending node and perihelion). The sum of the angles ☊ $+ \omega$ of

which the first is measured in the ecliptic and the second in the plane of the orbit, is called the *longitude of perihelion* π.

6. Period P (sidereal period measured in tropical years) or the daily motion μ ($\mu = 360 \times 60 \times 60''/P_{\text{days}}$).
7. Epoch E or the instant in time of perihelion passage T.

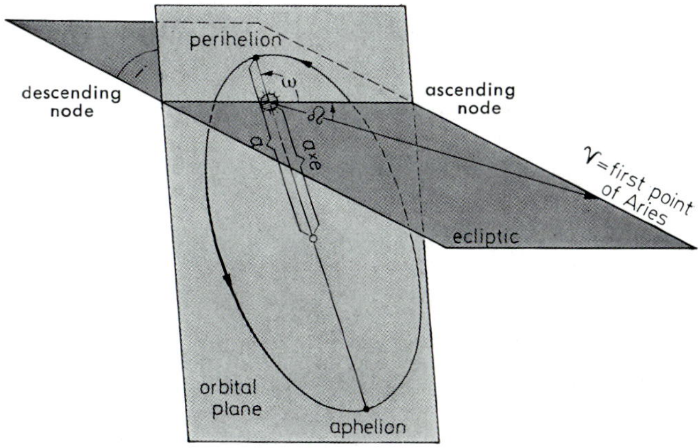

Fig. 5.4. Orbital elements of a planet or comet

The elements a and e determine the size and shape of the orbit (Fig. 3.3), i and ☊ the orbital plane, ω the orientation of the orbit in the plane. The course of the motion is determined by P and T, where, in accordance with Kepler's third law, the period P is determined by a, apart from small corrections (see below).

Table 5.1 contains the orbital elements of the planets that are of interest to us here (Fig. 5.5). In addition to the planets known to the ancients, we have included the brightest of the *minor planets* (Ceres) as well as Uranus, Neptune and Pluto. We shall later recount something of the interesting story of their discovery. To our planetary system there belong also the *comets* and *meteors*.

Nowadays we denote comets by the year of discovery and in each year we number them according to the sequence of their perihelion passages (see below); the name of the discoverer is often added. In ancient times and in the middle ages, according to the dogma of the immutability of the heavenly regions, the comets were relegated to the atmospheric surroundings of the Earth. *Tycho Brahe* first made exact observations of the comets of 1577 and 1585 and the parallaxes he derived showed that, for example, the comet of 1577 was at least six times as distant as the Moon. *Isaac Newton* recognized

Table 5.1. *Planetary system*

Planet	Symbol	Discovery	Some orbital elements				
			Sidereal period years	Semi-major axis of orbit Astron. units AU	Semi-major axis of orbit Million km 10^6 km	Eccen-tricity e	Inclination to ecliptic i
Inner planets							
Mercury	☿		0.241	0.387	57.9	0.206	7°0′
Venus	♀		0.615	0.723	108.2	0.007	3°24′
Earth	⊕		1.000	1.000	149.6	0.017	0°0′
Outer planets							
Mars	♂		1.880	1.524	227.9	0.093	1°51′
Minor planets		Ceres: { *Piazzi*, 1801; *Gauss* }	4.603	2.767	413.6	0.076	10°37′
Jupiter	♃		11.862	5.203	778	0.048	1°18′
Saturn	♄		29.457	9.539	1427	0.056	2°29′
Uranus	⛢	*W. Herschel*, 1781	84.015	19.182	2870	0.047	0°46′
Neptune	♆	*Leverrier* and *Galle*, 1846	164.788	30.057	4496	0.009	1°46′
Pluto	♇	*Lowell* and *Tombaugh*, 1930	247.7	39.5	5910	0.247	17°8′

that the comets move round the Sun in elongated ellipses or parabolas, that is in conic sections having eccentricity a little less than or equal to unity. His great contemporary, *Edmond Halley* improved the techniques for computing cometary orbits and in 1705 he was able to show that the comet of 1682 (since named "Halley's comet") has a period of about 76.2 years. According to Kepler's third law, the major axis of the orbit is therefore $2 \times 76^{2/3} = 36$ astronomical units, i. e. the aphelion lies somewhat beyond the orbit of Neptune. Halley's computed orbit also showed that the bright comet of 1682 was identical with those of 1531 and 1607, so he was also able to predict the return of the comet in 1758. Altogether 28 returns of Halley's comet since 240 B.C. have been witnessed.

In the main, cometary orbits fall into two groups: a) Nearly parabolic orbits with periods greater than 100 years, perihelion passages at about 1 AU yielding the highest probalitity of discovery. b) Elliptic paths of short-period comets. The aphelia are concentrated near the orbits of the major planets, particularly Jupiter. Such cometary families may have arisen by the "capture" of long-period comets by Jupiter or some other planet. Aver-

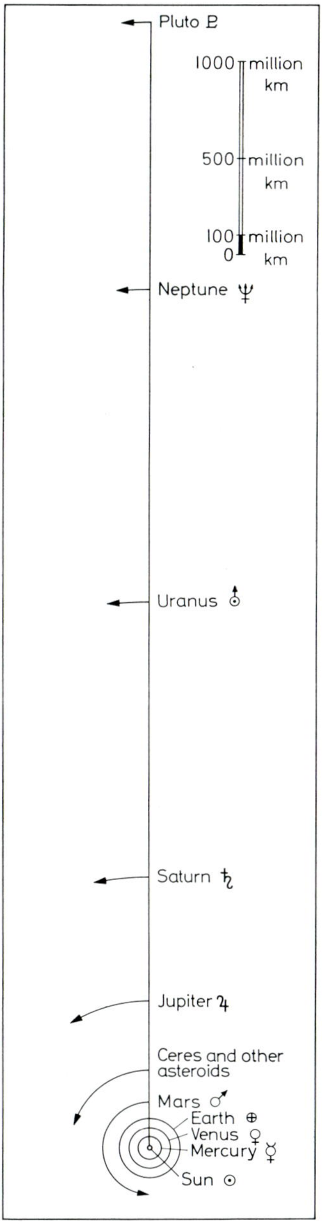

Fig. 5.5. *Mean orbital radii (semimajor axes) of the planets.* The lengths of the arcs correspond to the mean motion in one (terrestrial) year. During this time Venus makes 1.62 revolutions and Mercury 4.15

age values of arbital elements are about $a = 3.6$ AU, $e = 0.56$ (thus giving aphelion distance 5.6 AU almost equal to $a_{♃} = 5.2$ AU) and $i \approx 15°$.

Encke's comet has the shortest known period of 3.30 years. The comets *Schwassmann-Wachmann* 1925 II ($a = 6.4$ AU, $e = 0.132$) and Oterma 1943 ($a = 3.96$ AU, $e = 0.144$) have nearly circular orbits.

The swarms of shooting stars or *meteors,* that on certain dates in the year seem, as a matter of perspective, to emanate from their so-called *radiants* in the sky, are simply fragments of comets whose orbits almost intersect that of the Earth; this is shown by the periodicities of the occurrences. In many cases, the material seems to be fairly well concentrated in the path, so that from time to time depending on their periods specially lively displays of shooting stars are observed. For example there was the famous fall of Leonids (radiant RA 152°, $\delta + 22°$) that *A. von Humboldt* observed in 1799 November 11/12 in South America and that can be associated with the comet 1866 I of period 33 years. In addition, there are the so-called *sporadic meteors,* for which a periodicity cannot be discerned. That shooting stars are actually small heavenly bodies, which penetrate the Earth's atmosphere and so become incandescent, was first established in 1798 by two Göttingen students *Brandes* and *Benzenberg* by making corresponding observations from two sufficiently separated sites and then calculating the heights.

Even before that in 1794 *E. F. F. Chladni* had shown that *meteorites* are none other than (more massive) meteoric bodies that have reached the ground.

Neither amongst comets nor amongst meteors has anyone found *hyperbolic orbits,* i. e. orbits of objects that have entered the solar system from outside.

In Chapters 7 and 8 we shall consider the physical structure of planets, of their atmospheres and satellites as well as that of comets, meteors and meteorites.

Here we must again consider the important question as to how one can measure in kilometres the distance of the Earth from the Sun, or more precisely, the semi-major axis of the Earth's orbit, which we have defined as one astronomical unit (AU). Astronomers prefer to speak of the *solar parallax* $\pi_{\odot}$, the angle subtended by the Earth's equatorial radius $a = 6378$ km at the centre of the Sun. Unlike the case of the Moon, the solar parallax is too small for direct measurement. Consequently, astronomers first determine from observations made at several observatories in both the northern and the southern hemisphere the distance of, for instance, a planet or minor planet whose orbit comes sufficiently near the Earth. Formerly, they observed Mars in opposition or Venus at inferior conjunction. More recently, they have made extensive series of observations at opposition of the minor planet Eros, which is favourable for the purpose. There has now developed a *combination* of these astronomical methods with *radar* techniques. It is now possible to make direct measurements with high precision of the distance not only of the Moon but also of Venus (and other planets will soon be reached), by combining the time-lapse of reflected radio-signals with terrestrial measurements of light-speed. Using the detailed measurements, one computes the radius of the Earth's orbit basically using Kepler's third law, but in practice using very difficult celestial mechanics.

Instead of the radius of the Earth's orbit, one can also measure the velocity of the Earth in its orbit with the help of the doppler effect (Fig. 5.6). If a radiation-source moves with radial-velocity v (relative velocity-component in the line of sight) away from an observer, then the wavelength λ or the frequency $\nu = c/\lambda$ (c = light-speed) suffers a change

$$\Delta\lambda = \lambda\, v/c \quad \text{or} \quad \Delta\nu = \nu\, v/c\,. \tag{5.2}$$

A relative *recession* of the source from the observer (by definition, a positive radial velocity) produces an increase in the wavelength, i. e. a redshift of the spectral lines, and a *decrease* in the frequency, and conversely.

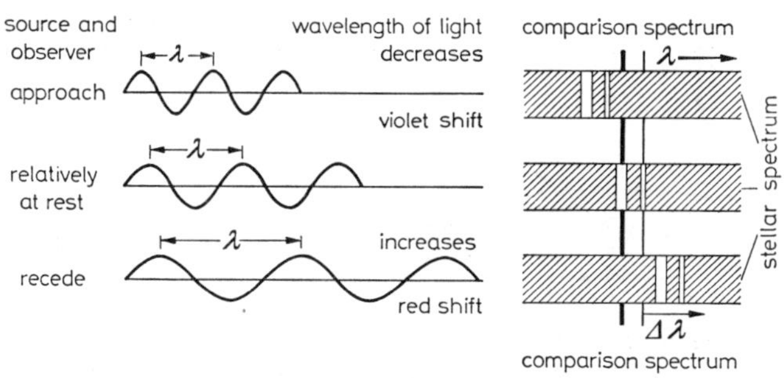

Fig. 5.6. Doppler effect $\Delta\lambda/\lambda = v/c$

In practice, one follows during a considerable fraction of a year either the radial velocity of a fixed star relative to the Earth, given by the doppler effect shown by its spectral lines, or else the relative velocity of, say, Venus and the Earth, given by the frequency shift of radar signals, the shift produced by a moving reflector being double the shift from a moving source.

The historically very significant first measurement of the speed of light by O. *Römer 1675* depends upon a corresponding concept. He determined the frequencies of revolution of the moon's of Jupiter from their passages before and behind Jupiter's disk. Because of the finite speed c of light-propagation, if the Earth was receding from Jupiter the frequencies diminished, if it was approaching they increased. Using the then available determination of the solar parallax, Römer derived a remarkably good estimate of the light-speed. That he thereby anticipated the doppler principle (5.2) some two hundred years before its first spectroscopic application is scarcely evident from the usual presentations.

The *aberration of light* discovered in 1725 by the Astronomer Royal J. Bradley in the course of his attempt to measure stellar

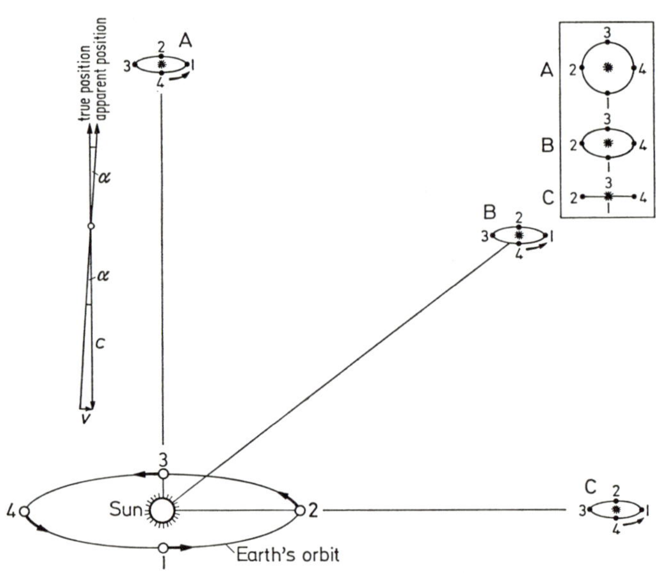

Fig. 5.7. *Aberration of light*. The star's light appears to be deflected by an angle v/c in the direction of the velocity vector of the Earth (see left of diagram), where v is the velocity component of the Earth perpendicular to the ray of light and c is the velocity of light. A star therefore describes a circle of radius $\alpha =$ Earth's velocity/velocity of light $= 20''.48$ at the pole of the ecliptic, a straight line of maximum extension $\pm\alpha$ *in* the ecliptic, and in between an ellipse (shown upper right). $*$ True position of star. An observer looking to the right sees the star at intervals of $\frac{1}{4}$ year in the positions 1—2—3—4

parallaxes also depends upon the finiteness of the light-speed (Fig. 5.7). If from the moving Earth we observe a star that, for an observer at rest, lies always in a direction at right angles to the motion, we have slightly to incline our telescope towards the direction of the Earth's velocity v by the angle of aberration v/c in order that it should point at the star. Thus, in the course of a year the star describes a small circle about the pole of the ecliptic (Fig. 5.7, upper right); a star in the ecliptic moves to and fro in a line; in between, a star describes the appropriate ellipse. Usually, one illustrates aberration by the analogy of an astronomer who hurries along carrying an umbrella through vertically-falling rain.

This demonstration, as well as our elementary explanation of the doppler effect, is open to criticism because it does not take account of the principle that the light-speed is independent of the motion of the source. A consistent interpretation of all effects, and the first to agree to order $(v/c)^2$ with the experiments of Michelson and others, was first given by A. Einstein's special relativity theory (1905).

In conclusion we recapitulate the quantities we have discussed and their mutual interrelations

Equatorial Earth-radius $a = 6378$ km
Solar parallax (equatorial horizontal parallax) $\pi_\odot = 8''.794$.
Astronomical unit = semi-major axis of the Earth's orbit
$A = a/\pi_\odot = 149.6 \times 10^6$ km
Light-speed $c = 299\,793$ km/sec
Light-time for 1 AU $= A/c = 498.5$ sec
Mean orbital speed of the Earth $v = 29.8$ km/sec
Equatorial rotational speed of the Earth 0.465 km/sec
Aberration constant $v/c \approx 20''.48$.

6. Mechanics and Theory of Gravitation

After protracted and hazardous beginnings with Kepler and Galileo, Isaac Newton in his *Principia* (1687) constructed the *mechanics* of terrestrial and cosmical systems. Using this in combination with his *law of gravitation,* in the same work he deduced Kepler's laws and many other regularities in the motions of the planets. Small wonder that the further development of *celestial mechanics* remained for almost two centuries one of the chief fields of endeavour for great mathematicians and astronomers.

We state Newton's three laws of motion in modern language:

Law I. *A body continues in its state of rest or of uniform motion in a straight line, except in so far as it is acted upon by external forces* (Law of inertia).

We represent a velocity in magnitude and direction by a vector $\boldsymbol{v}$ similarly a force by $\boldsymbol{F}$. The addition and subtraction of such quantities follows the well-known parallelogram law; they can be specified by their components in a rectangular coordinate system x, y, z, that is by their projections on the coordinate-axes, for example $\boldsymbol{v} = (v_x, v_y, v_z)$. If the moving body has mass m, *Newton* defined its *momentum* as the vector

$$\boldsymbol{p} = m\,\boldsymbol{v} \tag{6.1}$$

This important concept makes possible the statement of

Law II. *The rate of change of momentum of a body is proportional to the magnitude of the external force acting upon it and takes place in the direction of the force.*

In symbols we write for the case of a single body, denoting the time by t,

$$\frac{\mathrm{d}\boldsymbol{p}}{\mathrm{d}t} = \frac{\mathrm{d}}{\mathrm{d}t}(m\,\boldsymbol{v}) = \boldsymbol{F}. \tag{6.2}$$

Law I is obviously the special case of Law II when $\boldsymbol{F} = 0$. We can also express the velocity $\boldsymbol{v}$ as the rate of change of the position vector $\boldsymbol{r}$ with components x, y, z; so we write $\boldsymbol{v} = \mathrm{d}\boldsymbol{r}/\mathrm{d}t$ and consequently for the case of constant m,

$$m\,\mathrm{d}^2\boldsymbol{r}/\mathrm{d}t^2 = \boldsymbol{F}. \tag{6.3}$$

The statement, force = mass × acceleration, is valid however *only* for masses that do not change with time, while equation (6.2) remains true also in special relativity theory where the mass depends upon the speed according to the formula $m = m_0(1 - v^2/c^2)^{-1/2}$ in which we call $m_0 = m_{v=0}$ the rest-mass. If we have N bodies which we distinguish by the subscript $k = 1, 2, \ldots, N$, then corresponding to (6.2) we have the N vector equations

$$\frac{\mathrm{d}\boldsymbol{p}_k}{\mathrm{d}t} = \frac{\mathrm{d}}{\mathrm{d}t}(m_k\,\boldsymbol{v}_k) = \boldsymbol{F}_k \tag{6.4}$$

or the corresponding $3\,N$ equations for the components.

Newton's last law concerns the interaction of two bodies and asserts:

Law III. *The forces that two bodies exert upon each other are equal in magnitude and opposite in direction* (Law of action and reaction).

Then if $\boldsymbol{F}_{ik}$ is the force which body i exerts upon body k we have

$$\boldsymbol{F}_{ik} = -\boldsymbol{F}_{ki}. \tag{6.5}$$

As a simple example of Newton's laws of motion we consider (Fig. 6.1) a mass m that moves with uniform speed $\boldsymbol{v}$ in a (horizontal)

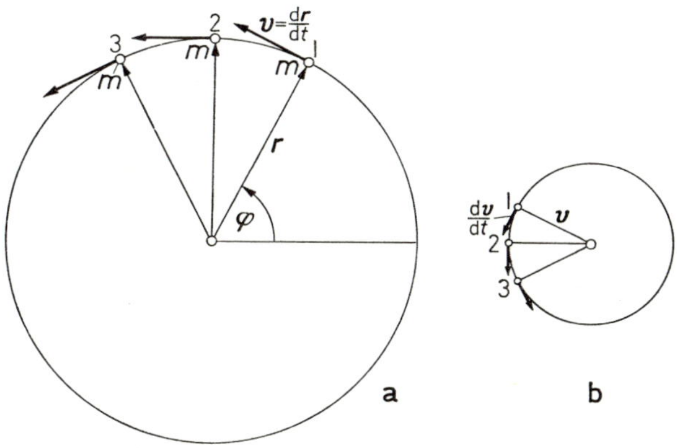

Fig. 6.1 a and b. *Calculation of centrifugal force.* a) *Circular orbit of the mass m.* Position vector $\boldsymbol{r}$ at the times 1, 2, 3 ... Velocity vector $\boldsymbol{v} = \mathrm{d}\boldsymbol{r}/\mathrm{d}t$ in the direction of the tangent to the orbit, and of magnitude $v = r\,\mathrm{d}\varphi/\mathrm{d}t$. b) *Hodograph.* Velocity vector at the times 1, 2, 3 ... Acceleration vector $\mathrm{d}\boldsymbol{v}/\mathrm{d}t$ in the direction of the tangent of the hodograph and therefore parallel to $-\boldsymbol{r}$. The magnitude of the acceleration is $v\,\mathrm{d}\varphi/\mathrm{d}t = v^2/r$

circle at the end of a string of length r. (The magnitude of a vector quantity, its so-called "modulus", is usually denoted by modulus signs, thus $|\boldsymbol{v}|$, or by the use of the corresponding italic letter, thus $v = |\boldsymbol{v}|$.) The angular speed $\mathrm{d}\varphi/\mathrm{d}t$, where φ is in circular measure, is then equal to v/r. If we draw successive velocity vectors from a point so as to construct the so-called *hodograph*, we read off the acceleration as having magnitude $|\mathrm{d}\boldsymbol{v}/\mathrm{d}t| = (v/r)v$ and as being directed towards the centre of the circle in Fig. 6.1 (a). So we derive the law of centrifugal force

$$F = m\,v^2/r \tag{6.6}$$

which had been discovered by Christian Huygens before Newton. Newton's third law then tells us that the string exerts the same force outwards on its point of attachment as that by which it pulls the moving body inwards.

If the detailed structure of a body is irrelevant for a problem in mechanics, we speak of it as a *point-mass* (particle). For instance, in the theory of planetary motion we treat the Earth as a point-mass; if we are interested in an atom, we treat its radiating electron as a point-mass, and so on.

Starting from Newton's three laws of motion of a particle, we proceed first to the equations of motion of a *system of particles.* Thence we derive the three *conservation laws* of mechanics, to which we shall repeatedly appeal.

1. Linear Momentum: Motion of Mass-Centre

In a system of point-masses m_k, denoted by subscripts i or k ($i,\ k=1,2,\ldots N$), we distinguish between the *internal force* $\mathbf{F}_{ik}$, which particle i exerts upon particle k, and the *external force* $\mathbf{F}_k{}^{(e)}$, which acts upon particle k from outside the system. The equation of motion (6.4) for particle k becomes

$$\mathrm{d}\boldsymbol{p}_k/\mathrm{d}t = \boldsymbol{F}_k{}^{(e)} + \sum_{i=1}^{N} \boldsymbol{F}_{ik}\,. \tag{6.7}$$

Using the law of action and reaction (6.5) and summing over all particles of the system, we get

$$\frac{\mathrm{d}}{\mathrm{d}t}\sum \boldsymbol{p}_k = \boldsymbol{F}_k{}^{(e)}$$

where the summation Σ is taken over all particles of the system from $k=1$ to $k=N$. If we consider the N particles as a single system we may define

$$\left.\begin{array}{ll} \text{total momentum} & \boldsymbol{P} = \sum \boldsymbol{p}_k \\ \text{resultant external force} & \boldsymbol{F} = \sum \boldsymbol{F}_k{}^{(e)}\,. \end{array}\right\} \tag{6.8}$$

Then the equation of motion becomes

$$\mathrm{d}\boldsymbol{P}/\mathrm{d}t = \boldsymbol{F} \tag{6.9}$$

like the equation for a single particle. If there is no external force ($\boldsymbol{F}=0$) then we obtain

$$\text{conservation of linear momentum}\quad \boldsymbol{P} = \sum \boldsymbol{p}_k = \text{constant.} \tag{6.10}$$

We can render the meaning of (6.9) and (6.10) a little more intuitive if we define for the system

$$\text{total mass}\ \ M = \sum m_k \tag{6.11}$$

and position vector $\boldsymbol{R}$ of mass-centre S, such that

$$M\boldsymbol{R} = \sum m_k\,\boldsymbol{r}_k\,. \tag{6.12}$$

Then equation (6.9) becomes the equation of motion of the mass-centre

$$M\,\mathrm{d}^2\mathbf{R}/\mathrm{d}t^2 = \boldsymbol{F} \tag{6.13}$$

analogous to that for a single particle. We infer that in the force-free case $\boldsymbol{F}=0$ the mass-centre (corresponding to Law I) must perform uniform motion in a straight line expressed by $\mathrm{d}\boldsymbol{R}/\mathrm{dt}=$ constant.

2. *Moment of Momentum: Conservation of Angular Momentum*

We consider first (Fig. 6.2) a particle m_k that can revolve about a fixed point O at the end of a rod $\boldsymbol{r}_k$. Let a force $\boldsymbol{F}_k$ act on m_k. This

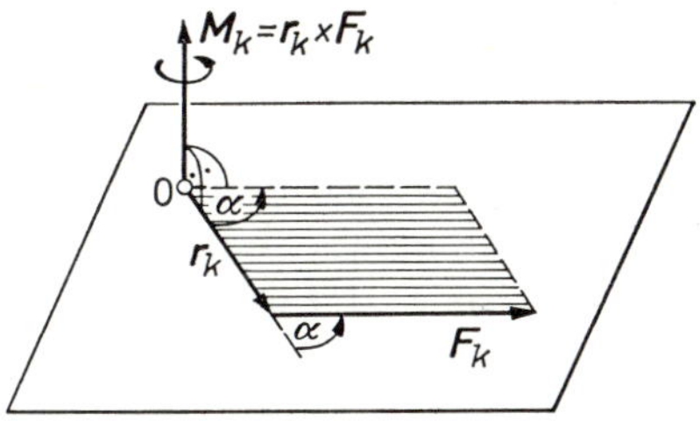

Fig. 6.2. *Couple* $\boldsymbol{M}_k = \boldsymbol{r}_k \times \boldsymbol{F}_k$. The magnitude of $\boldsymbol{M}_k$ is $|\boldsymbol{r}_k|\,|\boldsymbol{F}_k|\sin\alpha$ and is equal to the area of the parallelogram spanned by $\boldsymbol{r}_k$ and $\boldsymbol{F}_k$

"tries" to make m_k revolve about an axis through O perpendicular to the plane containing $\boldsymbol{r}_k$ and $\boldsymbol{F}_k$. For this, only the transverse component $|\boldsymbol{F}_k|\sin\alpha$ is effective, α being the angle between $\boldsymbol{r}_k$ and $\boldsymbol{F}_k$. The quantity, rod-length $|\boldsymbol{r}_k|$ times the effective force-component $|\boldsymbol{F}_k|\sin\alpha$ erected as a vector perpendicular to the plane of $\boldsymbol{r}_k$ and $\boldsymbol{F}_k$ is known mathematically as the *vector product* $\boldsymbol{r}_k \times \boldsymbol{F}_k$ (sometimes the "cross"-product) and physically as the *moment of the force* about O, or the turning moment $\boldsymbol{M}_k = \boldsymbol{r}_k \times \boldsymbol{F}_k$. Just as we define this moment in the case of a force, so in the case of the momentum $\boldsymbol{p}_k = m_k \boldsymbol{v}_k$ we define the *moment of momentum* or *angular momentum* $\boldsymbol{N}_k$ by

$$\boldsymbol{N}_k = \boldsymbol{r}_k \times \boldsymbol{p}_k = \boldsymbol{r}_k \times m_k \boldsymbol{v}_k\,.$$

The order of the factors in a produce like $\boldsymbol{M} = \boldsymbol{r} \times \boldsymbol{F}$ is such that a right-handed screw turning from $\boldsymbol{r}$ towards $\boldsymbol{F}$ will advance in the direction of $\boldsymbol{M}$.

From Newton's equation of motion (6.4) we now obtain by vector multiplication with $\boldsymbol{r}_k$ on the left

$$\boldsymbol{r}_k \times \frac{\mathrm{d}}{\mathrm{d}t}(m_k \boldsymbol{v}_k) = \boldsymbol{r}_k \times \boldsymbol{F}_k \quad \text{or} \quad \frac{\mathrm{d}}{\mathrm{d}t}(\boldsymbol{r}_k \times m_k \boldsymbol{v}_k) = \boldsymbol{r}_k \times \boldsymbol{F}_k\,. \tag{6.14}$$

(The second form follows because $\frac{\mathrm{d}}{\mathrm{d}t}(\boldsymbol{r}\times m\,\boldsymbol{v}) = \frac{\mathrm{d}\boldsymbol{r}}{\mathrm{d}t}\times m\,\boldsymbol{v} + \boldsymbol{r}\frac{\mathrm{d}}{\mathrm{d}t}(m\,\boldsymbol{v})$ and, since $\mathrm{d}\boldsymbol{r}/\mathrm{d}t = \boldsymbol{v}$, the first term on the right is the vector product of two parallel vectors which is zero by definition.)

For a system of particles we define the resultant moment of all the internal and external forces about the fixed point O, and also the total angular momentum of the system by the equations

$$\left.\begin{array}{ll}\text{resultant turning moment} & \boldsymbol{M} = \sum \boldsymbol{r}_k \times \boldsymbol{F}_k = \sum \boldsymbol{r}_k \times (\boldsymbol{F}_k^{(e)} + \sum_i \boldsymbol{F}_{ik}) \\ \text{total angular momentum} & \boldsymbol{N} = \sum \boldsymbol{r}_k \times \boldsymbol{p}_k = \sum \boldsymbol{r}_k \times m_k\,\boldsymbol{v}_k\end{array}\right\} \quad (6.15)$$

Then the equation of motion becomes

$$\mathrm{d}\boldsymbol{N}/\mathrm{d}t = \boldsymbol{M} \qquad (6.16)$$

or *the rate of change of angular momentum is equal to the sum of the moments of all the forces.*

If only *central forces* operate within the system, that is forces like gravitation that act only along the lines joining the particles, the contribution to $\boldsymbol{M}$ of the internal forces vanishes and on the right hand-side of (6.16) there remains only the moment of all the external forces.

If now all the external forces vanish, or at least if their resultant moment is zero, we have $\mathrm{d}\boldsymbol{N}/\mathrm{d}t = 0$. So we get the important law of conservation of angular momentum $\boldsymbol{N} = \sum \boldsymbol{r}_k \times m_k\,\boldsymbol{v}_k = \text{constant}.$ (6.17)

3. *Energy-law*

If a particle of mass m_k moves under the action of a force $\boldsymbol{F}_k$ through a small displacement $\mathrm{d}\boldsymbol{r}_k$ making angle α with $\boldsymbol{F}_k$ the work performed on the particle is

$$\mathrm{d}W = |\boldsymbol{F}_k|\,|\mathrm{d}\boldsymbol{r}_k|\cos\alpha = \boldsymbol{F}_k\cdot\mathrm{d}\boldsymbol{r}_k\,. \qquad (6.18)$$

This (scalar) quantity we call the scalar product of the two vectors and we denote it by the dot between the factors. To evaluate the work done in a finite displacement $A\to B$ we use Newton's equation of motion (6.4) with $\boldsymbol{v}_k = \mathrm{d}\boldsymbol{r}_k/\mathrm{d}t$ and we obtain

$$\int_A^B \boldsymbol{F}_k\cdot\mathrm{d}\boldsymbol{r}_k = \int_A^B \frac{\mathrm{d}}{\mathrm{d}t}(m_k\,\boldsymbol{v}_k)\cdot\mathrm{d}\boldsymbol{r}_k = \tfrac{1}{2}m_k\,\boldsymbol{v}_k^2\Big|_A^B\,. \qquad (6.19)$$

We call the quantity $\frac{1}{2}\,m_k\,\boldsymbol{v}_k^2$ the *kinetic energy* T_k of the particle and we can take its sum T over any number of particles. Further, if $\sum \boldsymbol{F}_k\cdot\mathrm{d}\boldsymbol{r}_k$ is a perfect differential $-\,\mathrm{d}V$, that is, if the work done by the forces is independent of the actual paths of the particles and depends only upon their initial and final states, then the sum of the kinetic energy T plus the potential energy V is constant. We have the further important law

$$\text{conservation of energy} \quad E = T + V = \text{constant.} \tag{6.20}$$

In a system of particles moving under gravitational forces (see below) and in many other cases T depends only on the velocities and V only on the positions of the particles.

4. Law of Gravitation: Celestial Mechanics

In order to obtain a theory of cosmical motions, Newton had to have in addition to his laws of mechanics also his law of gravitation (about 1665):

Two particles of mass m_i and m_k at distance r apart attract each other in the direction of the line joining them with a force

$$F = -G \frac{m_i m_k}{r^2}. \tag{6.21}$$

As a result of an integration that we shall not repeat here, Newton then showed that exactly the same law of attraction (6.21) holds good for two massive spheres (Sun, planets, ...) of finite extent, as for the two corresponding massive particles. He next verified the law of gravitation (6.21) assuming that free fall at the Earth's surface *(Galileo)* and the revolution of the Moon are both governed by the force of attraction of the Earth.

The acceleration (= force/mass) in the case of free fall can be measured by experiments with falling bodies, or more accurately by use of a pendulum. Its numerical value at the equator is 978.05 cm/sec^2, or taking account of the centrifugal acceleration of the Earth's rotation of 3.39 cm/sec^2, it is

$$g_{\oplus} = 981.4\ \text{cm/sec}^2. \tag{6.22}$$

Again the Moon moves on its circular path of radius r with velocity $v = 2\pi r/T$, when $T = 1$ sidereal month, and so experiences the acceleration (Fig. 6.1)

$$g_{☾} = v^2/r = 4\pi^2 r/T^2 = 0.272\ \text{cm/sec}^2 \tag{6.23}$$

using $r = 384\,400$ km $= 3.844 \times 10^{10}$ cm, $T = 27^{\text{d}}.32 = 27.32 \times 86\,400$ sec. The accelerations $g_{\oplus}$ and $g_{☾}$ are in fact inversely proportional to the squares of the radii of the Earth R and the lunar orbit r

$$\text{that is } g_{\oplus} : g_{☾} = \frac{1}{R^2} : \frac{1}{r^2} = 3620\,. \tag{6.24}$$

The value of the universal constant of gravitation G here appears only in combination with the hitherto unknown mass of the Earth M. Similarly, in other astronomical problems G occurs only along with the mass of the attracting celestial body. One cannot in principle determine a value for G from astronomical measurements; it has to be evaluated by terrestrial measurements.

First Maskelyne in 1774 made use of the deflection of a plumbline, that is the effect on the direction of a plumbline of the gravitational attraction of a mountain. Then in 1798 Henry Cavendish, after whom the Cavendish Laboratory in Cambridge is named, used a torsion balance, and in 1881 P. v. Jolly, the teacher of Max Planck, used a suitable form of lever-balance. The result of modern measurements is

$$G = (6.668 \pm 0.005) \times 10^{-8}\,\text{dyn cm}^2\,\text{g}^{-2}\,. \tag{6.25}$$

Since the acceleration due to gravity at the Earth's surface (for the time-being neglecting rotation and flattening) is related to the mass M and radius R by the formula already used, $g = GM/R^2$, we can now compute the *mass* M and *mean density* of the Earth $\bar{\varrho} = M/\frac{4}{3}\pi R^3$, and we obtain

$$M = 5.98 \times 10^{24}\,\text{kg} \qquad \bar{\varrho} = 5.52\,\text{g/cm}^3 \tag{6.26}$$

We shall in due course mention the geophysical significance of these values.

First we return to Newton's *Principia* and derive Kepler's laws from the laws of mechanics and of gravitation in order to gain a better understanding of Kepler's laws themselves and of the constants appearing in them.

Since clearly the mass of the Sun is so much greater than that of the planets, we first treat the Sun as at rest and measure the radius-vectors of the planets from the centre of the Sun. Also we first leave out of account the mutual attractions of the planets, their so-called perturbations.

The motion of a planet round the Sun then proceeds under the action of the *central force* $-G\mathfrak{M}m/r^2$, where $\mathfrak{M}$ is again the mass of the Sun and m ($\ll \mathfrak{M}$) that of the planet. Consequently, the law of angular momentum (6.17) applies, i. e. the angular momentum-vector

$$\boldsymbol{N} = \boldsymbol{r} \times m\boldsymbol{v} \tag{6.27}$$

where $\boldsymbol{v}$ is the velocity-vector of the planet, is constant in magnitude and direction. The vectors $\boldsymbol{r}$ and $\boldsymbol{v}$ remain therefore in the same plane perpendicular to $\boldsymbol{N}$, the *fixed plane* of the planet's motion. The magnitude $|\boldsymbol{r}\times\boldsymbol{v}| = r\,v\sin\alpha$ is twice the area traced out in unit time by the radius vector $\boldsymbol{r}$ of the planet (Fig. 6.3). The law of angular momentum is then identical with the assertion that each planet moves in a fixed plane with constant areal velocity (Kepler's second law and part of Kepler's first law).

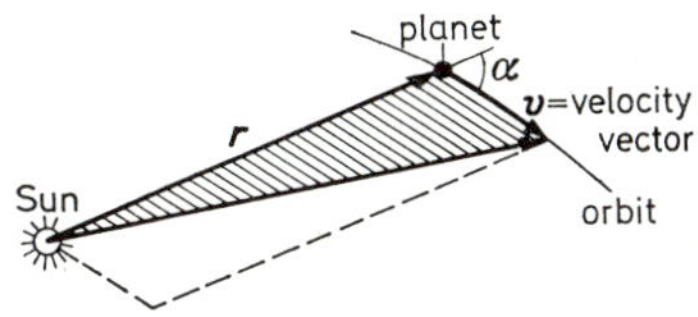

Fig. 6.3. *Rate of description of area by position-vector of a planet* ("areal speed") $\frac{1}{2}|\boldsymbol{r}\times\boldsymbol{v}| = \frac{1}{2}\,r\,v\sin\alpha$

Here we shall omit the somewhat formidable calculation which shows that the path of a particle (planet, comet, ...) under the action of a central force proportional to r^{-2} must be a *conic section*, i. e. circle (eccentricity $e=0$), ellipse ($0<e<1$), parabola ($e=1$) or hyperbola ($e>1$) with the Sun at a focus (Kepler's first law).

Instead, we readily write down the *energy-law* for the motions of the planets, etc. Consider a particle of mass m brought from infinity ($r\to\infty$) to rest at distance r from the Sun ($\mathfrak{M}$) under the action of the gravitational force $-G\,\mathfrak{M}\,m/r^2$. The work done is equal to its potential energy $V(r)$, where

$$V(r) = -G\int_r^\infty \frac{\mathfrak{M}\,m}{r^2}\,\mathrm{d}r = -G\frac{\mathfrak{M}\,m}{r}. \tag{6.28}$$

Since no work is done in any transverse motion, we see that the expression is independent of the choice of integration path. Referred to unit mass, we call $\varphi\,(r = -G\,\mathfrak{M}/r$ the *potential* at distance r from the Sun. This is one of the fundamental concepts of celestial mechanics as well as of theoretical physics.

The total energy of a planet $E=T+V$, that is

$$E = \tfrac{1}{2}\,m\,v^2 - G\,\mathfrak{M}\,m/r \tag{6.29}$$

or reckoned per unit mass

$$E/m = \tfrac{1}{2}\,v^2 - G\,\mathfrak{M}/r \tag{6.30}$$

thus remains constant in time. We again infer that the speed increases from aphelion to perihelion.

The complete calculation of planetary motion and the derivation of Kepler's third law we shall carry through only for circular orbits.

Nevertheless, with a view to subsequent generalization, we no longer require the planetary mass to be small compared with the solar mass. We therefore consider the motion of two masses in the first place about their mass-centre S, and in the second place the relative motion referred to, say, the more massive body. Let m_1, m_2 be the masses, a_1, a_2 their distances from the mass-centre S and $a = a_1 + a_2$ their distance from each other. From the definition of the mass-centre, we have then (Fig. 6.4)

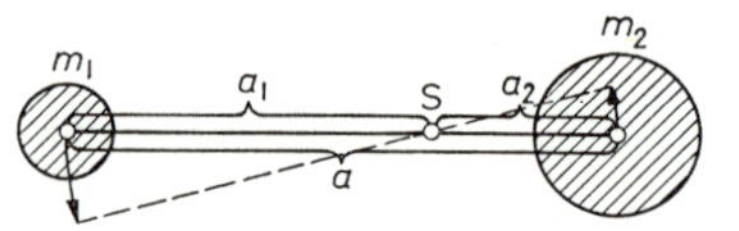

Fig. 6.4. *Motion of the masses m_1 and m_2 about their common-centre of gravity S, where $m_1 a_1 = m_2 a_2$*

$$a_1 : a_2 : a = m_2 : m_1 : m_1 + m_2 \tag{6.31}$$

$$\text{or} \qquad m_1 a_1 = m_2 a_2 = \frac{m_1 m_2}{m_1 + m_2} a \, .$$

For each mass the force of attraction $G m_1 m_2 / a^2$ must balance the centrifugal force. Denoting the period of the system by T, the latter force for the particle m_1 is

$$m_1 v_1^2 / a_1 = (2 \pi / T)^2 m_1 a_1 \tag{6.32}$$

and from the law of action and reaction, or from (6.31), the force for the particle m_2 has the same magnitude. Again using (6.31) and rearranging the factors, we obtain

$$a^3 / T^2 = G(m_1 + m_2) / 4 \pi^2 . \tag{6.33}$$

If we drop the assumption of circular orbits, the masses m_1, m_2 move round their mass-centre S in *similar conics*, also the relative orbit is another similar conic. The *semi-major axes*, which we also denote by a_1, a_2 and a, then replace the orbital radii, but we shall not give the proof here. Thus we obtain the generalized third law of Kepler (6.33). As we shall see, in the solar system the mass of even the greatest planet Jupiter is only about one-thousandth the solar mass. To this accuracy, we can replace the mass $m_1 + m_2$ by the solar mass $\mathfrak{M}$. Then by inserting numerical values for the motion of the Earth or another planet in (6.33) we derive the mass of the Sun. From the apparent semi-diameter 16′ in combination with the value of a we get the radius $R_\odot$. Then from $\frac{4}{3} \pi R^3 \bar{\varrho} = \mathfrak{M}$ we finally evaluate the mean density. The results are for the Sun

$$\begin{array}{lll} \text{Mass} & \mathfrak{M}_\odot & = 1.989 \times 10^{33}\,\text{g} \\ \text{Radius} & R_\odot & = 6.960 \times 10^{10}\,\text{cm} \\ \text{Density} & \bar{\varrho}_\odot & = 1.409\,\text{g cm}^{-3} . \end{array} \tag{6.34}$$

The uncertainty amounts to about $\pm$ one unit in the last place of decimals.

In corresponding manner, we compute the masses of the planets from the motions of their satellites (Table 7.1) again using Kepler's third law. If no satellite is available then, as is obviously much more difficult, we must have recourse to the mutual *perturbations* of the planets.

It is interesting to consider the Kepler problem further from the standpoint of the energy-law (6.29). For a circular orbit the centrifugal force is equal to the force of attraction of the two masses and so $m\,v^2/r = G\,\mathfrak{M}\,m/r^2$, or

$$\tfrac{1}{2}\,m\,v^2 = \tfrac{1}{2}G\,\mathfrak{M}\,m/r \qquad \text{(circular orbit).} \tag{6.35}$$

Thus the kinetic energy T is equal to one-half the potential energy with the sign reversed

$$T = -\tfrac{1}{2}V. \tag{6.36}$$

One can show that, averaged over time, this result holds good for any systems of point-masses held together by gravitational forces between them (obeying the inverse-square law). This is the *virial theorem* which is important, for example, in the theory of stellar systems.

We now consider the case of a *parabolic* orbit, for example that of a non-periodic comet. At infinity both the potential and kinetic energies are equal to zero. Thus $E = 0$ and from (6.29)

$$\tfrac{1}{2}m\,v^2 = G\,\mathfrak{M}\,m/r \quad \text{or} \quad T = -V. \tag{6.37}$$

Thus at the same distance from the Sun, the *parabolic speed* is $\sqrt{2}$ times the speed in a circular orbit. For example, the mean speed of the Earth is 30 km/sec, so the speed of a comet or meteor-stream that encounters us in parabolic orbit is $30\sqrt{2} = 42.4$ km/sec.

We also briefly consider the orbits of artificial satellites and space-vehicles, taking into account only the gravitational field of the Earth. A satellite having a circular orbit in the immediate vicinity of the Earth, disregarding the atmosphere, according to (6.35) must have speed $v_0 = 7.9$ km/sec. For larger orbital radius r, we have v proportional to $r^{-1/2}$. In order to send a non-self-propelled space vehicle to an infinite distance against the gravitational field of the Earth alone, we have to give it initial speed at least equal to the parabolic or *escape speed* $v_0\sqrt{2} = 11.2$ km/sec.

Besides the Kepler-problem, Newton solved many other problems of celestial mechanics.

We first discuss *precession,* at least in outline. The wandering of the Earth's axis round the pole of the ecliptic depends upon the same principle as the corresponding motion of a spinning-top under the effect of terrestrial gravitation. The equatorial bulge of the Earth is attracted towards the plane of the ecliptic by the Moon and the Sun, whose masses in the mean over such long time-intervals we can think of as being spread around their paths (and here we ignore the small inclination of the Moon's orbit) (Fig. 6.5). The turning moment

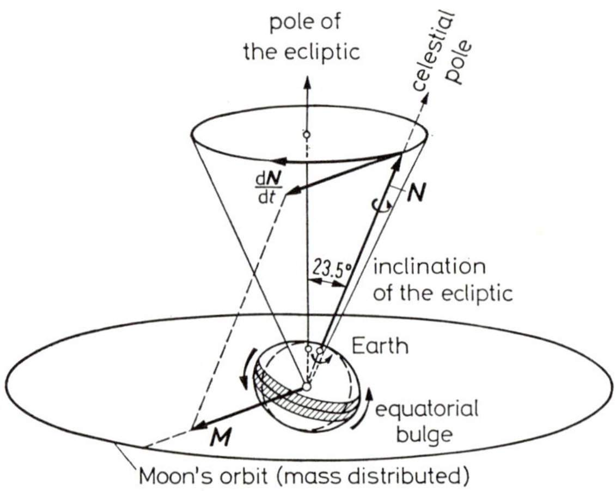

Fig. 6.5. *Lunar precession*

$\boldsymbol{M}$ so produced acts in accordance with equation (6.16) upon the Earth's angular momentum vector $\boldsymbol{N}$, which is practically along the axis of rotation. As we see at once from Fig. 6.5, the change of $\boldsymbol{N}$ corresponding to $\mathrm{d}\boldsymbol{N}/\mathrm{d}t = \boldsymbol{M}$ produces the familiar motion of $\boldsymbol{N}$, or of the Earth's axis, on a cone of fixed angle. Numerical calculation yields the correct period of luni-solar *precession* (Fig. 2.1).

Next we must briefly mention the old problem of the tides. These being known to Mediterranean peoples only by hearsay, in his time *Galileo* developed in controversy a wholly incorrect theory of the 12-hour alternation of ebb and flow, which contributed to bringing about his unfortunate trial by the Inquisition. It was again *Newton* who presented a statical theory of the tides (Fig. 6.6).

In the motion of the Earth and the Moon about their common mass-centre, the acceleration vector (which is the difference between gravitation due to the Moon and centrifugal acceleration due to this orbital motion) of a particle of water in the ocean acts *upwards* when the Moon crosses the meridian. So there the water is lifted and we have high water. In keeping with the apparent motion of the Moon (one "Moon-day" $= 24^h 51^m$) two high-water "hills" and two

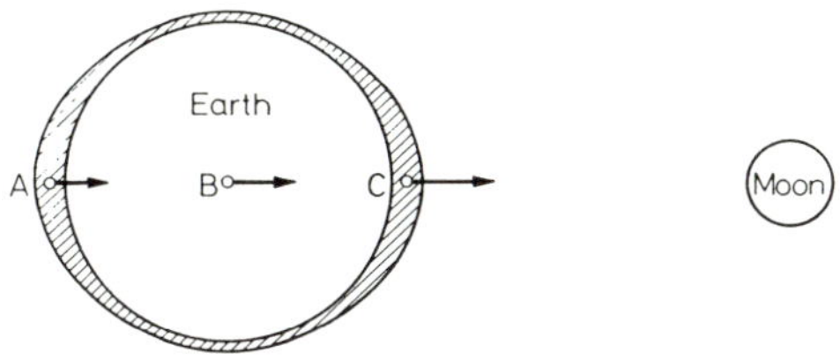

Fig. 6.6. *Static theory of the tides.* The acceleration towards the Moon of the three points shown is, corresponding to their various distances from it, $b - \Delta$ for point A (lower culmination of the Moon), b for point B (centre of gravity of the Earth) and $b+\Delta$ for point C (upper culmination of the Moon). The (rigid) Earth as a whole has the acceleration b. Consequently at A and C there is an excess acceleration Δ, which produces high water at both points

low-water "valleys" continually travel round the Earth, getting later each day by 51 minutes. The tidal force of the Sun is about half that of the Moon. At new moon and full moon the tidal forces of the Moon and the Sun act together and produce "spring" tides; at first and last quarters, they oppose each other and we have "neap" tides. Actually this statical theory explains only the crude features of the phenomena. The *dynamical* theory of tides investigates the way in which forced oscillations, corresponding to the different periods of the apparent motions of the Moon and the Sun, of the various oceanic basins are excited. Following *G. H. Darwin*, tidal prediction consists essentially in Fourier analysis and synthesis appropriate to the stated astronomical periods. As already mentioned in Chapter 3, *tidal friction* in narrow seas acts as a brake on the Earth's rotation and produces a *lengthening* of the day. In accordance with the conservation of angular momentum (6.17) the angular momentum that the Earth loses in this way must reappear as additional orbital angular momentum of the Moon. From Kepler's third law the angular momentum per unit mass is proportional to the orbital radius, and so the Moon must slowly recede from the Earth. The conjecture based upon this, that the Moon came out of the Earth (perhaps out of the Pacific Ocean), is shown by accurate calculation to be an unjustified extrapolation (see page 65).

We must not leave the theory of planetary motion without clarifying the decisive change from the Ptolemaic to the Copernican world-systems by looking back from a modern view-point (Fig. 6.7).

With the Sun as origin (heliocentric) let $\boldsymbol{r}_P$, $\boldsymbol{r}_E$ be the position vectors of a planet and of the Earth. Then the position vector of the

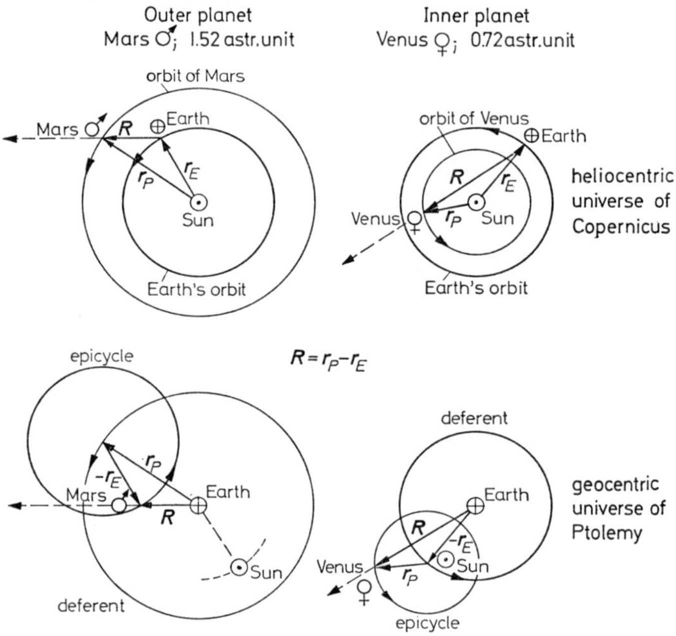

Fig. 6.7. *Motion of an outer planet (Mars) and an inner planet (Venus)* on the celestial sphere, represented heliocentrically and geocentrically. The broken arrow ←−−− shows the position of the planet in the sky. The monthly motions of the Earth and the planet are shown by arrow points on the circles. We have for

	Outer planets	Inner planets
deferent	$\boldsymbol{r}_P$	$-\boldsymbol{r}_E$
epicycle	$-\boldsymbol{r}_E$	$\boldsymbol{r}_P$

planet referred to the Earth (geocentric) is specified by the vector difference

$$\boldsymbol{R} = \boldsymbol{r}_P - \boldsymbol{r}_E . \tag{6.38}$$

Let us look back from here for a moment to the geocentric view of Ptolemy: (a) In the case of the outer planets, for example Mars ♂ we begin by drawing the vector $\boldsymbol{r}_P$ out from the Earth and letting it rotate in the same way as it originally rotated about the Sun. To $\boldsymbol{r}_P$ in accordance with (6.38) we add the vector $-\boldsymbol{r}_E$, which is the position vector of the Sun seen from the Earth, and so we obtain the position vector $\boldsymbol{R}$ of the planet as seen from the Earth. The immaterial circle that $\boldsymbol{r}_P$ discribes round the Earth with the sidereal period of the planet is the Ptolemaic *deferent*. The other circle that the planet at the end-point of the vector $-\boldsymbol{r}_E$ describes around the point $\boldsymbol{r}_P$ with the sideral period of the Earth is the Ptolemaic *epicycle*. (b) In the case of the inner planets it seemed better to Ptolemy first to let the greater vector revolve around the Earth as deferent with period equal to a sideral year, and then to let the smaller vector $\boldsymbol{r}_P$ *describe* about the end-point $-\boldsymbol{r}_E$ of the other the Ptolemaic epicycle with period equal to the sidereal period of the planet.

Thus far the geocentric construction expresses exactly the relation $\boldsymbol{R} = \boldsymbol{r}_P - \boldsymbol{r}_E$. The representation (b), applied to *all* the planets, expresses the world-system of Tycho Brahe.

Actually, however, we have not completed the transformation to the Ptolemaic system. So long as mankind could measure only the position of the planets in the sky i. e. only their directions and not their distances, they were concerned only with the direction of the vector $\boldsymbol{R}$ and not its magnitude. Thus one could change the scale of $\boldsymbol{R}$ for each planet. That is, the vectors

$$\boldsymbol{R}_P' = A_P \, \boldsymbol{R}_P \tag{6.39}$$

with a fixed but arbitrary numerical factor A_P for each planet would give, in accordance with the Ptolemaic system, a completely satisfactory description of the motions of the planets in the sky.

We now see clearly what is lost when we retreat step by step from the Copernican to the Ptolemaic system:

1. The change of reference system entails the renunciation of any simple mechanical explanation.

2. While the scale-factors A_P leave the position of the planets in the sky unchanged, we thereby lose the mutual relationships of the planets in space.

3. The fact that in the Ptolemaic system the year-period, corresponding to the motion of the vector $\boldsymbol{r}_E$, has to be introduced *independently* for each planet shows the artificiality of the older system.

It is important to make it clear, however, that the purely kinematic consideration of the motion of the planets in the sky did not permit a discrimination between the old and the new system. Essential progress came with Galileo's first observations with his telescope (1609): a) One could look upon Jupiter with its freely revolving moons as a "model" of the Copernican planetary system. b) The phases of Venus determined the relative positions of the Sun, Earth and Venus. The smallness of the phase angle of Jupiter, say, gave at any rate qualitative support for Copernicus. The *concept* of a celestial *mechanics* presupposed — and this should not be forgotten — the basic uniformity of cosmical and terrestrial matter and of its physics.

7. Physical Constitution of Planets and Satellites

In recent decades the study of planets and satellites has developed into one of the most difficult chapters in astrophysics. We can leave the discussion of the necessary *instruments* to Chapter 9. At the present time, the interpretation of the observations demands all the resources of *physical chemistry* (chemical equilibrium, phase-diagrams, and so on). Also the formation and evolution of our planetary system (cosmogony) can be usefully discussed only in the context of the evolution of the stars and of the Galaxy.

Here we must impose severe restrictions on ourselves and ask in the first place: What possibilities have we got at the present time for the study of the planets?

a) We need not go further into the measurement of the apparent and true *diameters* of the planets, their *masses* and the derived *mean densities*. Numerical values of these and other parameters are collected in Table 7.1. Fig. 7.1 gives a graphic representation of the true dimensions of the planets and the Sun.

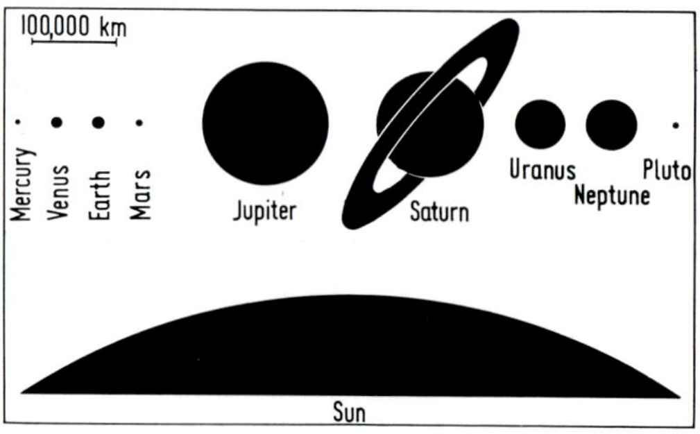

Fig. 7.1. *Relative sizes of the planets and Sun drawn to scale*

b) The rotation-period can be got from the telescopic observation of any sufficiently permanent surface-features. Another possibility is provided by the doppler effect in the Fraunhofer lines in reflected sunlight or of the absorption lines of the planetary atmosphere itself.

c) The reflective power of a planet, etc. is described by the value of the *albedo,* defined as the ratio of the light reflected or scattered in all directions to the incident light. So far as it is measurable, the brightness as a function of phase-angle, or the surface brightness as a function of the directions in which light comes on and off, as well as measurements of the polarization of the light, provide rather extensive information.

d) In the *spectrum* of a planet, the presence of at any rate many gases can be verified from their absorption bands. It is hoped that the many telluric bands of H_2O, CO_2, O_3, ... that hamper observations made through the Earth's atmosphere may be avoided by observations from artificial satellites. Blurred absorption bands in the infra-red permit us to check, for instance, the presence of solid ice (H_2O) in the polar caps of Mars.

e) Comparison of the intensity of the planet's own infra-red thermal radiation with the reflected solar radiation in the visible region (Pettit and Nicholson) permits the measurement of the *surface temperature.* Difficulties in interpreting such measurements can arise because the atmospheric gases, which may have a different temperature from the ground, make a significant contribution to the infra-red radiation. Another possibility for measuring the temperature of the Moon and the planets is offered by the absolute intensity of thermal *radio* radiation in the centimetre and decimetre wavelength region.

The results for the individual planets of the means of observation mentioned in a)—e) are collected in Table 7.1; we have included the numbers of presently known satellites. We now supplement this quantitative summary by a short description of individual planets and their satellite-systems:

Mercury ☿ is very difficult to observe since it is never more than $\pm 28°$ from the Sun. Recent radar measurements combined with older visual observations show that, contrary to what was formerly believed, Mercury's period of rotation is not the same as its period of revolution (88 days) but is 58.4 ± 0.4 days.

Venus ♀ possesses a dense atmosphere; light-scattering produced by it is evident at the edge of the planet's shadow. Infrared absorption bands show that the atmosphere consists to a great extent of carbon dioxide CO_2. Thanks to the cloudbelt (whose chemical structure is not known) a solid surface is unlikely to be detected. Recent radar measurements give the wholly unexpected result that Venus has

Table 7.1. *Physical features of planets*

Planet	Radius $R_{\oplus}=1$	Mass $M_{\oplus}=1$	Density: average g cm^{-3}	Rotation period day (d), hour (h), minute (m)	Satellites: number	Temperature:** average surface T °K	Atmosphere: constituents detected
Terrestrial planets							
Mercury	0.38	0.054	5.46	$58^d\ 10^h$	0	442	—
Venus	0.961	0.814	5.06	247^d	0	229	CO_2, H_2O
Earth	1.000	1.000	5.52	$23^h\ 56^m$	1	246	N_2, O_2, CO_2
Mars	0.523	0.107	4.12	$24^h\ 37^m$	2	216	CO_2, H_2O
Ceres *	0.055	~0.0001	3.3 ?	$9^h\ 05^m$	0	160	—
Major planets							
Jupiter	10.97	317.45	1.33	$9^h\ 50^m$	12	93	CH_4, NH_3
Saturn	9.03	95.06	0.71	$10^h\ 15^m$	10	68	CH_4, NH_3
Uranus	3.72	14.50	1.55	$10^h.8$	5	47	CH_4, H_2
Neptune	3.43	17.60	2.41	$15^h.8$	2	38	CH_4, H_2
Pluto	0.45	0.05 ?	3 ?	$6^d.4$	0	43	

* Largest asteroid: total mass all asteroids $\sim 3 \cdot 10^{-4} M_{\oplus}$

** Calculated, taking account of albedo.

retrograde rotation with period 247 ± 5 days. Venus and Mercury possess no satellites; their masses are calculated from perturbations.

Mars ♂ shows a vivid red colour, such as might be due to iron compounds. We see at the poles white polar caps that retreat in the Martian summer and advance in the Martian winter. A long-wavelength absorption band shows that they consist of hoar frost (solid H_2O). In the rarefied atmosphere there is a certain amount of H_2O and CO_2; Variations betoken sand-storms. Pictures televised by the space-probe Mariner IV of 1965 July 15 from distances of 17 000 to 20 000 km from the centre of Mars reveal numerous *craters* with

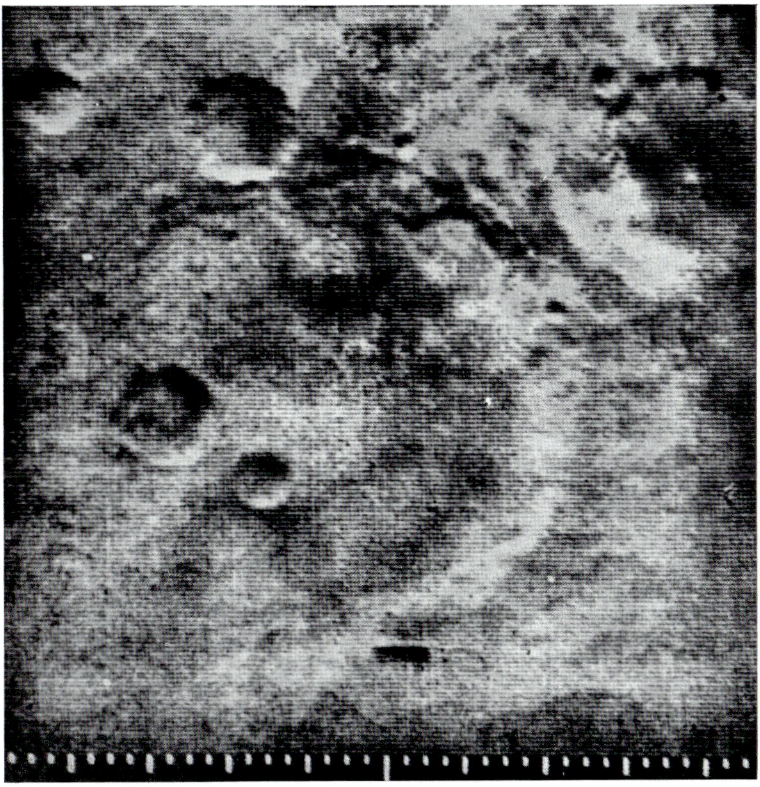

Fig. 7.2. *Surface of Mars,* photographed from the space probe Mariner IV through a green filter from a distance of 12 000 km. The region covered by the picture measures approximately 240 × 272 km

diameters ranging from the limit of 3 to 4 km of instrumental resolution up to about 120 km. These craters are in many ways similar to those on our Moon. Measurements by Mariner IV showed also that

Mars possesses no permanent magnetic field as much as 10^{-4} to 10^{-3} times the Earth's field. The observations of the so-called *canals* on Mars depend upon a physiological-optical contrast-phenomenon. Fig. 7.2 shows part of the surface photographed from Mariner IV.

Asteroids or minor planets. Kepler had conjectured the existence of a heavenly body in the "gap" between Mars and Jupiter (Fig. 5.5). On 1801 January 1 Piazzi in Palermo discovered the first known asteroid Ceres, but by mid-February it was lost to sight in the vicinity of the Sun. By October of the same year the 24-year old K. F. Gauss had computed its orbit and ephemerides, so that Zach was able to recover it. As a sequel to this brilliant performance, the "princeps mathematicorum" in his *Theoria Motus* 1809 solved the general problem of orbit-determination, that is the evaluation of all the orbital elements of a planet or comet from three complete observations. To-day over 2000 asteroids are known, most of them between Mars and Jupiter. Objects with known orbits are given a number and a name. 1 Ceres, the largest and brightest asteroid, has a diameter about 700 km; the smallest are less than 1 km in diameter. 433 Eros, whose orbit has the exceptional eccentricity $e = 0.233$, at its opposition of 1931 approached to within 0.17 AU of the Earth and made possible a favourable measurement of the solar parallax. The periodic luminosity changes of many asteroids (for instance, Eros with a $5^{h}\,16^{m}$ period) show that they have irregular shapes.

Jupiter ♃ (Fig. 7.3) the largest and most massive of the planets (about $^{1}/_{1000}$ solar mass) has a dense atmosphere with pronounced stripes parallel to the equator, similar to the circulation-system of the Earth. The so-called red spot has been seen for many years. *Galileo* discovered the four brightest satellites I—IV; with ever larger telescopes, up to the present twelve have been discovered.

Saturn ♄ (Fig. 7.4) resembles Jupiter in many ways. In 1659 *Christian Huygens* discovered Saturn's rings (of which *Galileo* had observed some indications) and Titan, the brightest of Saturn's satellites; today we know altogether 10 satellites. Keeler's measurement (1895) of the speed of rotation of Saturn's rings from the doppler effect in reflected sunlight showed that the various zones of the ring-system revolve according to Kepler's third law and thus that they are composed of small particles. As Roche has shown, a larger satellite in the position of the rings would be disrupted by tidal action. The infra-red spectrum shows that the rings consist at least partly of ice. The gaps in the rings are produced by *resonance* because the orbital periods of particles in the gaps would be rational fractions of the periods of the inner satellites. The "gaps" in the asteroidal belt arise in analogous manner by resonance with the revolution of Jupiter. As shown by numerous absorption bands in the spectra, the atmo-

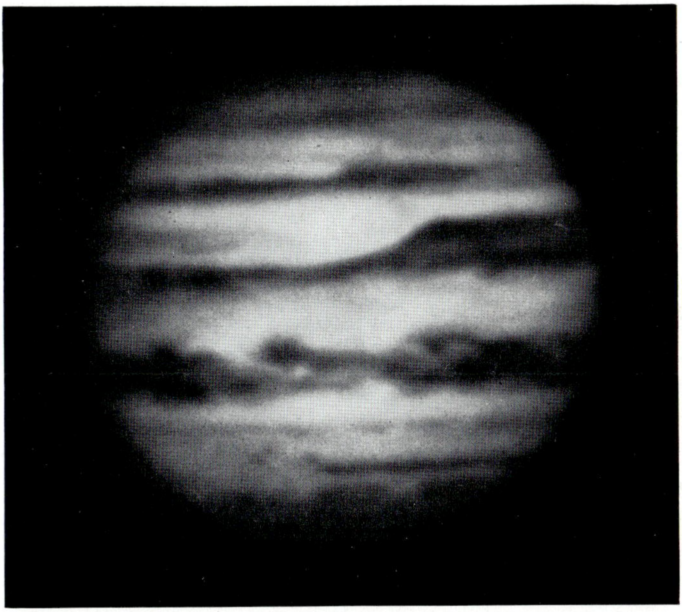

Fig. 7.3. *Jupiter*. Equatorial radius 71 350 km and mass 1/1047 solar mass. Pic du Midi, 60 cm refractor (photograph by B. Lyot and H. Camichel)

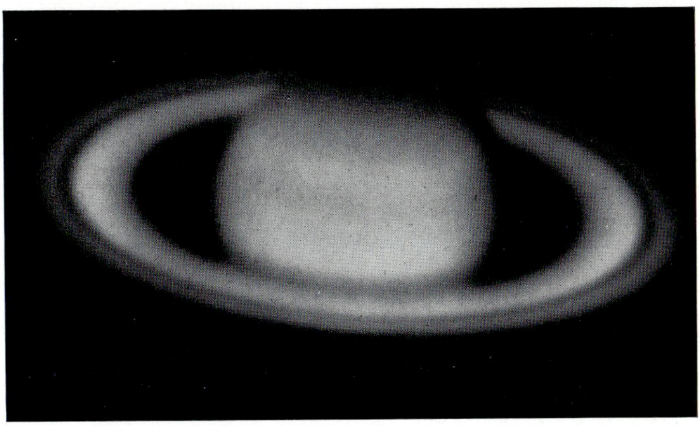

Fig. 7.4. *Saturn*. Equatorial radius 60 400 km and mass 1/3498 solar mass. Pic du Midi, 60 cm refractor (photograph by H. Camichel)

spheres of the major planets Jupiter and Saturn, and also those of Uranus and Neptune, consist essentially of hydrogen H_2, methane CH_4 and ammonium NH_3, that is *hydrides*, in contrast to the oxides (CO_2, H_2O, ...) of the inner planets. Saturn's satellite Titan surprisingly possesses its own atmosphere in which CH_4 is detected.

Uranus ♅ was fortuitously discovered by W. Herschel in 1781. We know five satellites. As with Neptune, particular features in the disk are scarcely recognizable.

Neptune ♆. From perturbations of the orbit of Uranus, Adams and Leverrier inferred the existence of a planet with a longer period. Galle 1846 found Neptune near to the computed position. We know two satellites.

Pluto ♇. Perturbations of the orbits of Uranus and Neptune led to the inference of a trans-neptunian planet. C. Tombaugh at the Lowell Observatory discovered Pluto in 1930. This small body permits the recognition of no atmosphere and no surface structure. Its orbital elements, mass, etc. fall so markedly out of sequence with those of the other outer planets that it is thought possibly to be a "captured" body.

We shall amplify these particulars with two theoretical comments:
a) The absolute temperature T at the surface of a planet — supposing it to have no significant energy source of its own — is such that the radiation it gives out in the infra-red (according to the Stefan-Boltzmann law of radiation this is proportional to the surface area times T^4) is equal to the radiation it takes in from the Sun (at distance 1 AU this is given by the *solar constant* (Chapter 12)). The fact that the albedo is less than unity is easily allowed for. Without further discussion, however, the "greenhouse" effect of the atmosphere is not so easily taken into account. In the case of fast rotation, the temperature tends to be the same over the dark and light hemispheres; in the case of slow rotation the balance between incoming and outgoing radiation applies to the sunlit hemisphere by itself. These two cases provide upper and lower bounds to the surface temperature of the planet; the measured values in general lie between them (Table 7.1).
b) We ask further, to what extent can a planet, satellite, etc., retain an atmosphere? According to the kinetic theory of gases, the molecules of a gas of molecular weight μ and absolute temperature T have (most probable) speed $\bar{v} = (2\,\mathcal{R}\,T/\mu)^{1/2}$, where $\mathcal{R} = 8.314 \times 10^7$ erg deg^{-1} mol^{-1} denotes the gas constant. From (6.36) a molecule with speed v can escape from a body of mass M, radius r, if $\frac{1}{2}\,v^2 \geqq G\,M/r$. Considering

the Maxwell-Boltzmann distribution of molecular speeds for $v > \bar{v}$, we infer that Mercury, the Moon, and most of the other satellites can have practically *no* atmosphere, but that Titan, the largest satellite of Saturn, with a temperature calculated in accordance with (a), can actually retain a CH_4-atmosphere for a long time.

To attempt here a comprehensive view, we take note: The asteroids obviously mark the separation between the terrestrial planets Mercury, Venus, Earth, Mars and the quite differently constituted major planets Jupiter, Saturn, Uranus, Neptune. (Pluto as a "foundling" must remain out of consideration here. The asteroids with a total mass about 1/3500 Earth mass may have come from the fragmentation of one or more larger bodies.)

The first set have masses less than or about equal to the Earth-mass and density from 3.8 to 5.52 g/cm³. So they are obviously composed mainly of solid material. In chemical language, their atmospheres are *oxidizing;* they contain O_2, CO_2, H_2O, N_2, . . . On the other hand, the outer planets have masses between 14.6 and 318 Earth-masses and mean densities from 0.7 to 2.2 g/cm³. The fact that they have *reducing* atmospheres with H_2, CH_4, NH_3 . . . makes us think that they may be composed mainly of hydrogen and hydrides of the lighter elements. We shall see that they thus resemble the original solar material much more than do the inner planets which have retained almost only compounds of heavy elements.

As a sample of the inner planets we consider our Earth somewhat more in detail; obviously it is not our intention to give a conspectus of geophysics.

In consequence of its rotation the Earth is, to a good approximation, an oblate spheroid with

$$\begin{array}{ll} \text{equatorial radius} & a = 6378.2 \text{ km} \\ \text{polar radius} & b = 6356.9 \text{ km} \end{array}$$

and the flattening is

$$\frac{a-b}{a} = \frac{1}{298}.$$

The flattening and the centrifugal force result in gravity at the equator being $^1/_{190}$ weaker than at the poles.

We already know the mean density of the Earth $\bar{\varrho} = 5.52$ g/cm³. The density of the Earth's crust (granite, basalt) is 2.6 to 3 g/cm³. The moment of inertia inferred from the motion of the spinning Earth gives at any rate a summary of information about the increase of density with depth. Further knowledge comes from the propagation of earthquake-waves. In an earthquake, longitudinal and transverse elastic waves are set up in the comparatively superficial epicentre. These then propagate themselves through the Earth's interior and are there refracted, reflected and transformed into one another in accordance with the depth dependence of the elastic constants. By an exact study of seismic waves, about 1906 E. Wiechert in Göttingen discovered that there are in the Earth's interior several surfaces of discontinuity where the elastic constants and the density change abruptly. The Earth's crust ($\varrho = 2.6$ to $3.0\,\mathrm{g/cm^3}$) has a thickness of about 30 km beneath plain-lands. Under the continents, the thickness increases up to about 60 km and under the oceans it decreases to a few kilometres. Its lower boundary forms the Mohorovičič-discontinuity (which it has been proposed to reach by boring a "Mohole" in the bed of the Pacific Ocean). Then, down to a depth of 2900 km there follows the *mantle* having $\varrho = 3.3$ to $5.7\,\mathrm{g/cm^3}$ which may consist mainly of silicates. In the deep interior, from 2900 km to 6370 km we have the *core* with $\varrho = 9.4$ to about $17\,\mathrm{g/cm^3}$. No transverse waves are propagated in the core. In this sense we may think of the outer core as *fluid*, but of excessively great viscosity. However, more extensive investigations suggest that the so-called *inner-core* below a depth of 5000 km is again *solid*.

We can only indirectly infer the chemico-mineralogical composition of the Earth's core. Laboratory experiments at high pressures and temperatures, along with the geophysical data, yield an acceptable phase-diagram if we assume that the core, like iron meteorites, consists of 90 per cent *iron* and 10 per cent *nickel*.

We can fairly accurately calculate the inward increase in *pressure* from the hydrostatic equation; for the centre* we find $p \approx 3.5 \times 10^6$ atmospheres $= 3.5 \times 10^{12}$ dynes/cm². By comparison, Bridgman could produce about 4×10^5 atmospheres in the laboratory.

The increase of *temperature* with depth can be measured in deep bore-holes and a geothermal gradient of about 30°/km is obtained. The temperature distribution at greater depths is determined on the other hand by the generation of heat by the radioactive substances U^{238}, Th^{232} and, to a lesser extent K^{40}, and on the other hand by the

* Elementary estimate: With mean gravity $\frac{1}{2}g = 5 \times 10^2\,\mathrm{cm/sec^2}$ and mean density $\varrho = 5.5\,\mathrm{g/cm^3}$ a column of length $R = 6.37 \times 10^8$ cm (Earth's radius) exerts a pressure $p = \frac{1}{2}\,g\,\varrho\,R \approx 1.8 \times 10^{12}\,\mathrm{dyn/cm^2}$ which agrees with the exact calculation to within a factor 2.

slow outward heat-transport by conduction and convection in the magma. Thus the temperature in the core must be at least some thousands of degrees, but almost certainly not as much as 10 000° K.

The Earth's magnetic field, the *geomagnetic field* together with its *secular variations (rapid variations* caused by the Sun will be considered later) may be accounted for in the following way, according to W. M. Elsasser and Sir Edward Bullard:— In connexion with the heat transport by convection mentioned above, the *magma,* that is fluid stony material, forms great eddies. If in such eddies in conducting material some traces of a magnetic field are present, then these can be intensified as in the self-exciting dynamo of W. v. Siemens. The details of such a self-exciting dynamo in the Earth's interior have not yet been tidied up, but on the whole this theory offers the best chance of being correct.

Thanks to the circumstance that when certain minerals were being formed the magnetic field present at the time was, so to say, frozen into them (P. M. S. Blackett and others), it is possible to reconstruct the geomagnetic field of past geological epochs. Such paleomagnetic measurements have shown that the field-vectors of the past can best be put in order by resorting to A. Wegener's hypothesis of *continental drift* as inferred from the map of the world. The continents, that is pieces of the crust, were fractured as a result of disturbances in the mantle (and core?) and were pressed together in places where there are the great mountain-folds. We cannot here go into the arguments for this view from animal- and plant-geography and from paleoclimatology.

The best information concerning the duration of the various geological ages and the age of the Earth, defined as the time since the formation of the Earth's crust, is to-day given by the methods of *radioactive age-determination.* One employs the *radioactive decay* of the following isotopes, where we give only the stable end-products of each series:

	Half-life $T_{1/2}$
$U^{238} \rightarrow Pb^{206} + 8\,He^4$	4.49×10^9 years
$U^{235} \rightarrow Pb^{207} + 7\,He^4$	0.713×10^9 years
$Th^{232} \rightarrow Pb^{208} + 6\,He^4$	13.9×10^9 years
$Rb^{87} \rightarrow Sr^{87} + \beta^-$	61×10^9 years
$K^{40} \nearrow A^{40} + K(\gamma)$, $K^{40} \searrow Ca^{40} + \beta^-$	1.3×10^9 years

In all applications we somehow determine the ratio of the end-product to the initial isotope. If this ratio was zero to begin with ($t=0$) then after time t it must be equal to $2^{t/T}-1$.

Table 7.2 shows in brief the most important geological strata and their absolute dating. (The author is indebted for this to Professor K. Krömmelbein, Kiel.)

The age of the Earth since the separation of its materials or the formation of the crust (through a mechanism by no means clearly understood!) can be fairly exactly set at

$$4.5 \pm 0.3 \times 10^9 \text{ years} \tag{7.1}$$

As we saw, the planets Jupiter, Saturn, Uranus and Neptune are distinguished from the terrestrial planets by their low mean density and the composition of their atmospheres out of H_2, CH_4, NH_3 . . . The value of the moment of inertia as known from celestial mechanics, along with the known values of the mass and radius, shows that the density of any of these planets increases inwards very considerably.

All this shows that the major planets consist to a great extent of *hydrogen*. Partly experimentally and partly theoretically, the equation of state is to some degree known. Above about 2 million atmospheres the electrons are largely free and it forms a metallic phase. As regards the temperature distribution in the interior, we have to rely upon estimates. Exact calculations, particularly those of R. Wildt and de Marcus, have shown that pure hydrogen planets of the right mass would be too small. Satisfactory agreement with observation can be got by assuming an admixture of helium (and some heavier elements) in the ratio H:He$\geqq$14 for Jupiter, $\geqq$11 for Saturn, by numbers of atoms. As we shall see, this would mean that the major planets are made of essentially the same mixture of elements as the Sun. For the case of Jupiter an order-of-magnitude representation of the density as a function of distance from the centre $r/r_{♃}$ is the following:

$r/r_{♃}$	1.0	0.9	0.8	0.7	0.6	0.5	0.4	0.3	0.2	0.1	0
$\mathrm{g\,cm^{-3}}$	0.0002	0.48	1.1	1.6	2.1	2.7	3.1	3.6	4.1	19	31

The structure of planetary atmospheres is determined simply by hydrostatic equilibrium

$$\mathrm{d}p = -g\,\varrho\,\mathrm{d}h\,. \tag{7.2}$$

That is to say, the weight of an element of a layer of thickness $\mathrm{d}h$ is supported against gravity g by the pressure difference between its faces. Further, the pressure p is related to the density ϱ and the mean molecular weight μ by the equation of state of an ideal gas

$$p = \varrho \mathcal{R} T/\mu . \tag{7.3}$$

where $\mathcal{R} = 8.314 \times 10^7$ erg/degree is the universal gas-constant. Substitution in (7.2) gives

$$\frac{dp}{p} = -\frac{g\,\mu}{\mathcal{R}T}\,dh \quad \text{or} \quad \frac{dp}{p} = -\frac{dh}{H} \text{ where } H = \frac{\mathcal{R}T}{g\,\mu} \tag{7.4}$$

If the equivalent height or so-called scale-height H is constant, we can integrate and obtain for a considerable range of height the barometric formula

$$\ln p - \ln p_0 = -\frac{h}{H} \quad \text{or} \quad p = p_0\, e^{-h/H}, \tag{7.5}$$

when p_0 is the pressure on the "ground" $h = 0$.

The structure of a planetary atmosphere depends on (a) the acceleration due to gravity on the planet g, (b) the mean molecular weight μ, that is, on the chemical composition and state of dissociation and ionization of the gases, and (c) the temperature distribution $T(h)$. This last is in turn determined by the mechanism of heat transport, i. e. the input and output of heat-energy in each layer h to $h + dh$. For instance, in the Earth's atmosphere, in the lowest layer, the *troposphere,* the absorbed solar heat is carried away by convection, and this leads to a uniform decrease of temperature upwards. Above the so-called *tropopause* at a height of about 10 km, radiation takes over the transport of energy, and we have next the almost isothermal *stratosphere.* Here we find a different mechanism of absorption of solar radiation and also the long wavelength re-emission into space. Related to the formation of ozone O_3, we encounter a warm layer at height about 25 km. After a short temperature-decrease, above about 90 km the temperature rises again to more than 1000 °K as a result of the dissociation and ionization of the atmospheric gases N_2 and O_2. The electrically conducting layers of the *ionosphere* arise from the ionization (the maximum electron density of the E-layer occurring at about 115 km and of the F-layer at 300—400 km), and make possible the propagation of not too short radio waves around the globe.

The conditions in the atmospheres of the planets need be in no way any simpler. The study of their ionospheres, analogous to that of the Earth, by the methods of radio- and radar-astronomy and with the aid of space vehicles is only now beginning.

Table 7.2. *History of the Earth*

Era	Times of principal mountain-building	Epochs
CENOZOIC	Alpine	Snails-mussels; Mammals
MESOZOIC		Goniatites Ammonites
PALAEOZOIC	Hercynian (Variscan) Caledonian Assyntic	Goniatites Ammonites; Brachiopods; Armoured fish; Graptolites; Trilobites
PROTEROZOIC	Algomian	
AZOIC	Laurentian	

Oldest rocks known (according to absolute age-determinations) 3300—3400 millions years.
Origin of Earth's crust about 4500 million years ago.

Fauna	Period (age in millions of years from the present)	Flora
Earliest man	Quaternary 1	NEOPHYTIC Age of angiosperms
→	Tertiary 70 ± 2	
→ Extinction of ammonites and dinosaurs	Cretaceous 135 ± 5	
→ Earliest birds	Jurassic 180 ± 5	MESOPHYTIC Age of conifers
→ Earliest mammals	Triassic 225 ± 5	
→ Extinction of many groups of palaeozoic animals	Permian 270 ± 5	PALAEOPHYTIC Age of ferns
→ Earliest reptiles	Carboniferous 350 ± 10	
→ Earliest amphibians	Devonian 400 ± 10	
	Silurian 440 ± 10	EOPHYTIC Age of algae
Earliest vertebrates	Ordovician 500 ± 15	
First appearance of many groups of invertebrates →	Cambrian 600 ± 20	
ALGONKIAN ARCHEAN	} PRE-CAMBRIAN	

Let us now take a look at the *satellites* of the planets. The satellites close to any planet all have nearly circular orbits with small inclination to the planetary equators. Only the outer, highly perturbed and

Fig. 7.5. *Moon.* The crater Posidonius (lower right) with a diameter of 100 km and the smaller crater Chacornac (lower left) on the edge of Mare Serenitatis (above); in Mare Serenitatis a 180 m higher well-defined mountain range; numerous small craters scattered everywhere. Lick Observatory, 1962 March 25, 120-in reflector

maybe captured satellites show high inclinations and large eccentricities. The absolute values of the masses of the larger satellites lie in the range 50 to 150×10^{24} g. In regard to the *mass-ratio* of satellite:

planet, our Moon with 1:81.30 is altogether exceptional; for example, the four Galilean moons of Jupiter all have mass-ratios less than $1:10^4$.

With mean density $\bar{\varrho} = 3.34\ \mathrm{g\ cm^{-3}}$ our Moon is evidently composed of material like that of the outer parts of the Earth. Selenographic studies with large telescopes (Fig. 7.5) and with space-probes (Figs. 7.6 and 7.7) have familiarized us with the formations on the Moon's surface. Today there can be little doubt that the circular forms of the great *maria* (naturally the reference to "seas" is of only historic relevance) and the craters down to pits less than 1 metre across have all been caused by the impact of meteorite-like bodies. The heights of the lunar mountains vary with the diameters of the craters in the same way as in the case of corresponding features of terrestrial explosion-craters. The greatest heights are limited by the solidity of the material, and so are of the same order of magnitude as on the Earth, as Galileo long since inferred from the variation in the illumination at the edge of the dark part, i. e. the terminator. Also the long-believed thesis that the Moon had been pulled out of the Earth, perhaps out of the Pacific Ocean, has been shown to be mechanically untenable, while it is possible that the whole Moon has been produced by the falling-together of small solid pieces. However, in many places, e. g. in the floors of the maria, there are signs of secondary fusion. In particular, the question as to which lunar features are due to impacts and which to vulcanism is still frequently debated. As is shown by the slow equalization of temperature between the surface layers and the deeper layers in the Earth's shadow during a lunar eclipse, and as pictures taken by Ranger VII (Fig. 7.6) confirm,

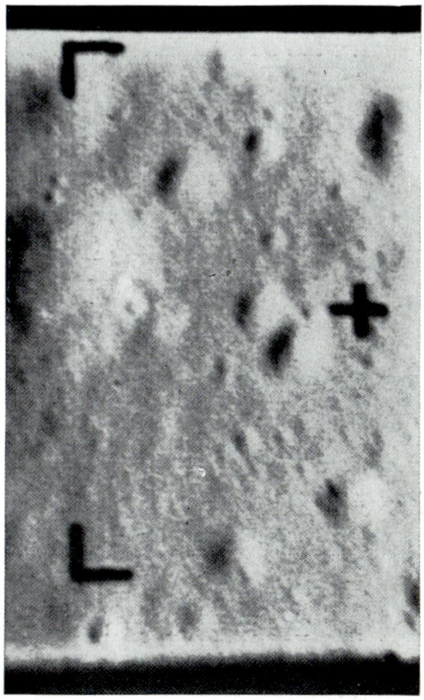

Fig. 7.6. *Last photograph taken by the American space probe Ranger 7 before striking the lunar surface, 1964, July 31.* The picture was taken from a height of approximately 300 metres with an $f/1$ television camera and shows a region measuring approximately 50×30 metres. The smallest craters measure barely a meter across

the Moon's surface is covered by porous rocks of very low thermal conductivity. The light streamers radiating, for example, from the crater Copernicus evidently consist of material thrown out in the

Fig. 7.7. *Photograph of the lunar surface on the eastern edge of Oceanus Procellarum,* transmitted from the first "soft-landing" Soviet space probe Luna 9, on 1966 February 3. In the foreground objects as small as 1—2 mm can be discerned

impact that produced the crater. In this context we must remember that gravity on the Moon is only one-sixth that on the Earth.

For the satellites of other planets one can in some cases determine at least the mass, from perturbations, and the diameter, using a telescope or an interferometer (see below). The resulting mean densities are mostly of the same order as for the Moon and terrestrial rocks, about 2 to 4 g cm^{-3}. Some satellites show periodic fluctuations in brightness, which indicate an irregular shape; this, as well as the CH_4 atmosphere of the largest satellite of Saturn, Titan, speaks for a "cold" origin. We shall return to this in connexion with the cosmogony of the solar system.

8. Comets, Meteors and Meteorites, Interplanetary Dust; Structure and Composition

We begin with a short review of cometary orbits and some immediate inferences.

a) The orbits of long-period comets with periods of the order 10^2 to 10^6 years show randomly distributed inclinations i, direct and retrograde motions being about equally likely. The eccentricities e are a little less than or nearly equal to 1 so that we have to do with elongated ellipses, or parabolas as limiting cases. Hyperbolic orbits ($e>1$) are produced only occasionally as derivative consequences of perturbations by the large planets. Since at a large distance from the Sun the velocity is very small, the comets must come from a cloud of such bodies that accompanies the Sun in its journey through the stellar system.

b) The short-period comets mostly move in direct elliptic orbits of small inclination ($i \approx 15°$) whose aphelia lie near the path of one of the large planets. The cometary families of Jupiter, Saturn, .. evidently result from the capture of long-period comets. Since in the

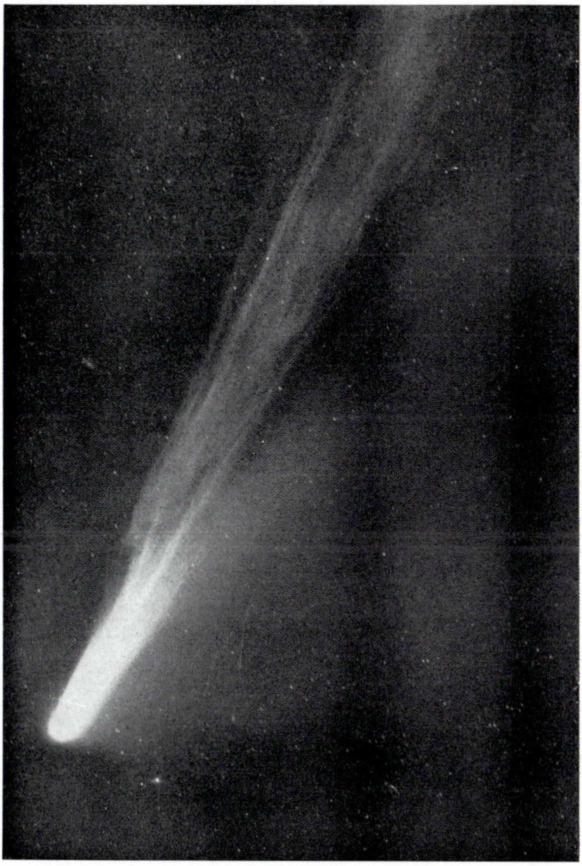

Fig. 8.1. *Comet Mrkos, 1957 d.* 1957 August 23. Mt. Wilson and Palomar Observatories 48-in Schmidt camera

course of time such comets are dissipated by breaking-up and evaporation of their material, the swarm of short-period comets must be continually replenished by further captures.

Photographs with suitable exposure-times (Fig. 8.1) show that a comet in the first place possesses a *nucleus* (not always detectable) of only a few kilometres diameter. As a diffuse nebulous envelope around this, often in the form of parabolic shells, sometimes as rays emanating from the nucleus, there is then the *coma*. The nucleus and the coma together are called the *head* of the comet; its diameter is about 2×10^4 to 2×10^5 km. Somewhat inside the distance of Mars, the comet developes the well-known *tail*, which may have a visible length from 10^7 km up to occasionally even 1.5×10^8 km ($=1$ AU).

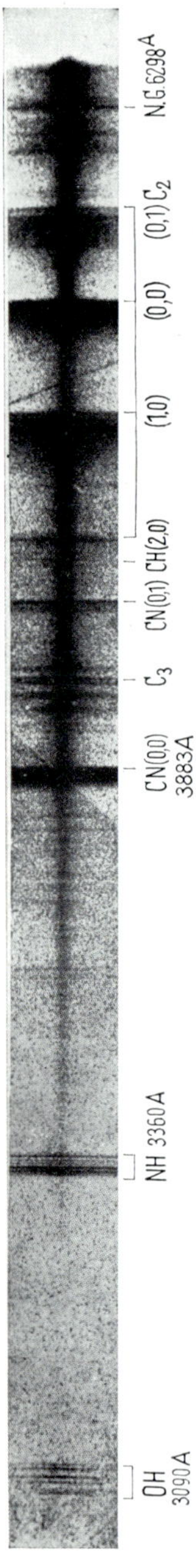

Fig. 8.2. *Spectrum of head of the comet 1941 I Cunningham* (0.87 AU away from the Sun). In the centre is the continuous spectrum of the reflected sunlight. On the outside are the emission bands of the molecules OH, NH, CH, CN, C_2, C_3 .... N.G. (nightglow) denotes the night sky line [OI] 6300 Å

The spectrum of the *head* of a comet (Fig. 8.2) shows partly sunlight whose intensity-distribution indicates scattering by particles of the dimensions of the wavelength of visible light (about $0.6\,\mu = 6000$ Å). In addition there are emission bands of the molecules, radicals and radical-ions

$$\mathrm{CH,\ NH,\ OH,\ CN,\ C_2,\ NH_2,\ C_3,\ OH^+,\ CH^+} \qquad (8.1)$$

and, when the comet is near the Sun, the atomic spectral lines of

Na and occasionally Fe, Ni, Cr, Co, K, Ca II, [OI].

Spectra of cometary *tails*, corresponding to their lower density (making recombination difficult, see below), show chiefly the ions of radicals and molecules

$$\mathrm{N_2^+,\ CO^+,\ OH^+,\ CO_2^+}\ \text{and again CN.} \qquad (8.2)$$

It is seen that all the cometary molecules, etc., are compounds of the cosmically abundant light elements H, C, N, O.

The development of a comet and its spectrum can be pictured in the following way:

At a great distance from the Sun, only the nucleus is present; it contains smaller and larger fragments of stone and nickel-iron — like meteorites (see below) — mixed with compounds of the light elements mentioned above, especially hydrides, forming a sort of "ice" (F. Whipple), chemically comparable to the major planets. On approaching the Sun, substances like H_2O, NH_3, CH_4, etc., evaporate. These parent molecules, streaming away at about 1 km/sec, when in the coma are further reduced by photochemical processes, and the molecules, etc., listed in (8.1) are excited to flourescence by solar radiation. This is shown by the fact that gaps in the solar spectrum caused by the crowding together of Fraunhofer lines are observed to be reproduced also in the band-spectra. In the cometary tail, the molecules and radicals are further ionized by the short-wavelength solar radiation; because of the low density, the rate of recombination of positive ions with electrons is small.

The characteristic forms and motions of cometary tails can be explained, as Bessel, Bredichin and others long since realized, by the hypothesis of some force of repulsion emanating from the Sun, often much stronger than solar gravitation.

As shown by their spectra, the broad diffuse tails of so-called type II consist in the main of colloidal particles of size about the wavelength of light. Radiation pressure on such particles may in fact attain a value many times the force of gravity, as the observations require (every absorbed or scattered photon $h\nu$ imparts an impulse of order $h\nu/c$).

The narrow, elongated tails of so-called type I, on the other hand, as their spectra show, consist mainly of molecular ions like CO^+, etc. Here the computed radiation pressure no longer suffices to account for the large values of the ratio radiational-acceleration: gravity. According to L. Biermann these *plasma-tails* are to a greater extent blown away from the Sun by an ever-present corpuscular radiation, the so-called *solar wind* (see below). At the distance of the Earth, this consists of a stream of ionized hydrogen, i. e. protons and electrons, with 1 to 10 such particles per cm^3 and with velocity about 400 km/sec. Maybe in this way we may account for the often, but not always, observed influence of solar activity on comets. Investigations with the aid of space-vehicles have already provided interesting information about the solar wind, and in future may well make possible more definite assertions on the subject.

As has become apparent in recent times, the *meteors* or shooting stars represent only a selection of the aggregate of the small bodies in the solar system. We now distinguish rather sharply between a meteor, the short-lived luminous object in the sky — ranging from the fireball making things bright as day to the telescopic shooting-star — and the parent body, the meteorite. Since cosmic bodies reaching the vicinity of the Earth in almost parabolic orbits have velocity 42 km sec^{-1} and since the velocity of the Earth in its orbit is 30 km sec^{-1}, then according to the direction of approach they arrive with speeds between 12 and 72 km sec^{-1} in the evening and morning respectively. The bodies become heated on entering the Earth's atmosphere. In the case of larger portions of matter the heat cannot penetrate the interior sufficiently quickly, the surface becomes pitted by fusion or burns off, and they reach the ground as *meteorites.* The largest known meteorite is "Hoba West" in Southwest Africa of about 50 tons. The impact of much more considerable masses must have produced meteoritic craters as on the Moon. For example, the famous crater of Canyon Diablo in Arizona has a maximum diameter of 1300 metres and at the present time a depth of 174 metres. According to geological indications, it must have come some 20 000 years ago from the impact of an iron meteorite of about 2 million tons. Its kinetic energy corresponds to about a 30-megaton hydrogen bomb! Small meteors burn away in the atmosphere, being the common shooting stars at a height about 100 km. In their flight through the atmosphere they ionize the air in a cylindrical volume. In the case of the great meteor-streams they make a contribution to the ionosphere, the so-called anomalous E-layer at about 100 km height. Also, such a conducting cylinder scatters electromagnetic waves as does a wire, predominantly at right angles to its own direction. The great strides made in meteor-astronomy since the application of radar techniques by Hey, Lovell and others depend upon this. On the radar screen one sees only the larger bodies themselves, but one easily obtains the direction normal to the ionized trail even for shooting-stars that are below the limit of visual detection. One can also measure the velocities by radar methods. The outstanding advantage above visual observation is that these methods are independent of time of day and of cloudiness, so that the falsification of statistics from such causes is avoided. The most exact results for the brighter meteors at night are got from photographic observations, using wide-angle cameras of high light-gathering power, made as near as possible simultaneously at two stations suitably far apart. In this way one obtains the position of the path in space. Rotating sectors are used to interrupt the exposure and so to make it possible to compute the velocity in the path. The strength of the image shows the brightness and its often rapid fluctuations. Using an objective prism (see below) one can obtain spectra of the brightest meteors and so gain insight into the mechanism of their radiation and into their chemical composition.

Since the air restistance is proportional to the cross-section $\sim$(diameter)^2 while the weight is proportional to the mass $\sim$(diameter)^3, one sees easily that for ever smaller particles the resistance so predominates that they are no longer made to glow and they simply drift undamaged to the ground. These *micrometeorites* are smaller than a few microns (10^{-4} cm). Using suitable collecting devices they are found in large quantities on the ground and also in deep-sea sediments; naturally there is difficulty in distinguishing them from terrestrial dust. In recent times, investigations with rockets (from which instruments are recovered) and satellites have yielded important new results. With the former micrometeorites have been captured, and under an electron microscope these are identified as:

a) fluffy particles,

b) compact, but not fused, particles of irregular shape,

c) globules.

As the only cosmic material immediately accessible to us, meteorites have been closely studied first by the methods of mineralogy and petrology and more recently for the traces of radioactive elements and anomalous isotopes.

In the first place, we distinguish between *iron meteorites* (density ≈ 7.8 g cm^{-3}), the nickel-iron crystals of which in their characteristic Widmannstetter etched figures (Fig. 8.3) preclude confusion with terrestrial iron, and *stony meteorites* (density ≈ 3.4 g cm^{-3}). The latter resemble more terrestrial silicate rocks, but the frequent inclusions of up to pea-sized chondrules indicate considerable difference of origin. The detailed classification of meteorites, into which we cannot enter, show at anyrate that their origin was bound up with complicated separation-processes in magmas. On the other hand some parts of certain meteorites retain enclosed rare gases, the isotope-ratios of which show that they originated in solar material.

While it was originally believed that in the chemical analyses of many meteorites by V. M. Goldschmidt and the Noddacks we should have before us the *cosmical abundances* of the elements and their isotopes*, we now seek rather to use them along with the quantitative analysis of the Sun to obtain clues to the early history of the meteorites and of our planetary system.

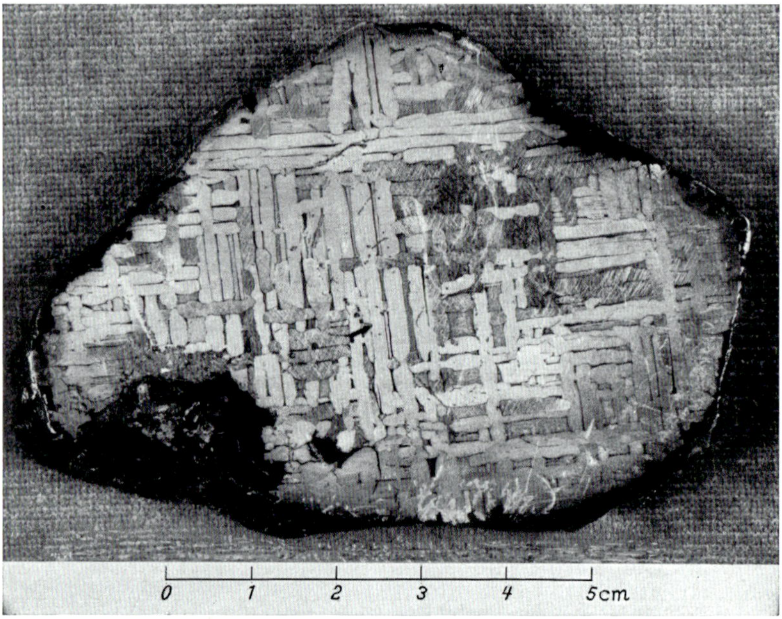

Fig. 8.3. *The Toluca iron meteorite*—named after its place of discovery. The polished and etched cross section shows the *Widmannstetter figures*. These are formed by Fe-Ni plate crystals of Kamazite (7 percent Ni) and Taenite (with larger nickel content) which are juxtaposed parallel to the four surface-pairs of an octahedron; such meteorites are called Octahedrites

The radioactive age-determination yields for meteorites a maximum age that agrees with that of the Earth to within possible errors, namely 4.5×10^9 years. In addition, one can measure how long a meteorite in space has been exposed to irradiation by energetic cosmic-ray protons (assuming constant flux). By the splitting of heavy nuclei (spallation) — down to a certain depth of penetration in larger bodies — this produces all possible stable and radioactive isotopes from the quantities of which one can calculate an *irradiation age*. The irradiation times for iron meteorites work out at some 10^8 to 10^9 years; for stony meteorites only about 10^7 to 10^8 years. The latter may well give

* A serious difficulty arises from the fact that iron meteorites predominate amongst the larger bodies, and stony meteorites amongst the smaller ones.

the time since the body examined was broken off a larger body in a collision.

Following G. S. Hawkins, Fig. 8.4 shows a conspectus of the flux densities of all "particles" present in the vicinity of the Earth *, with

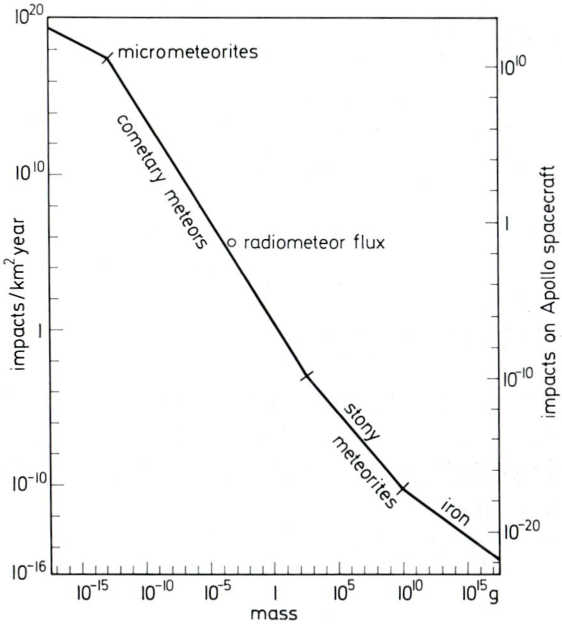

Fig. 8.4. *Number of meteorites, etc. of mass greater than m grams which strike in one year a surface of 1 km² outside the Earth's atmosphere.* The scale on the right gives the probability that an Apollo spacecraft will be hit by a body of mass greater than m during a 10-day journey. It is equal to unity for particles of 0.01 mg; these will pass through aluminium sheet up to half a millimetre thick

masses from 10^{-16} g to 10^{16} g corresponding to diameters of order of magnitude from 10^{-5} cm to 10^{5} cm. In all, a mass of about 1.2×10^{9} g, predominantly in the form of micrometeorites, falls on to the Earth each day. Especially from radar measurements, we know from the orbits that a considerable proportion of meteorites are of cometary origin; another fraction, the *sporadic meteors,* move on statistically distributed ellipses having eccentricities almost equal to unity. Hyperbolic paths or velocities do *not* occur. The larger bodies may more closely resemble asteroids; it should be added that we know very little about possible connections between comets and asteroids.

* We shall discuss the interplanetary plasma, as well as the zodiacal light, in connexion with the physics of the Sun (Chapter 20).

The *zodiacal light* arises from the reflection or scattering of sunlight by interplanetary dust. One can observe it as a cone-shaped illumination of the sky in the region of the zodiac shortly after sunset or shortly before sunrise. During solar eclipses one observes near the Sun an extension of the zodiacal light produced by strong forward scattering (Tyndall scattering) by interplanetary dust. This is seen as an outer part of the solar corona. It is called the F- or *Fraunhofer corona,* because its spectrum, like that of the zodiacal light, contains the dark Fraunhofer lines of the solar spectrum. In both cases the scattered light is partially polarized.

Interlude

9. Astronomical and Astrophysical Instruments

Great scientific advances are often bound up with the invention or introduction of new kinds of instruments. The telescope, the clock, the photographic plate, photometer, spectrograph and finally the entire arsenal of modern electronics each signalizes an epoch of astronomical investigation. Equally important, however — as we must not forget — is the conception of new ideas and hypotheses for the analysis of the observations. Fruitful scientific achievements depend almost every time on the interplay of new sets of concepts and new instrumental developments, that can succeed in penetrating new domains of reality only in combination with each other. "Wonder en is gheen wonder" as we must say with Simon Stevin (1548—1620) ("A marvel is never a marvel").

The passage from classical astronomy to astrophysics in so far as the distinction has any meaning — forms perhaps the most convenient place at which to start considering some astronomical and astrophysical instruments and techniques of measurement.

Fig. 9.1 recalls the principles of Galileo's telescope (1609) and Kepler's (Dioptrice, 1611). In both the magnification is determined by the ratio of the focal distances of the objective and of the eyepiece. Galileo's arrangement gave an upright image and so became the prototype of opera glasses. Kepler's tube permitted the insertion of cross-wires in the common focal plane of the objective and eyepiece — so it was useful for the exact setting of angles, e. g. in the meridian circle. If we replace the cross-wires by micrometer threads (Fig. 9.1) we can measure visually the relative positions of double-stars, the diameters of planetary disks, etc.

By the invention of *achromatic lenses* in 1758, J. Dollond and others overcame the troublesome colour-fringes (chromatic aberration)

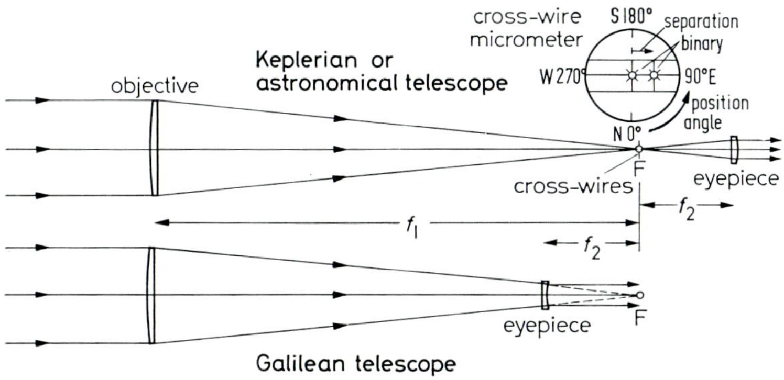

Fig. 9.1. *Keplerian and Galilean telescope.* The eyepiece is a convex lens in the former case, and a concave lens in the latter. F denotes the common focus of objective and eyepiece. The magnification M is equal to the ratio of the focal lenghts of the objective f_1 and eyepiece f_2; $M = f_1/f_2 = 5$ in the diagram. A Keplerian telescope can be used for binary measurements if a cross-wire micrometer is placed at the focus F. The curvature of the lenses is shown exaggerated

of a telescope with simple lenses. An achromatic convergent lens, for example, a telescopic objective, consists of a convex lens (convergent lens, positive focal length) made of crown glass, of which the dispersive power is relatively small in comparison with its refractive power, together with a concave lens (divergent lens, negative focal length) made of flint glass, of which the dispersive power is large in comparison with its refractive power. Precisely speaking, with a two-lens objective we can succeed in making the change of focal length with wavelength vanish, that is $df/d\lambda = 0$, only at a single wavelength λ_0. In the case of a visual objective we choose $\lambda_0 \approx 5290$ Å corresponding to the maximum sensitivity of the eye; in the case of a photographic objective we choose $\lambda_0 \approx 4250$ Å corresponding to the maximum sensitivity of the ordinary photographic (blue) plate.

We consider more exactly the image of a region of the sky made by a telescope on a photographic plate in the focal plane of the objective. (In the case of visual observation, the focal plane would be regarded as being enlarged by the eyepiece, as by magnifying glass.) The operation of converting the plane wave coming from "infinity" into a convergent spherical wave is achieved by the lens by virtue of the fact that in glass (refractive index $n > 1$) the light travels n times more slowly and so the light-waves are n times shorter than in the

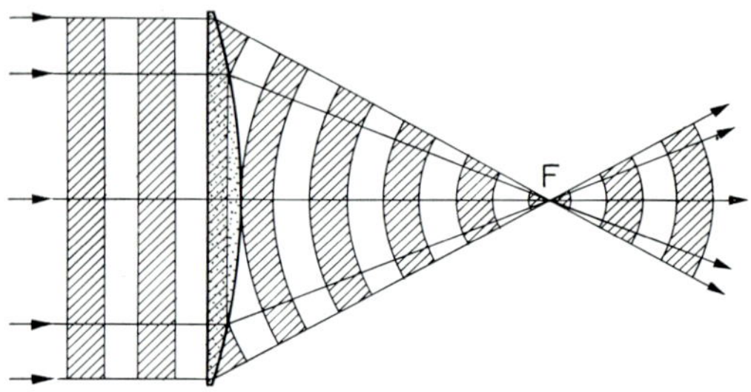

Fig. 9.2. *Formation of an image by a plano-convex lens.* Plane wavesurfaces arrive from a star on the far left, the rays being perpendicular to these surfaces. The velocity of light is n times smaller (where n is the refractive index) in the lens, so that the wave surfaces are bent into spherical form. These spherical surfaces first converge towards and then diverge away from the focus F

vacuum. Consequently the wave surface of the light behind the objective is held back in the middle (Fig. 9.2). This account, the mathematical formulation of which by means of the "eikonal" we owe to H. Bruns, W. R. Hamilton, and K. Schwarzschild, often very considerably simplifies the understanding of optical instruments compared with the direct application of Snell's law of refraction.

What the lens-telescope or *refractor* achieves by the insertion of layers of different thickness with $n > 1$ in the optical path is done by the mirror-telescope or *reflector* (Isaac Newton c. 1670) by means of a concave mirror. This starts with the advantage that it can have no colour-error. As a simple geometrical consideration shows, a spherical mirror (Fig. 9.3 a) focuses a parallel beam near the axis at a focal distance f equal to half the radius of curvature R. Rays further away from the axis after reflection meet the optical axis at a smaller distance from the apex of the mirror. The resulting image-error is called *spherical aberration.* A *parabolic mirror* (Fig. 9.3 b) achieves the exact focussing of a beam parallel to the axis at a single focus. We see this at once, if we regard the paraboloid as the limiting case of an ellipsoid when one focus is removed to infinity. Unfortunately, however, a parabolic mirror gives a good image only in the immediate neighbourhood of the optical axis. At larger aperture-ratios, the usable diameter of the image-field is very meagre on account of the rapid increase outwards of the image-errors of an oblique beam. For example, the reflector of the Hamburg-Bergedorf Observatory with a mirror of diameter 1 metre and focal length 3 metres and so an aperture-ratio F/3 has an image-field of only 10′ to 15′ diameter.

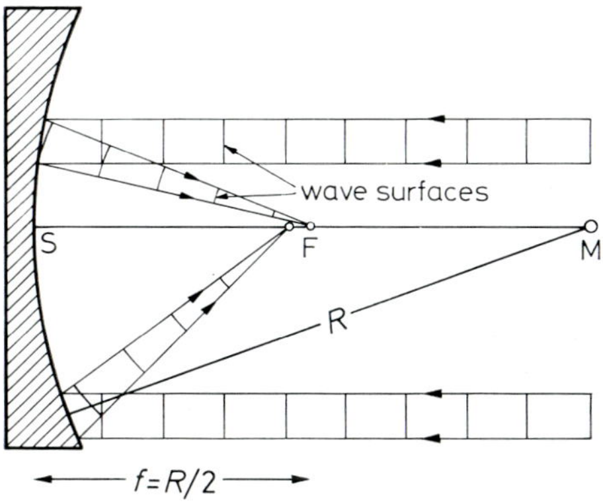

Fig. 9.3 a

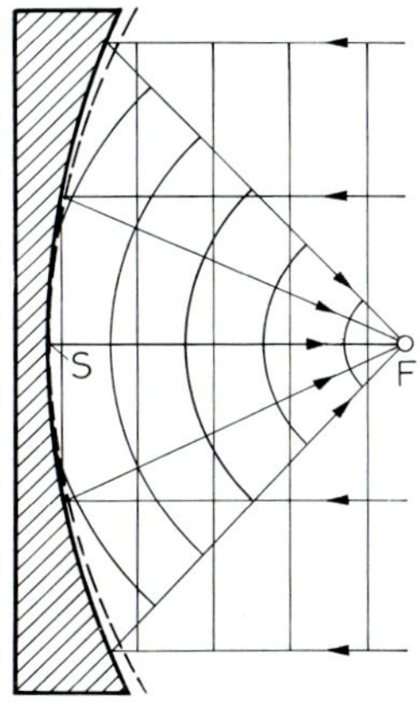

Fig. 9.3 b

Fig. 9.3 a. *Spherical mirror.* A bundle of rays close to the axis (see upper part of diagram) is brought to a focus *F*, whose distance from the axial point *S* of the mirror is equal to the focal length $f = k/2$, where k is the radius of curvature of the mirror. A bundle of rays further away from the axis (lower part of diagram) is brought to a focus nearer to *S*. This is known as spherical aberration. The plane waves incident from the right are reflected as convergent spherical waves

Fig. 9.3 b. *Parabolic mirror.* This brings all rays parallel to the axis to a single focus *F*, i. e. an incident plane wave is reflected as a convergent spherical wave. This has the same curvature as the paraboloid at the axial point *S* of the mirror

The ingenious construction of the Schmidt camera (1930/31) satisfied the desire of astronomers for a telescope with a large image field *and* a large aperture-ratio (light-intensity). Bernhard Schmidt (1879 to 1935) first remarked that a spherical mirror of radius *R* focuses narrow parallel beams that reach the vicinity of the centre of the sphere *from any direction* on to a concentric sphere of radius $\frac{1}{2}R$, corresponding to the known focal length $\frac{1}{2}R$ of the spherical mirror. With a small aperture-ratio one can thus obtain a good image over a large angular region upon a curved film, if one simply furnishes the spherical mirror with an entry stop at the centre of curvature, that is at twice the focal length from the mirror (Fig. 9.4). If one wishes to attain high light-intensity and opens the entry stop wider, spherical aberration will make itself noticeable by the blurring of stellar images. B. Schmidt overcame this by introducing into the entry aperture a thin aspherically figured corrector-plate such that by means of appropriate glass-thickness and a small displacement of the focal surface one compensates the optical path-differences which

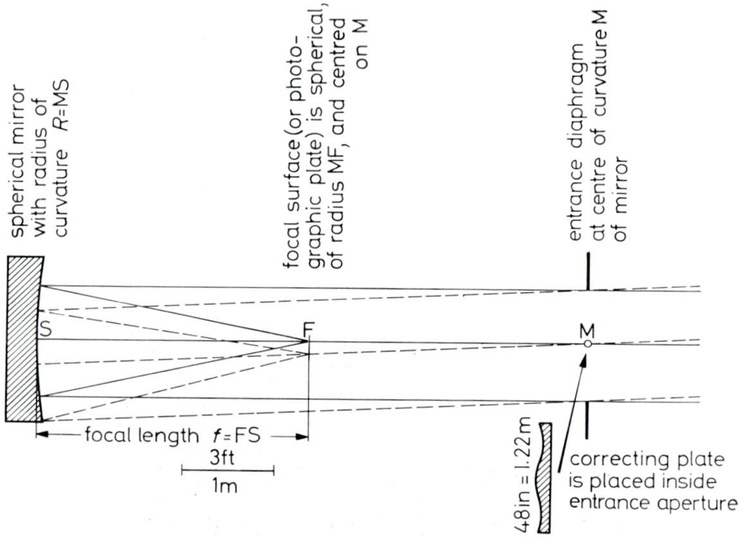

Fig. 9.4. *Schmidt camera* (after Bernhard Schmidt). This employs a spherical mirror with radius of curvature $R=MS$. The entrance diaphragm is located at the centre of curvature M of the mirror. This ensures that all parallel bundles of rays admitted by the diaphragm-aperture (some of them appreciably inclined to the optical axis) are brought together under similar conditions on a spherical surface (the focal surface) centred on M and of radius $\frac{1}{2}R=MF$. The focal length is thus $f=FS=\frac{1}{2}R$. A thin aspherical correcting plate is placed in the entrance aperture in order to remove spherical aberration. (Dimensions shown for Mt. Wilson and Palomar 48-inch Schmidt telescope)

correspond to the separation between the paraboloid and the spherical surface in Fig. 9.3 b. On account of the smallness of these differences, this is possible for a large range of angles of entry and without chromatic errors.

The mounting of a telescope has the purpose of enabling it to follow the daily motion with the best attainable accuracy. Therefore such an equatorial or parallactic mounting has a *polar axis* parallel to the Earth's axis, which is driven by a sidereal clock, and perpendicular to this the *declination axis*, both having the appropriate graduated circles. The astronomer corrects small errors in the clockwork, refraction etc., by guiding using electrically controlled fine adjustments.

Refractors, of which the aperture ratio is in the range F/20 to F/10, are mostly given the so-called Fraunhofer or "German" mounting as for example the largest instrument of this kind with a

Fig. 9.5. *Yerkes Observatory 40-inch (1 m) refractor.* Fraunhofer (German) mounting

1 metre objective and 19.4 metre focal length at the Yerkes Observatory of the University of Chicago (Fig. 9.5).

Reflectors mostly have aperture ratio F/5 to F/3 and one employs either one of the various sorts of *fork-mountings* (the declination axis passes through the centre of gravity of the tube and is reduced to two pivots on the sides of the tube) or the *English* mounting in which the north and south ends of the polar axis rest on separate piers. The largest reflector at present is the Hale telescope of the Mt. Wilson and Palomar Observatories with a main mirror of 200 inches $\approx$ 5 metres diameter and 16.8 metres focal length (Fig. 9.6). With many of the newer reflectors one can take photographs at the prime focus of the main mirror (Fig. 9.7). However, one can also interpose a convex mirror and produce the image at the Cassegrain focus behind a hole drilled through the main mirror. With both arrangements one can also throw the image sideways out of the tube using a 45° plane mirror. Finally, by means of a complicated mirror-system one can lead the light through a hollow polar axis to form the image of a star, for example, at the coudé focus on the slit of a large fixed spectrograph (Fig. 9.7 c).

Fig. 9.6. *Mount Wilson and Palomar Observatories 200-inch (5 m) Hale reflector*

Schmidt telescopes have, as a rule, aperture ratios F/3.5 to F/2.5 but they can go down to F/0.3. The Palomar 48-inch Schmidt telescope (F/2.5) has a corrector-plate of diameter 48 inches = 122 cm; in order to avoid vignetting, the spherical mirror has to have a greater diameter of 183 cm. The famous *Sky Survey* was made with this instrument: some 900 fields $7^\circ \times 7^\circ$ each with one plate in the blue and one in the red to limiting magnitudes 21^m and 20^m cover the whole northern sky down to $-32°$ declination. The Schmidt mirror of the Schwarzschild Observatory at Tautenburg in Thuringia is somewhat larger, with a corrector-plate of 134 cm.

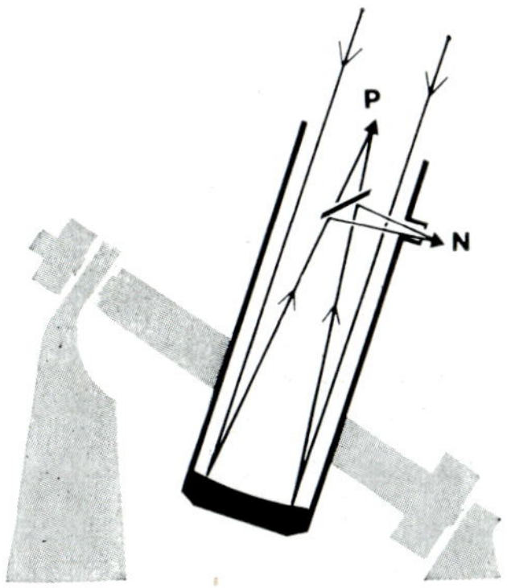

a

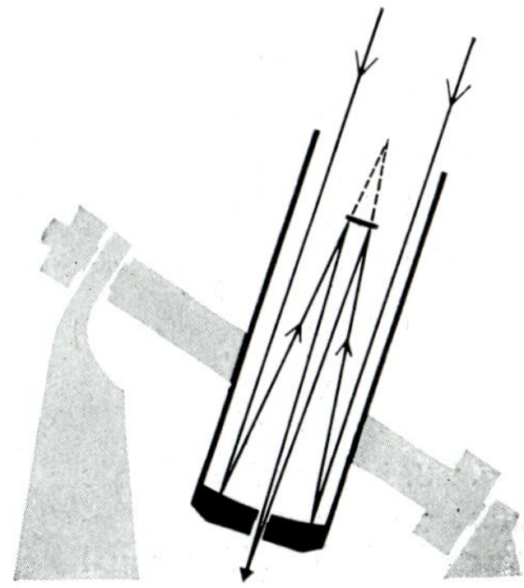

b

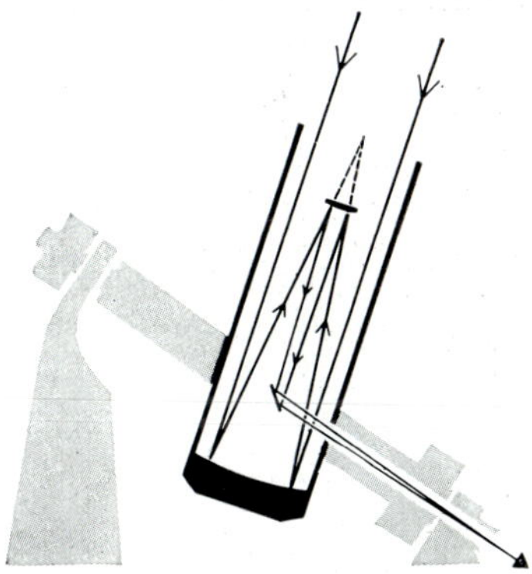

c

Fig. 9.7 a—c. *Mount Wilson and Palomar Observatories 100-inch (2.5 m) Hooker reflector*

	focal length	focal ratio
a) Newtonian or prime focus	42 feet	$F/5$
b) Cassegrain focus	133 feet	$F/16$
c) Coudé focus	250 feet	$F/30$

Among the special instruments of positional astronomy we must mention at least the *meridian circle* (O. Römer 1704). The telescope can be moved in the meridian about an east-west axis. Using a pendulum clock of high precision or a quartz clock, the right ascension is determined by the time of passage of the star through the meridian at right angles to threads in the focal plane. The simultaneous determination of the culmination and so of the declination of the star is made possible by a horizontal thread in the field of view and the divided circle fixed to the axis. Modern determinations of position reach an accuracy of few hundredths of a second of arc. On a divided circle of 1 metre radius 0.1 second of arc corresponds to half a micron!

We now seek to form an idea of the effectiveness of different telescopes for one task or another. The visual observer first asks about *magnification*. As we have said, this is simply the ratio of the focal lengths of the objective and eyepiece. The diffraction of light at the aperture of entry sets a limit to the perception of ever smaller objects. The smallest angular separation of two stars, of a double star, for example, that one can discern is called the *resolving power*. A square aperture of side D (this being easier to discuss than a circular aperture) with parallel light as from a star forms a diffraction-image that is bright in the centre; on either side we first have darkness produced by interference where the illumination of both halves (Fresnel zones) cancels out. According to Fig. 9.8 this corresponds to an angle in circular measure of λ/D. We approximate to a circle of radius R by a square that lies between the circumscribed square with $D=2R$ and the inscribed square with $D=\sqrt{2}R=1.41R$.

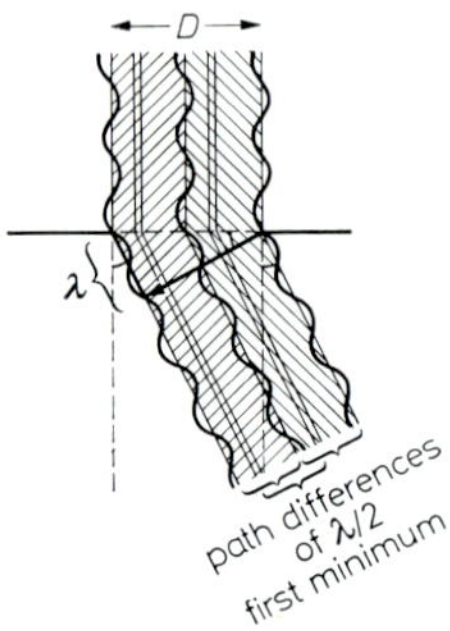

Fig. 9.8. *Diffraction of light at a slit or rectangular aperture of width D.* Two equal bundles of rays a distance $\frac{1}{2}D$ apart and having a path difference of half a wavelength i.e. of $\frac{1}{2}\lambda$ will produce through *interference* a first diffraction minimum at ange λ/D with their original direction

Taking the mean, we obtain the radius in circular measure of the diffraction disk for an aperture of radius R as $\varrho=0.58\,\lambda/R$. A more exact calculation gives

$$\varrho=0.610\,\lambda/R. \tag{9.1}$$

At this separation ϱ the diffraction disks of two stars half overlap and are just separable. Thus (9.1) gives the resolving power of the telescope. Following astronomical usage, if we reckon ϱ in seconds of arc and the aperture of the telescope $2R$ in inches (1 inch = 2.54 cm), with $\lambda=5290$ Å we obtain

$$\text{theoretical resolving power } \varrho=5''.2/2R_{\text{inch}}. \tag{9.2}$$

For double-star observing Dawes found empirically $\varrho=4''.5/2R_{\text{inch}}$.

This limit is attained with first-class refractors in exceptionally good "seeing" conditions. With larger reflectors the *thermal deformation* of the very sensitive mirror usually produces larger images. In photographs *scintillation* alone produces image-diameters of order $0''.5$ to $3''$.

The theoretical resolving power of a telescope is determined by the interference of peripheral rays. A. A. Michelson achieved a somewhat greater resolving power by his *stellar interferometer* in

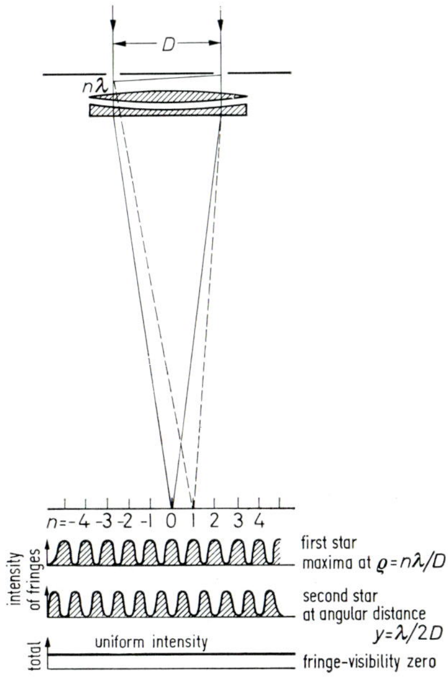

Fig. 9.9 a

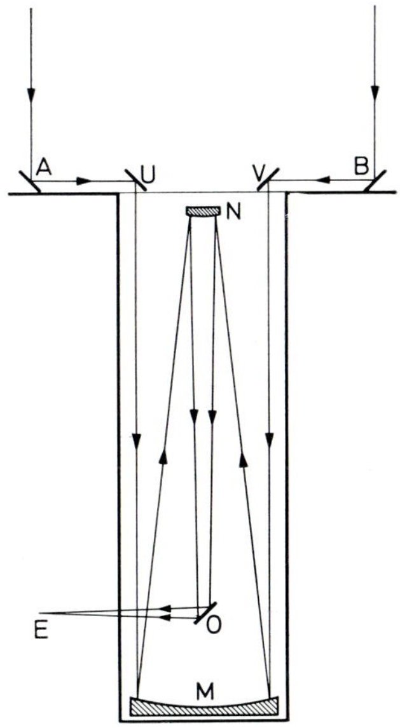

Fig. 9.9 b

Fig. 9.9 a. *Stellar interferometer (after A. A. Michelson).* A point-source star produces a system of interference fringes, whose distances $\varrho = n\,\lambda/D$ from the optical axis are given by $n = 0, \pm 1, \pm 2 \ldots$. The fringe systems of two (equally bright) stars become superposed and produce uniform intensity i. e. *fringe-visibility* zero at angular distances $\gamma = \frac{1}{2}\,\lambda/D, \frac{3}{2}\lambda D \ldots$

Fig. 9.9 b. *20-foot (6 m) interferometer (Mount Wilson and Palomar Observatories).* A steel beam over the aperture of the 100-inch reflector carries two fixed inner mirrors U and V and two moveable outer mirrors A and B (all inclined at 45°). The separation AB, whose maximum value is 20 feet (6 m), corresponds to the distance D between the two slits in Fig. 9 a. Observations are made visually at the Cassegrain focus E

which he placed two slits at separation D in front of the object glass (Fig. 9.9 a). Then a "point-source" star yields a system of interference fringes at angular separation

$$\varrho = n\,\lambda/D \quad (n = 0, 1, 2 \ldots). \tag{9.3}$$

If one now observes a double star whose components have angular separation y in the direction of a line joining the two slits, the fringe-systems of the two stars are superimposed. One has maximum fringe-visibility if $y = n\,\lambda/D$. In between, the fringe-visibility drops to zero if the component stars are equally bright, otherwise it passes through a minimum. Conversely, if the two slits in front of the object-glass are slowly moved apart, we get

maximum fringe-visibility for $y = 0,\ \lambda/D,\ 2\,\lambda/D, \ldots$
minimum fringe-visibility for $y = \frac{1}{2}\lambda/D,\ \frac{3}{2}\,\lambda/D, \ldots$

If one observes a small disk of angular diameter y', a precise calculation shows that this is effectively equivalent to two point-sources at separation $y = 0.41\,y'$ and one obtains the first visibility-minimum at slit-separation D_0 corresponding to $0.41\,y' = \frac{1}{2}\lambda/D_0$, or

$$y' = 1.22\,\lambda/D_0\,. \tag{9.4}$$

Since $D_0 \leqq 2\,R$ in (9.1), it seems that little is gained as compared with the ordinary use of the telescope; actually, however, the judgment of fringe-visibility is less impaired by poor seeing than is measurement with a thread-micrometer. Thus Michelson and others were first able to measure the diameters of Jupiter's satellites, close double-stars, etc.

Later, however, Michelson placed in front of the 100-inch reflector a mirror-system as in a stereotelescope so that he could make $D_0 > 2\,R$ (Fig. 9.9 b). Thereby it was possible to measure directly the angular diameters of some red giant stars, the greatest being about 0".04.

In the Michelson interferometer both rays must be brought together correct in phase. This difficulty, which makes it impossible to construct larger instruments, is overcome by the correlation interferometer of R. Hanbury Brown in the following way: — Two concave mirrors collect the light of a star on to one photomultiplier each. The correlation of the current fluctuations shown by the two photomultipliers for a given frequency-interval is measured. As the theory shows, this correlation-measure is connected with D and y or y' in exactly the same way as the visibility of the fringes in the Michelson interferometer. The first measurements of the angular diameters of Sirius and Vega make the new method appear very promising.

The effectiveness of an instrument for the photography of sources of weak surface-brightness (e. g. gaseous nebulae) depends, as in a camera, first upon the aperture-ratio and secondly on absorption and reflexion losses in the optics. In both respects the Schmidt mirror and its variants are unrivalled.

The question as to how faint a star can be photographed is much more complicated. The limiting brightness for a telescope is obviously determined by the requirement that the small stellar disk, as affected by scintillation, diffraction, plate-grain, etc., should be in recognizable contrast to the plate-background (plate-haze) as affected by night-sky illumination and other disturbances. It follows that one reaches faint stars first with larger aperture; then with given aperture one goes further with small aperture ratio, i. e. long focal-length. In practice one must take account also of the exposure times.

If we seek to delimit the domains of competence of telescopes using lenses or mirrors relative to each other, apart from the difference in aperture ratio, we must remember that for refractors the quality of the image is less affected by temperature-fluctuations, and the theoretical resolving power is closely approached, while a mirror is highly temperature-sensitive. Therefore the refractor is primarily appropriate for visual binary stars, planetary surfaces, trigonometric parallaxes, . . ., the reflector for spectroscopy, direct photography at any rate of faint objects . . . Photoelectric photometry and other tasks may be undertaken with both types of instrument.

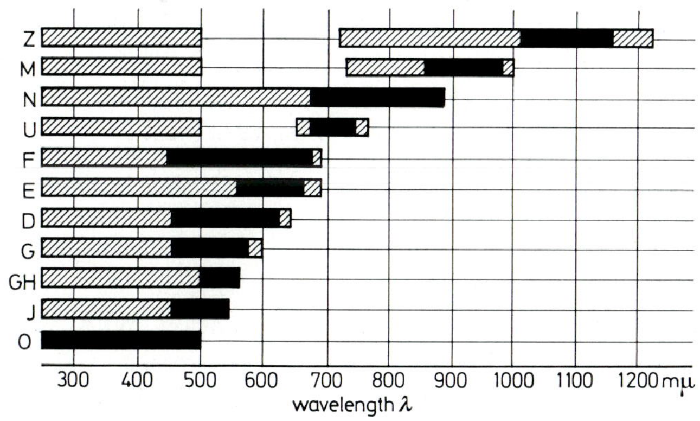

Fig. 9.10. Spectral sensitivity ranges of Kodak plates with varying sensitisation. ▬ spectral region for which the sensitizing class considered is especially valuable. ▨ total range of sensitivity. Wavelengths are given in mμ

The auxiliary equipment for registering and measuring the radiation of stars, nebulae, etc., is just as important as the telescope.

To-day *visual observations* play a part only where we have to do with the rapid discernment of small angles or small details near to the limit permitted by scintillation, thus for the observation of visual binaries, planetary surfaces, solar granulation, etc.

The *photographic plate* is still one of the most important tools of the astronomer. Figure 9.10 shows the sensitivity curves of certain frequently-used Kodak plates. The most highly sensitive blue-plates O are those used most for astronomical photography. *D* and *F* are sensitized emulsions; in combination with suitable colour-filters they serve for photography in the visual and red parts of the spectrum. *N* and *Z* plates for infra-red photography have to be hypersensitized with ammonia.

The relation between light-intensity and plate-darkening can only be determined individually and empirically for each plate. Photographic photometry must therefore operate so far as possible differentially. Either one measures the deflection of a microphotometer as a measure of the blackening at the centre of the star-image, or using an iris-photometer one closes an iris-stop round the small stellar image (in particular that got with a Schmidt camera) until the deflexion of the photometer takes a suitable pre-selected value. In both cases the interpretation of the deflections must be carried out in accordance with an otherwise determined brightness-scale. This can be obtained, for example, using two exposures on the same plate, one of which is made with a measured reduction-ratio by means of a neutral filter ("half-filter method"). Or else one places in front of the instrument a coarse grating (made of metal rods or the like) that produces on either side of the star diffraction images with a known reduction-ratio. In recent times these methods have been almost superceded by the more exact and simpler photoelectric measurement of magnitude-sequences (Chapter 13).

Photoelectric photometry employs as measuring device either photocells or photomultipliers in combination with appropriate electronic amplifiers and registering galvanometers, and at the present time also with automated electronic data-reduction. The spectral sensitivity can be adjusted by choice of a suitable photocathode for measurements from the ultraviolet (at the limit of transparency of our atmosphere $\lambda \sim 3000$ Å) down to the infrared ($\lambda \sim 12\,000$ Å). At somewhat shorter wavelengths there starts the region of use of lead-sulphide cells and similar semi-conducting systems. In astronomy, if one speaks of measuring the total radiation with a thermo-element or a bolometer one must remember that for terrestrial observations the atmosphere and maybe also the telescope completely absorb considerable parts of the spectrum.

It is important to be clear as to the relative advantages and disadvantages of photographic and photoelectric photometry and as to their effectual cooperation. Photoelectric measuring devices are of higher constancy and can be used easily and precisely along with laboratory methods over great ranges of brightness, while a single photographic plate covers only a very restricted range of brightness, about 1 : 20. The instrumental errors of a photoelectric measurement are about ten times smaller than those of photographic photometry. On the other hand, a single photographic exposure includes an incredible number of stars (e. g. in clusters) while the observer working photoelectrically must set and measure each single star for itself. By and large, the present division of tasks is therefore:

Photoelectric photometry: brightness-scales, exact light curves of individual variable stars, exact colour-indices (see below.).

Photographic photometry: photometry of larger star-fields (clusters, Milky Way, . . .), surveys of certain kinds of stars etc., connected with photoelectrically measured scale-fields.

The success of photoelectric photometry on the one hand and the development of television techniques on the other suggest that even for direct pictures the photographic plate will be augmented or replaced by photoelectric devices. In the *image-converter* of Lallemand the electrons leaving the photocathode receive an additional acceleration; in this way one can achieve a sensitivity that considerably surpasses that of the photographic plate. The astronomical application of devices in the domain of television-techniques has been considered by several workers.

The further analysis of cosmic light-sources is taken over by *spectroscopy.* Here we treat first its instrumental side while we somewhat artificially defer consideration of its fundamental concepts and its applications.

The familiar laboratory prism- or grating-spectrograph can be fixed to a telescope the only purpose of which is to throw the radiation of a cosmic source, such as a star, through the slit of the spectrograph. Thence it follows: (1) The aperture ratio of the collimator must be equal to that of the telescope, or at any rate not smaller, for the illumination of the prism or the grating. (2) In good seeing the slit must admit most of the star-image. (3) Since an infinitely sharp spectral line would give an image of the slit on the plate (diffraction effects being generally negligible in stellar spectrographs), one arranges for the sake of economy that the image of the slit on the plate should be suited to the resolving power of the photographic emulsion (~ 0.05 mm). If one wishes to secure spectra of stars of a certain limiting magnitude within a given exposure-time (times exceeding five hours being inconvenient in practice) using a telescope of given size, then the focal length of the camera and the dispersion (Å per mm) are determined. Finally it is a great advantage if the image-field of the spectrograph-camera is so little curved that one can work with *plates* (bent if necessary): the subsequent measuring of films brings with it many technical inconveniences.

Nowadays one uses almost only *grating-spectrographs* (Fig. 9.11) since it is possible so to shape the grooves in the diffraction grating that a particular angle of reflexion ("blaze" angle) is strongly favoured. One chooses the collimator focal length to be as large as the dimensions of the available grating permit. One constructs the camera on the principle of the Schmidt camera, the advantages of which we already know—large field of view, small absorption and reflexion losses, small curvature of the image-field, small chromatic errors. Figure 9.11 c shows the large fixed coudé spectrograph (Th. Dunham Jr.) at the 100-inch Hooker reflector of the Mt. Wilson Observatory. The collimator lies in the direction of the polar axis of the telescope, its aperture-ratio is suited to the large focal length of the coudé system. After reflexion at the grating the light can be led at choice into Schmidt cameras of focal lengths 8, 16, 32, 73 and 114 inches. For example, the most-used 32-inch camera gives in the second order in the photographic region (3200—4900 Å) a dispersion of about 10 Å per millimetre; with this combination one can readily reach stars of magnitude 7 (see below).

Important auxiliary devices are:— (a) An iron arc, i. e. an electric arc between electrodes made of iron, the spectrum of which appearing above and below that of the star provides standards for the measurement of wavelengths and in particular for doppler effects. (b) An auxiliary light-entrance with a stepped slit provides continuous spectra of a filament-lamp with precisely known intensity-ratios. This makes it possible to assign the blackening-curve of the plate for each wavelength, that is in effect the relation between the microphotometer reading and the intensity at a given wavelength, and thus to evaluate the spectrum photometrically.

Fast spectrographs for the study of faint nebulae, stars, etc., with low dispersion are operated at the prime focus of the main mirror.

The *objective prism* is used for spectral surveys of whole star-fields, i. e. a prism is placed in front of the telescope in the position of

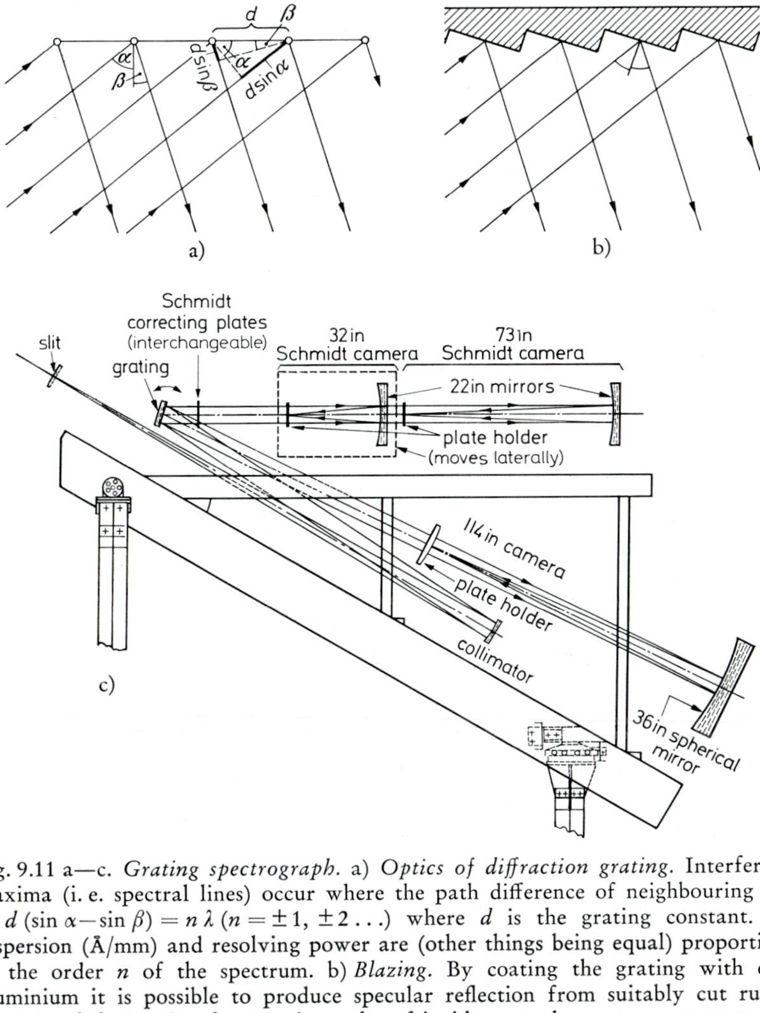

Fig. 9.11 a—c. *Grating spectrograph.* a) *Optics of diffraction grating.* Interference maxima (i. e. spectral lines) occur where the path difference of neighbouring rays is $d(\sin\alpha - \sin\beta) = n\lambda$ ($n = \pm 1, \pm 2 \ldots$) where d is the grating constant. The dispersion (Å/mm) and resolving power are (other things being equal) proportional to the order n of the spectrum. b) *Blazing.* By coating the grating with e. g. aluminium it is possible to produce specular reflection from suitably cut rulings or steps of the grating for certain angles of incidence and emergence so as to concentrate most of the light into the spectrum of one particular order. This leads to greater intensity of spectra. c) *Coudé spectrograph* at Mount Wilson. The coudé optical system of the 100-inch telescope forms the stellar image on the slit of the spectrograph. The collimator mirror forms the incident light into a parallel beam and directs this on the grating. The spectrally decomposed light is now ready to be formed into an image by one of the interchangeable Schmidt cameras (focal lengths 32, 73, 114 inches; also 8 and 16-inch cameras—not shown). The grating acts as the entrance stop of the camera. Correcting plates are only necessary for the short-focus cameras with greater focal ratio

minimum deflexion and so a spectrum of each star is obtained on a plate in the focal plane. In this way, for example, E. C. Pickering and A. Cannon at the Harvard Observatory produced the *Henry Draper Catalogue,* which along with position and magnitude contains the spectral type of something like a quarter of a million stars. In recent times the combination of a Schmidt mirror with a thin prism, e. g. at Warner and Swasey and Tonantzintla Observatories (dispersion ~ 320 Å/mm) has proved to be very productive. The objective prism does not permit the actual measurement of wavelengths or radial velocities. This is made possible by the *Fehrenbach prism.* It is a direct-view prism so arranged that light of a mean wavelength of the spectrum goes through as if it were a plane plate. One then makes two exposures one after the other on the same photographic plate, the prism being rotated through 180° about the optical axis between the exposures. For each star one thus obtains two spectra running in opposite senses, the mutual displacement of which makes possible the measurement of the doppler effect.

Radio astronomy opened up entirely new possibilities for observation in the region of

Wavelength λ	1 mm	10 cm	10 m	300 m
Frequency $\nu = c/\lambda$	3×10^{11}	3×10^{9}	3×10^{7}	10^{6} sec^{-1} or Hz
	300 GHz	3 GHz	30 MHz	1 MHz

The region is bounded on the short wavelength side at about $\lambda \approx 1$ to 5 mm by absorption particularly that of atmospheric oxygen, and on the long wavelength side at about $\lambda \approx 50$ m by reflection at the ionosphere.

K. G. Jansky in 1931 discovered the radio emission of the Milky Way in the metre-wavelength region. Using the subsequently improved receivers of radar devices, J. S. Hey and J. Southworth during World War II (c. 1942) discovered the radio emission of the disturbed and of the quiet Sun. After H. C. van de Hulst had predicted it, various investigators in Holland, U.S.A. and Australia discovered almost simultaneously in 1951 the 21-centimetre line of interstellar hydrogen. The doppler effect on this line opened up enormous possibilities for investigating the motion of interstellar matter in our Milky Way system and in other cosmic structures. Here we shall not yet pursue the well-nigh explosive development of radio astronomy but only summarize its most important types of instrument:

1. The *radio telescope* (Fig. 9.12) with a parabolic mirror made of sheet metal or wire-netting (mesh $\leqq \lambda/5$). As the focus the radiation is received by a dipole (of which the side away from the mirror is

Fig. 9.12. *Australian National Radio Astronomy Observatory, 210-foot (64 m) radio telescope (Parkes, New South Wales)*

shielded by a reflector-dipole or a plate etc.), a horn or the like. The high-frequency energy is amplified and rectified; finally its intensity is indicated by a registering galvanometer, for example, or it is digitalized for automatic data processing. Even for the largest paraboloid (Jodrell Bank: 80 m diameter), according to equation (9.1) which is valid also here, the resolving power is small compared with that of Galileo's first telescope!

2. In the procedure of M. Ryle (Cambridge), high resolving power is attained with a *radio interferometer* that corresponds exactly to Michelson's stellar interferometer; signals in correct phase from two radio telescopes are brought together and further amplified. Also the principles of the linear *diffraction grating* and of the two-dimensional *cross-grating*, here using a fixed wavelength, have been successfully taken over into antenna techniques in order to achieve high angular resolution. In its optical counterpart, we have already become

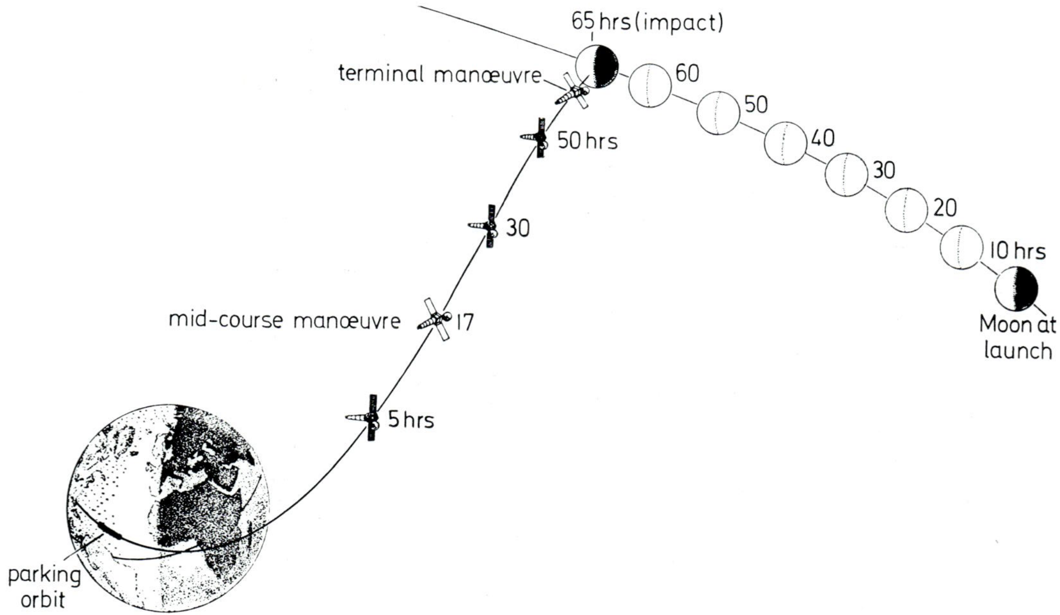

Fig. 9.13. *Journey of a Ranger space-probe to the Moon (approxymately 240,000 miles (400,000 km) in 65 hours).* After the first stage (an Atlas rocket) has burned out and fallen away, the second—an Agena—places the Ranger satellite in a "parking orbit". It remains in this orbit for several hundred miles until it has reached a suitable starting point. The Agena is then fired a second time and accelerates the Ranger and itself to the velocity needed for escape from the Earth's gravitational field. The Agena is then discarded and the Ranger lined up with respect to the Sun and the Earth. When the Ranger has reached the approximate half-way point on its orbit to the Moon, a correction is made by firing gas from suitably arranged jets. About 1 hour before reaching the Moon the Ranger is orientated with respect to the Moon itself, and approximately 15 minutes before hitting the surface, the television cameras are switched on. Thus, 5814 pictures were received (by the 85-foot radio antenna in Goldstone, Texas) from Ranger IX before it struck the lunar surface

acquainted with the principle of the *correlation-interferometer*, which was first applied by R. Hanbury Brown and others in radio astronomy. M. Ryle has further shown how, according to the principle of *aperture synthesis*, instead of the information got from one large instrument during one time interval, we can equally well use the information got successively from several smaller antennae in suitable pre-arranged positions. Thus also in radio astronomy resolving powers down to $0\overset{''}{.}1$ have been reached.

3. Here we cannot concern ourselves with amplification-techniques, although we must at least mention the enormous increase in precision of measurement using *masers* and *parametric amplifiers*. With such "almost noise-free" amplification one can push on to significantly weaker sources and so, other things being equal, to greater distances in space than by the sole use of conventional valve-amplifiers.

Observation from outside the atmospheric envelope of our Earth with *rockets* (the instrument-load of which can return to the ground) or *satellites* and *space-vehicles* (the results of whose measurements are transmitted by telemetry) opens to astrophysics, and particularly to the study of the Sun, the whole spectral range that is completely absorbed by the atmosphere:— Gamma-rays, the X-ray region and the short-wave ultraviolet. As we know, the transparency of atmospheric ozone O_3 begins only about 2850 Å. At the other end of the spectrum, successful radio observations down to 0.4 MHz have also been made outside the ionosphere. It is hoped with the specially stabilized OAO = Orbiting Astronomical Observatory to improve upon the limit of angular resolving power of about a second of arc set by atmospheric scintillation. Since the Moon has already been successfully televized both from the far side and from close quarters, (Fig. 9.13) it is now hoped that the same may be done for the planets. Extensive preparatory work has been done for the project of landing on the Moon.

Part II

Sun and Stars

Astrophysics of Individual Stars

10. Astronomy + Physics = Astrophysics

Historical Introduction

In Chapter 1 we sought to give an introduction to classical astronomy with some historical remarks. In the same spirit we now turn to the astrophysical investigation of the Sun and stars. The latter we consider in the first place as individuals. Part III will deal with the internal constitution and evolution of stars and with stellar systems, galaxies, etc.

At the end of chapter I we recalled the first measurements of trigonometric parallaxes by F. W. Bessel and F. G. W. Struve in 1838. They denoted a final vindication of the Copernican world-system, the correctness of which nobody doubted any more. Above all one had thereby gained a sure foundation for all cosmic distance-determinations. Bessel's parallax of 61 Cygni $p = 0\overset{''}{.}293$ asserted that this star had a *distance* of $1/p = 3.4$ parsecs or 11.1 light years. Thence we could, for example, compare the luminosity of this star directly with that of the Sun. However, trigonometric parallaxes first became an effective aid to astrophysics only when F. Schlesinger in 1903 with the Yerkes refractor and later at the Allegheny Observatory developed their photographic measurement to an incredible degree of precision ($\sim 0\overset{''}{.}01$).

Double stars give information about the *masses* of the stars. Sir William Herschel's observations of Castor (1803) left no doubt that here two stars are moving round each other in elliptical orbits under the influence of their mutual attraction. As long ago as 1782 J. Goodricke had observed in Algol (β Persei) the first eclipsing variable.

The work of H. N. Russell and H. Shapley (1912) is the exploitation of the many-sided information that these double stars have to offer. Pickering in 1889 discovered in the case of Mizar the first spectroscopic binary (motions measured by the doppler effect).

After its beginnings in the eighteenth century (Bouguer 1729, Lambert 1760 and others), *stellar photometry* or the measurement of the apparent luminosities of the stars gained a secure basis about a hundred years ago. For one thing, N. Pogson in 1850 introduced the definition that 1^m (one magnitude) should denote a decrease of the logarithmic brightness by 0.400, that is a luminosity-ratio $10^{0.4} = 2.512$. For another thing, J. C. F. Zöllner in 1861 constructed the first *visual stellar photometer* (with two Nicol prisms to give measurable light-diminution) with which *stellar colours* were measured as well.

About the same time the great star-catalogues of magnitudes and positions brought about an immense enlargement of our knowledge of the stellar system: In 1852/59 there appeared the *Bonner Durchmusterung* by F. Argelander and others with about 324 000 stars down to about $9^m\!.5$; later for the southern sky came the *Cordoba-Durchmusterung.*

Karl Schwarzschild initiated *photographic photometry* with the *Göttinger Aktinometrie* 1904/08. He recognized at once that the colour index = photographic minus visual magnitude forms a measure of the colour and so of the temperature of a star. Soon after the invention of the photocell by Elster and Geitel 1911, H. Rosenberg and then P. Guthnick and J. Stebbins began to develop *photoelectric photometry,* the possibilities of which were greatly enlarged by the replacement of the filar electrometer by electronic amplifiers and registering galvanometers, and then also by the invention of the photomultiplier. Thus to-day one can measure stellar luminosities with a precision of a few thousandths of a magnitude in usefully selected wavelength-intervals from the ultraviolet down to the infrared and derive the corresponding colour-indices. We must mention at least the six-colour photometry of J. Stebbins and A. E. Whitford (1943) and the internationally adopted system of UBV-magnitudes (*U*ltraviolet, *B*lue, *V*isual) of H. L. Johnson and W. W. Morgan (1951).

The *spectroscopy of the Sun and stars* developed in parallel with stellar photometry. In 1814 J. Fraunhofer discovered the dark lines in the solar spectrum that are named after him. With extremely modest equipment, in 1823 he succeeded in seeing similar lines in the spectra of certain stars and he noted their differences. Astrophysics proper, that is the investigation of the stars by physical methods began when in 1859 G. Kirchhoff and R. Bunsen in Heidelberg discovered

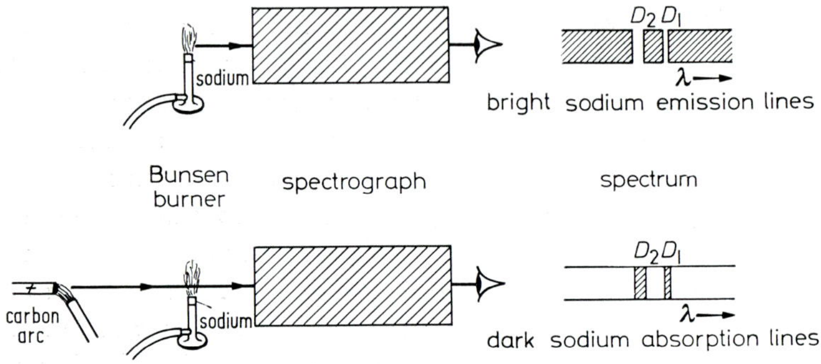

Fig. 10.1. Illustrating the basic experiment in spectral analysis (after G. Kirchhoff and R. Bunsen, 1859). A bunsen flame containing a trace of sodium (table-salt) shows bright sodium emission lines in its spectrum. However, when light from the positive crater of a carbon arc (whose temperature is appreciably higher than of the flame) is allowed to pass through the sodium flame, it produces a continuous spectrum crossed by dark sodium absorption lines, similar to those in the solar spectrum

spectral analysis as well as the meaning of the Fraunhofer lines in the solar spectrum (Fig. 10.1) and in 1860 Kirchhoff formulated the foundations of radiation-theory, in particular Kirchhoff's law, which lays down the relation between the emission and absorption of radiation in thermodynamic equilibrium. This law together with the doppler principle ($\Delta\lambda/\lambda = v/c$) formed for forty years the entire conceptual equipment of astrophysics. The spectroscopy of the Sun and stars first turned its attention to the following exercises:

1. Photographing the spectra and measuring the wavelengths for all the elements *in the laboratory*. Identification of the lines in stars and other cosmic light-sources (Sir William Huggins, F. E. Baxandall, N. Lockyer, H. Kayser, Charlotte E. Moore-Sitterly, and many others).

2. Photography and ever more exact measurement of the spectra of stars (H. Draper 1872; H. C. Vogel and J. Scheiner 1890, and others) and of the Sun (H. A. Rowland, production of good diffraction gratings, photographic map of the normal solar spectrum 1888, Preliminary Table of Solar Spectrum Wavelengths 1898 with about 23 000 lines).

3. Classification of stellar spectra, first in a one-dimensional sequence, essentially according to decreasing temperature. After

preparatory work by Huggins, Secchi, Vogel etc., E. C. Pickering with A. Cannon from 1885 onwards produced the Harvard-classification and the Henry Draper Catalogue. Later advances brought: The discovery of luminosity as a second parameter for the classification and thence the determination of *spectroscopic parallaxes* by A. Kohlschütter and W. S. Adams in 1914, and much later with a modern point of view the "Atlas of Stellar Spectra" (1943) by W. W. Morgan, P. C. Keenan, and E. Kellman giving the MKK-classification.

4. For a long time the measurement of *radial velocities* of stars, etc., using the doppler principle, claimed the predominant attention of astronomers. After visual attempts by W. Huggins in 1867, H. C. Vogel in 1888 succeeded in making the first useful photographic measurements of radial velocities, the rotation of the Sun having been verified spectroscopically shortly before that. W. W. Campbell (1862—1938) at the Lick Observatory, in particular, contributed to the furtherance of the technique of radial-velocity measurement. We shall discuss the motions and dynamics of stars and stellar systems in Part III.

The discovery in 1913 of the *Hertzsprung-Russell diagram* was the culmination of this epoch in astrophysics. About 1905 E. Hertzsprung had already recognized the distinction between "giant" and "dwarf" stars. On the basis of improved measurements of trigonometric parallax, carried out partly by himself, H. N. Russell constructed the famous diagram, with spectral type as abscissa and absolute magnitude (see below) as ordinate, which showed that most stars in our neighbourhood fall in the narrow strip of the *main sequence* (Fig. 15.2) while a smaller number occupy the region of the *giant stars.* Russell first associated with this diagram a theory of stellar evolution (start as red giants, compression and heating to reach the main sequence, cooling along the main sequence) which had, however, to be given up some ten years later. We shall be able to pursue such problems when we come to Part III.

What astrophysics most needed at the beginning of our century was an enlargement of its physical or conceptual basis. The theory of "cavity" radiation or, as it is called, "black-body" radiation, i. e. the radiation-field in thermodynamic equilibrium, begun by G. Kirchhoff in 1860, had been brought to a conclusion by M. Planck in 1900 with the discovery of quantum theory and of the law of spectral energy-distribution in black-body radiation. Astronomers then set themselves to estimate the temperatures of the stars by the application of Planck's law to their continuous spectra. However, at the same time the gifted Karl Schwarzschild (1873—1916) erected one of the pillars of a future theory of the stars, the theory of the *stationary radiation-field*. He showed in 1906 that in the Sun's photosphere (the layers that send out most of the radiation) the transport of energy outwards from the interior is performed by *radiation*. He calculated the increase of temperature with (optical) depth on the assumption of radiative equilibrium and showed that one thereby derives the correct centre-to-limb darkening of the solar disk. K. Schwarzschild's work on the solar eclipse of 1905 August 30 is a masterpiece of insight into observation and theory. In 1914 he investigated theoretically and by spectrophotometric measurements the radiative exchange in the broad H- and K-lines (3933, 3968 Å) of the solar spectrum. He clearly saw that the further prosecution of his undertakings required an atomistic theory of absorption coefficients i. e. of the interaction of radiation and matter. So he turned with great enthusiasm to the quantum theory of atomic structure founded by N. Bohr in 1913. There then resulted the famous work on the quantum theory of the Stark effect and of band spectra. In 1916 Schwarzschild died all too soon at the age of 43 years.

The combination of the theory of radiative equilibrium with the new atomic physics was presented on the one hand by A. S. Eddington during 1916—1926 within the scope of his theory of the internal constitution of the stars (see Part III). On the other hand, with his theory of thermal ionization* and excitation M. N. Saha in 1920 created a point of departure for a physical interpretation of the spectra of the Sun and stars. Thus stimulated, the whole of the fundamentals of the present day theory of stellar atmospheres and of solar and stellar spectra were soon developed. We must mention the work of R. H. Fowler, E. A. Milne and C. H. Payne on ionization in stellar atmospheres (1922-25), the measurement of multiplet intensities by L. S. Ornstein, H. C. Burger, H. B. Dorgelo and their calculation by R. de L. Kronig, A. Sommerfeld, H. Hönl, H. N. Russell, followed by the important work of B. Lindblad, A. Pannekoek, M. Minnaert and many others. By welding together the meanwhile further-developed theory of radiative energy-transport and the quantum theory of

* This had been developed by J. Eggert in 1919 in regard to stellar interiors.

line- and continuum-absorption coefficients, one could by 1927 begin in earnest to construct a rational theory of the spectra of the Sun and stars (M. Minnaert, O. Struve, A. Unsöld). This made it possible to infer from their spectra the *chemical composition* of the outer parts of the stars and so to study empirically the *evolution of the stars* in relation to the generation of energy by nuclear processes in stellar interiors.

With the discovery of the hydrogen convection zone, in 1930 A. Unsöld founded the theory of *convective streaming* in stellar atmospheres and especially in the solar atmosphere. A little later H. Siedentopf and L. Biermann put forward the connection with hydrodynamics (mixing-length theory), shortly after S. Rosseland had called attention to the astrophysical significance of turbulence.

We know that ionized gases — to-day we often speak of plasma — in stellar atmospheres and other cosmic structures possess high electrical conductivity. In the 1940s T. G. Cowling and H. Alfvén remarked that consequently cosmic magnetic fields could be dispersed by the ohmic dissipation of the related currents only in the course of very great time-intervals. Magnetic fields and flow-fields are in perpetual interaction; one must combine the fundamental equations of electrodynamics and of hydrodynamics to get those of magnetohydrodynamics (or hydromagnetics). In 1908 G. E. Hale, the founder of the Mt. Wilson Observatory, using the Zeeman effect in Fraunhofer lines discovered magnetic fields in sunspots up to about 4000 gauss. Using much more sensitive equipment, in 1952 H. W. Babcock was first able to measure the much weaker fields of a few gauss over the rest of the solar surface. Sunspots, the neighbouring "plages faculaires" (facula-surfaces), flares (eruptions), prominences and many other solar manifestations are statistically bound up with the 2×11.5-year cycle of *solar activity*. According to our current interpretation, all this — and it is what we least understand at present — belongs to the domain of *magnetohydrodynamics*. The study of streaming and of magnetic fields in the Sun also leads to some understanding of the heating of the corona, the outermost envelope of the Sun, to a temperature of 1 to 2 million degrees. Extremely complex processes occur here that lead to the production of the variable part of the radio-emission of the Sun as well as the various kinds of *corpuscular radiation* and of the *solar wind* already mentioned in connection with comets.

Again deferring consideration of the internal constitution and of the evolution of stars, we can briefly sum up the development of the physics of a single star with the help of the following key terms:

1. Radiation theory, interaction of radiation and matter.
2. Thermodynamics and hydrodynamics of streaming-processes.
3. Magneto-hydrodynamics and plasma physics.
4. Cosmic corpuscular radiation of super-thermal energy; astrophysics of cosmic rays, X-rays and of γ-rays — which are expected in the near future to comprise a scarcely less exciting domain of investigation than the rest.

11. Radiation Theory

In regard both to the radiation-fields within stellar atmospheres and stellar interiors and to the emitted radiation whose analysis gives us information about the structure and composition of the atmospheres, we have to do with the basic concepts of radiation theory.

In the radiation field to be studied — we make no special assumptions about it at this stage — we take a surface-element $d\sigma$ with normal n and we consider the radiant energy flowing per unit time through $d\sigma$ within a small solid angle $d\omega$ about a direction defined by polar angles ϑ, φ, where ϑ is the inclination to n (Fig. 11.1). In the spectral distribution of this energy we consider that within the frequency-interval ν to $\nu + d\nu$ and write it as

$$dE_\nu(\vartheta, \varphi) = I_\nu(\vartheta, \varphi)\, d\nu \cos\vartheta\, d\sigma\, d\omega \tag{11.1}$$

the cross-section of our radiation-bundle being $\cos\vartheta\, d\sigma$.

Accordingly the *intensity* $I_\nu(\vartheta, \varphi)$ denotes the energy-flow in unit time per unit frequency-interval per unit solid angle about the direction ϑ, φ across unit area perpendicular to this direction. The usual units are: time, second; frequency, one cycle per second, Hertz (Hz); solid angle, steradian; area, square cm. Instead of frequency, we can refer the spectral distribution to wavelength (with 1 cm as unit). Since $\nu = c/\lambda$ we have $d\nu = -(c/\lambda^2)\, d\lambda$. So from the defining property $I_\nu\, d\nu = -I_\lambda\, d\lambda$ we have

$$I_\lambda = (c/\lambda^2)\, I_\nu \tag{11.2}$$

or in symmetric form

$$\nu I_\nu = \lambda I_\lambda .$$

The intensity of total radiation I is got by integrating over all frequencies or wavelengths

$$I = \int_0^\infty I_\nu \, d\nu = \int_0^\infty I_\lambda \, d\lambda. \tag{11.3}$$

As a simple application we calculate the energy dE_ν that passes in unit time through the element $d\sigma$ and also through a second element $d\sigma'$ at distance r. The normals to $d\sigma$, $d\sigma'$ make angles ϑ, ϑ' (say) with the line joining $d\sigma$, $d\sigma'$ (Fig. 11.2). The element $d\sigma'$ subtends at $d\sigma$ the solid angle $d\omega = \cos\vartheta' \, d\sigma' / r^2$. We have therefore

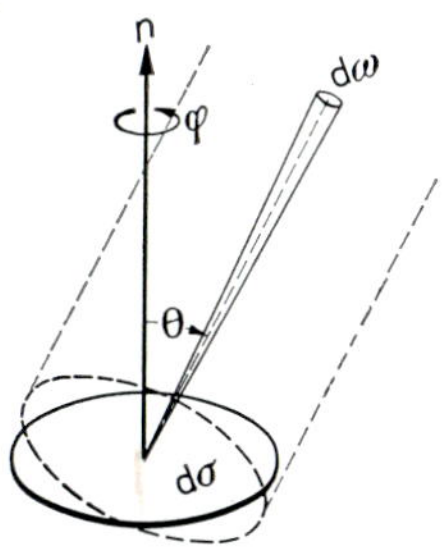

Fig. 11.1. Definition of the intensity of radiation

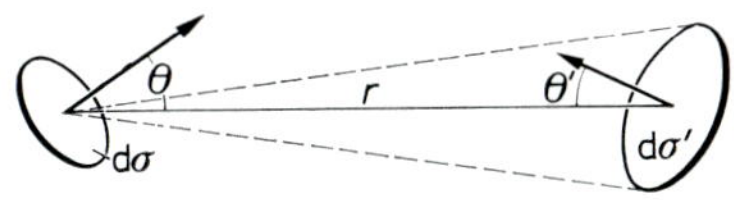

Fig. 11.2. Mutual radiation of two surface elements

$$dE_\nu = I_\nu \, d\nu \cos\vartheta \, d\sigma \, d\omega \quad \text{or} \quad dE_\nu = I_\nu \, d\nu \, \frac{\cos\vartheta \, d\sigma \cos\vartheta' \, d\sigma'}{r^2} \tag{11.4}$$

Also the element $d\sigma$ subtends at $d\sigma'$ the solid angle $d\omega' = \cos\vartheta \, d\sigma / r^2$. Therefore from (11.4) we can write

$$dE_\nu = I_\nu \, d\nu \cos\vartheta' \, d\sigma' \, d\omega'. \tag{11.5}$$

Thus the intensity I_ν along any ray of sunlight, for example, in accordance with the definition (11.1), has the same value in the immediate vicinity of the Sun and anywhere else outside it in space.

What in everyday language is rather vaguely called the "strength" of, say, sunlight corresponds more to the exact concept of *radiation-flux*. We define the flux πF_ν, in the direction n by writing the total energy of ν-radiation crossing the element $d\sigma$ in unit time as

$$\pi F_\nu \, d\sigma = \int_0^{2\pi} \int_0^{\pi} I_\nu(\vartheta, \varphi) \cos\vartheta \, d\sigma \sin\vartheta \, d\vartheta \, d\varphi \tag{11.6}$$

where from Fig. 11.1 we have used $d\omega = \sin\vartheta \, d\vartheta \, d\varphi$. In an isotropic radiation field I_ν is independent of ϑ and φ and so $F_\nu = 0$. It is often useful to separate πF_ν into

$$\text{outward flux} \quad (0 \leqq \vartheta \leqq \tfrac{1}{2}\pi) \qquad \pi F_\nu^+ = \int_0^{2\pi}\int_0^{\frac{1}{2}\pi} I_\nu \cos\vartheta \sin\vartheta \, \mathrm{d}\vartheta \, \mathrm{d}\varphi$$

and (11.7)

$$\text{inward flux} \quad (\tfrac{1}{2}\pi \leqq \vartheta \leqq \pi) \qquad \pi F_\nu^- = -\int_0^{2\pi}\int_{\frac{1}{2}\pi}^{\pi} I_\nu \cos\vartheta \sin\vartheta \, \mathrm{d}\vartheta \, \mathrm{d}\varphi$$

whence $\qquad F_\nu = F_\nu^+ - F_\nu^-.$

Analogously to (11.3) we define

$$\text{total radiation-flux} \quad \pi F = \int_0^\infty \pi F_\nu \, \mathrm{d}\nu = \int_0^\infty \pi F_\lambda \, \mathrm{d}\lambda. \tag{11.8}$$

Consider next the radiation from a spherical star (Fig. 11.3). The intensity I_ν of the radiation emerging from its atmosphere will depend only on the angle of emergence ϑ reckoned from the normal at the point P concerned, that is $I_\nu = I_\nu(\vartheta)$. As seen from Fig. 11.3, the same angle ϑ is also the angle between the sight-line and the radius through P. The distance of P from the centre of the star as seen in projection on a plane perpendicular to the sight-line is $\sin\vartheta$ in units of the radius.

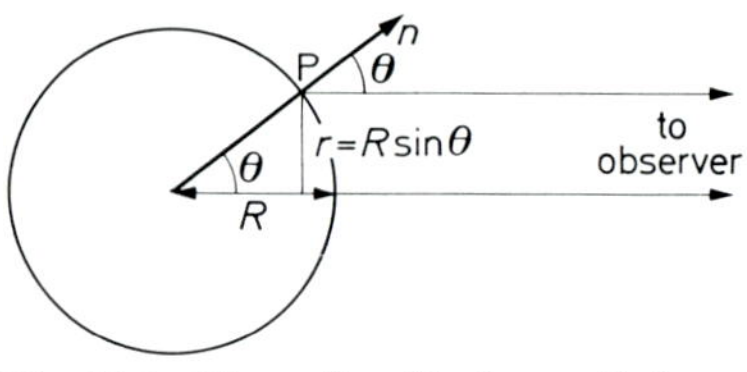

Fig. 11.3. Flux of radiation πF_ν from a star. The mean intensity is $\bar{I}_\nu = F_\nu$

The *mean intensity* $\bar{I}_\nu$ of the radiation emitted towards the observer from the visible disk, using (11.1), is given by

$$\pi R^2 \bar{I}_\nu = \int_0^{2\pi}\int_0^{\frac{1}{2}\pi} I_\nu(\vartheta) \cos\vartheta \, R^2 \sin\vartheta \, \mathrm{d}\vartheta \, \mathrm{d}\varphi. \tag{11.9}$$

If we cancel R^2 and compare with (11.7) we learn that

$$\bar{I}_\nu = F_\nu^+. \tag{11.10}$$

Thus the mean intensity of radiation over the apparent disk is equal to $1/\pi$ times the radiation flux at the surface. If the star of radius R is at a great distance r from the observer, then he sees the disk as subtending solid angle $\mathrm{d}\omega = \pi R^2/r^2$ and consequently he infers from (11.7)

$$\text{radiation-flux} \quad \pi F_\nu^* = \bar{I}_\nu \, \mathrm{d}\omega = F_\nu^+ \pi R^2/r^2. \tag{11.11}$$

Thus the theory of stellar spectra must have in view the calculation of the radiation-flux. As K. Schwarzschild was the first to remark, the advantage of observing the Sun is that in this case we can measure I_ν directly as a function of ϑ.

In the first instance phenomenologically, we must describe the emission and absorption of radiation: A volume-element dV emits per second in solid angle $d\omega$ and in frequency-interval ν to $\nu + d\nu$ the quantity of energy

$$\varepsilon_\nu \, d\nu \, dV \, d\omega \, . \tag{11.12}$$

The *emission-coefficient* ε_ν depends in general upon the frequency ν as well as the nature and state of the material (chemical composition, temperature and pressure) and also upon the direction. The total energy-output of an isotropically radiating volume-element dV per second is

$$dV \, 4\pi \int_0^\infty \varepsilon_\nu \, d\nu . \tag{11.13}$$

We set the emission against the energy-loss through *absorption* suffered by a narrow beam of radiation of intensity I_ν when it traverses a layer of matter of thickness ds. This is

$$dI_\nu/ds = -\varkappa_\nu I_\nu \, . \tag{11.14}$$

We call $\varkappa_\nu$ the *absorption-coefficient;* again it depends in general upon the nature and state of the material, and also upon direction. In (11.14) we have referred the absorption to a layer of thickness 1 cm. Instead of this, we can consider a layer that contains 1 g of material per square cm of the surface. Then we obtain the *mass absorption-coefficient* $\varkappa_{\nu,\mathrm{M}}$. If ϱ is the density of the material, then $\varkappa_\nu = \varkappa_{\nu,\mathrm{M}} \varrho$. Finally, if we reckon the absorption-coefficient per atom, we have the *atomic absorption-coefficient* $\varkappa_{\nu,\mathrm{at}}$. With n atoms per cm^3, we obtain $\varkappa_\nu = \varkappa_{\nu,\mathrm{at}} n$. As regards dimensions, we have

		Dimensions
Absorption-coefficient	$\varkappa_\nu$	$[cm^{-1}]$ or $[cm^2/cm^3]$
Mass absorption-coefficient	$\varkappa_{\nu,\mathrm{M}} = \varkappa_\nu/\varrho$	$[cm^2/g]$
Atomic absorption-coefficient	$\varkappa_{\nu,\mathrm{at}} = \varkappa_\nu/n$	$[cm^2]$
= absorption cross-section of the atom		(11.15)

If our pencil of radiation traverses a non-emitting layer of thickness ds then from (11.14)

$$\mathrm{d}I_\nu/I_\nu = -\varkappa_\nu\,\mathrm{d}s \tag{11.16}$$

and the intensity I_ν of the radiation after passage through the absorbing layer is related to the intensity $I_{\nu,0}$ of the incident radiation by

$$I_\nu/I_{\nu,0} = e^{-\tau_\nu} \tag{11.17}$$

where the dimensionless quantity

$$\tau_\nu = \int_0^s \varkappa_\nu\,\mathrm{d}s = \int_0^s \varkappa_{\nu,\mathrm{M}}\,\varrho\,\mathrm{d}s \tag{11.18}$$

is called the *optical thickness* of the layer traversed. For example, a layer of optical thickness $\tau_\nu = 1$ reduces a ray to $e^{-1} = 36.8$ percent

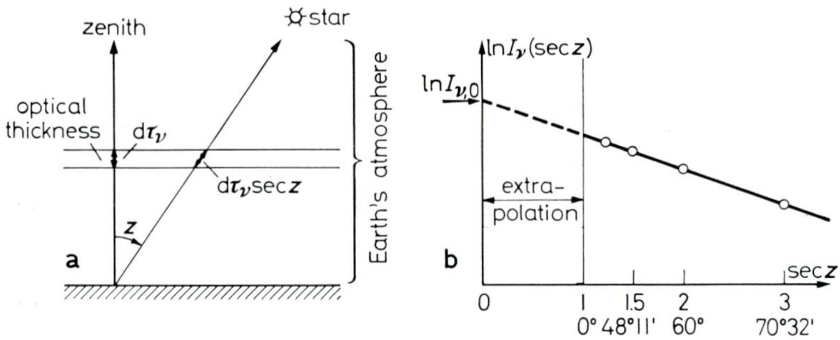

Fig. 11.4 a and b. a) Atmospheric extinction of radiation (frequency ν) of a star at zenith distance z. b) Extrapolation to sec $z = 0$ gives the intensity of radiation $I_{\nu,0}$ outside the Earth's atmosphere

of its original intensity. The *extinction* of the ν-radiation of a star (Fig. 11.4) at zenith distance z corresponds to a reduction-factor

$$I_\nu/I_{\nu,0} = e^{-\tau_\nu \sec z} \tag{11.19}$$

where τ_ν is the optical thickness, measured vertically, of the Earth's atmosphere at frequency ν. Provided τ_ν is not too large, we can derive the intensity outside the atmosphere by linear extrapolation of the values of $\ln I_\nu$, measured for different values of the zenith distance, for (formally) sec $z \to 0$.

We encounter particularly instructive relations if, following G. Kirchhoff (1860), we consider a radiation-field that is in *thermodynamic equilibrium*, or temperature-equilibrium, with its surroundings. We obtain such a radiation-field if we immerse an otherwise arbitrary cavity in a heat-bath of temperature T. Since by definition in such conditions all objects have the same temperature, it seems justifiable to speak of cavity-radiation of temperature T. (Here we always assume a refractive index of unity for the cavity; it would be easy to modify this hypothesis.) In an isothermal enclosure, every body and every surface element emit and absorb equal quantities of radiant energy per unit time. Starting from this we can easily show that the intensity I_ν of cavity-radiation is independent of the material contents and of the constitution of the walls of the cavity and of direction (the radiation is isotropic). We need consider only a cavity with two different chambers H_1 and H_2 (Fig. 11.5). Were $I_{\nu,1} \neq I_{\nu,2}$

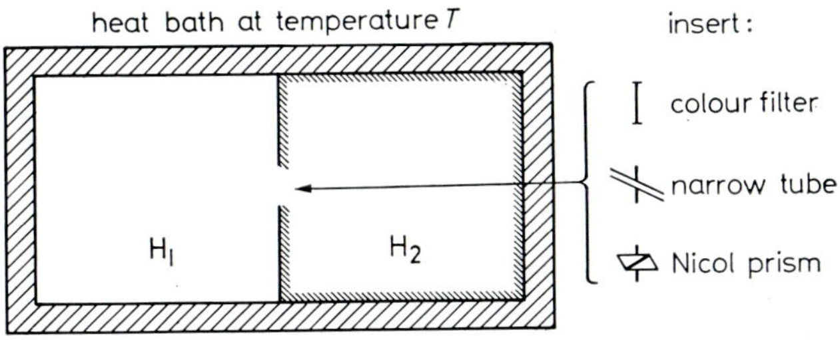

Fig. 11.5. Cavity radiation is unpolarized and isotropic. Its intensity is a universal function of ν and T, the Kirchhoff-Planck function B_ν (ν, T)

we could gain energy at an opening through which the chambers are put in communication by using say, a radiometer (the familiar "light-mill" of opticians' shop-windows) and we should have perpetual motion of the second kind. That $I_{\nu,1} = I_{\nu,2}$ must hold good for all frequencies, directions of propagation and of polarization we can readily show by introducing into the opening between H_1, H_2 in succession a colour filter, a tube of adjustable direction or a nicol prism. Also, in accordance with the second law of thermodynamics, we know that perpetual motion of the second kind is not possible. Thus the intensity of cavity radiation I_ν is a universal function $B_\nu(T)$ of ν and T.

We can produce and measure this quantity by "tapping" a cavity of temperature T through a sufficiently small opening. The important function $B_\nu(T)$, whose existence G. Kirchhoff recognized in 1860 and which M. Planck first explicitly calculated in 1900, we call the *Kirchhoff-Planck* function $B_\nu(T)$.

In thermodynamic equilibrium, i. e. in a cavity at temperature T, the emission and absorption by an arbitrary volume-element must be equal to each other. We write them down for a flat volume-element of base-area $d\sigma$ and height ds for the ν-radiation in solid angle $d\omega$ about the direction normal to $d\sigma$. Using (11.12), (11.14) we obtain

$$\begin{aligned} \text{emission per second} &= \varepsilon_\nu \, d\nu \, d\sigma \, ds \, d\omega \\ \text{absorption per second} &= \varkappa_\nu \, ds \, B_\nu(T) \, d\sigma \, d\omega \, d\nu . \end{aligned} \tag{11.20 a}$$

Equating these quantities we have Kirchhoff's law

$$\varepsilon_\nu = \varkappa_\nu B_\nu(T). \tag{11.20}$$

This asserts that *in a state of thermodynamic equilibrium the ratio of the emission-coefficient ε_ν to the absorption-coefficient $\varkappa_\nu$ is a universal function $B_\nu(T)$ of ν, T which at the same time expresses the intensity of cavity-radiation.*

As an important application we calculate the intensity I_ν of the radiation emitted by a layer of material of constant temperature T (e. g. the plasma in a gas-discharge) which has thickness s in the direction of the sight-line:

For a volume-element that, as seen by the observer, has cross-section 1 cm^2 and that extends from x to $x + dx$ in the direction of the sight-line ($0 \leqq x \leqq s$) the value of the radiation emitted per second per steradian is $\varkappa_\nu B_\nu(T) \, dx$. Up to its escape from the layer, this portion is reduced by the factor $\exp(-\varkappa_\nu x)$. Thus we obtain

$$I_\nu(s) = \int_0^s \varkappa_\nu B_\nu(T) \, e^{-\varkappa_\nu x} \, dx$$

or, introducing again the *optical depth* by $d\tau_\nu = \varkappa_\nu \, dx$ or $\tau_\nu = \varkappa_\nu s$

$$I_\nu(s) = B_\nu(T) \left(1 - e^{-\tau_\nu}\right) \approx \begin{cases} \tau_\nu B_\nu(T) & \text{for } \tau_\nu \ll 1 \\ B_\nu(T) & \text{for } \tau_\nu \gg 1 . \end{cases} \tag{11.21}$$

As is easily seen these formulae hold good also when $\varkappa_\nu$ depends upon x. Thus the emission by an isothermal optically thin layer ($\tau_\nu \ll 1$) is equal to its optical depth times the Kirchhoff-Planck function. The radiation intensity from an optically thick layer ($\tau_\nu \gg 1$), on the other hand, approximates to that of a black-body and cannot exceed this. Applied to a spectral line, this gives the familiar appearance of self-absorption: While two equally broad spectral lines from an optically thin layer show an intensity ratio $\sim \varkappa_1 : \varkappa_2$ in the case of an optically thick layer the ratio approaches unity. (See further Chapter 19.)

Instead of the emission and absorption by a volume-element, we can consider also those of a surface-element. We denote the emissive power by E_ν and the absorptive power A_ν (the reflective power is then $R_\nu = 1 - A_\nu$, then analogously to (11.20 a) we have

$$\begin{aligned} \text{emission per second} &= E_\nu \, d_\nu \, d\sigma \cos\vartheta \, d\omega \\ \text{absorption per second} &= A_\nu \, B_\nu(T) \, d\nu \, d\sigma \cos\vartheta \, d\omega \end{aligned} \tag{11.22 a}$$

and the Kirchhoff law receives the somewhat different form

$$E_\nu = A_\nu \, B_\nu(T) \, . \tag{11.22}$$

Thus *in thermodynamic equilibrium the ratio of emissive to absorptive powers is equal to the intensity of cavity-radiation $B_\nu(T)$.*

For a body that completely absorbs all frequencies ($A_\nu = 1$), a so-called *black body*, we have $E_\nu = B_\nu(\mathrm{T})$. Consequently we call cavity-radiation also *black-body radiation*, or *black radiation* for short. A small opening into a cavity swallows up all the radiation entering it, by multiple reflexion and absorption in the interior, whence we may again infer that cavity-radiation and black radiation are the same.

As regards the actual calculation of the function $B_\nu(T)$ Max Planck set about it in 1900 in the following way: He considered the interaction between the cavity radiation-field and, for instance, an electron bound elastically to an equilibrium position, a *harmonic oscillator*. Its vibrations are agitated by the electromagnetic waves of the radiation field until, in thermodynamic equilibrium, emission and absorption are equal to each other. Further, however, according to the Boltzmann law of equipartition, in thermodynamic equilibrium the mean energy of the harmonic oscillator is equal to kT, where k is the Boltzmann constant ($k = 1.38 \times 10^{-16}$ erg/degree). Clearly one can thence work backwards to compute the intensity $B_\nu(T)$ of the cavity-radiation in thermodynamic equilibrium at temperature T. Planck then recognized that he could correctly reproduce the measured intensities of cavity-radiation only if he extended the basic laws of classical mechanics and electrodynamics to meet the known requirements of *quantum theory*. Thus there resulted the *Planck radiation-formula* that we write on the ν- and λ-scales as

$$B_\nu(T) = \frac{2h\nu^3}{c^2} \frac{1}{e^{h\nu/kT} - 1} \quad \text{and} \quad B_\lambda(T) = \frac{2hc^2}{\lambda^5} \frac{1}{e^{hc/k\lambda T} - 1} \tag{11.23}$$

with the two important limiting cases

$$\frac{h\nu}{kT} \gg 1 \qquad B_\nu(T) \approx \frac{2h\nu^3}{c^2} e^{-h\nu/kT} \qquad \text{Wien's law} \tag{11.24}$$

$$\frac{h\nu}{kT} \ll 1 \qquad B_\nu(T) \approx \frac{2\nu^2 kT}{c^2} \qquad \text{Rayleigh-Jeans law.} \tag{11.25}$$

In the Rayleigh-Jeans law the characteristic quantity h of quantum theory has disappeared. Quite generally, quantum theory passes over into classical theory for light-quanta $h\nu$ whose energy is considerably smaller than the thermal energy kT (Bohr's correspondence principle). The radiation constant that appears in the exponent in the radiation law (11.23) is written as c_2. We have

$$c_2 = hc/k = 1.438_5 \text{ cm degree.} \tag{11.26}$$

We obtain the total radiation of a black body by integrating (11.23) over all frequencies $B(T) = \int_0^\infty B_\nu(T)\,d\nu$. The total flux, that is the emission of a black body from 1 cm² into free space is $\pi F^+ = \pi B(T)$. If we carry out the integration of Planck's formula (using the variable $x = h\nu/kT$) we derive the Stefan-Boltzmann radiation law

$$\pi F^+ = \pi B(T) = \sigma T^4 \tag{11.27}$$

with the radiation-constant

$$\sigma = \frac{2\pi^5 k^4}{15 c^2 h^3} = 5.67 \times 10^{-5} \text{ erg/cm}^2 \text{ sec degree}^4. \tag{11.28}$$

This law was found experimentally by J. Stefan in 1879 and derived theoretically by L. Boltzmann in 1884 using an ingenious calculation of the entropy of cavity-radiation.

12. The Sun

We first collect together some known results for the Sun. These will then often provide the most useful units for dealing with the stars.

Using the solar parallax of 8″.794 we first derived the mean Sun-Earth distance, or the astronomical unit,

$$1 \text{ AU} = 149.6 \times 10^6 \text{ km} = 23\,456 \text{ equatorial Earth-radii} \tag{12.1}$$

The corresponding apparent radius of the solar disk is

$$15'59''.63 = 959''.63 = 0.004\,652_4 \text{ radian.} \tag{12.2}$$

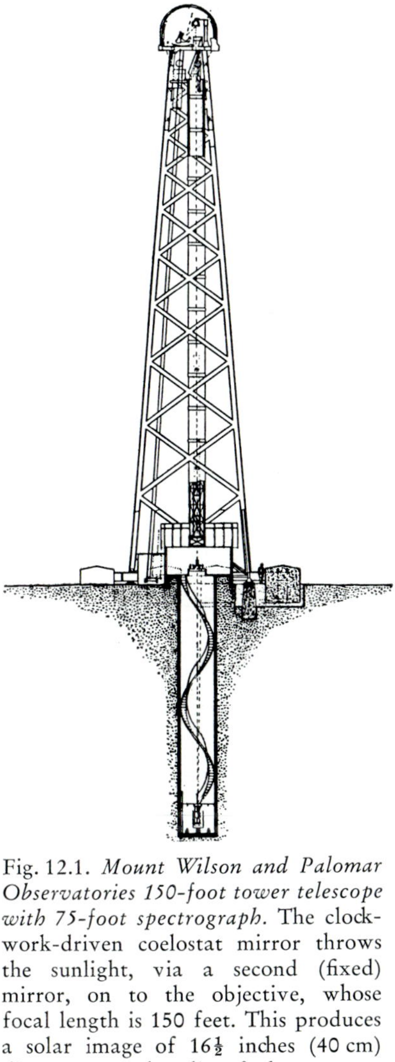

Fig. 12.1. *Mount Wilson and Palomar Observatories 150-foot tower telescope with 75-foot spectrograph.* The clockwork-driven coelostat mirror throws the sunlight, via a second (fixed) mirror, on to the objective, whose focal length is 150 feet. This produces a solar image of 16½ inches (40 cm) diameter on the slit of the spectrograph at about ground level. The grating and lens of the spectrograph are located underneath the tower in a temperature-constant well 75 feet below ground level. The spectrum can be photographed or observed visually near the slit

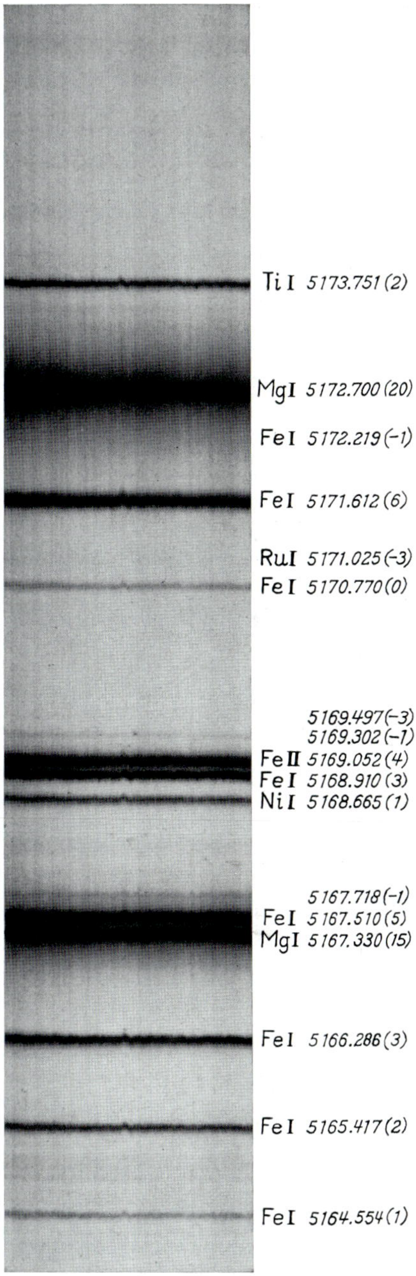

Fig. 12.2. *Solar spectrum (centre of disk),* λ 5164—5176 Å, 5th order, taken with the vacuum grating spectrograph of the McMath-Hulbert Observatory. Wavelength, identification and (estimated) Rowland intensity of the Fraunhofer lines are marked at the side

Thence we obtain

$$\text{solar radius} = 696\,000 \text{ km.} \tag{12.3}$$

Thus on the Sun

$$1' = 43\,500 \text{ km} \qquad 1'' = 725 \text{ km.} \tag{12.4}$$

On account of atmospheric scintillation the limit of resolution is about 500 km.

The astronomical unit also corresponds to 215 solar radii. A flattening of the Sun resulting from its rotation is not detectable ($< 0\rlap{.}''1$).

Using Kepler's third law we obtain

$$\text{solar mass } \mathfrak{M}_\odot = 1.989 \times 10^{33} \text{ g.} \tag{12.5}$$

Thence we find the mean density of the Sun $\bar{\varrho} = 1.409$ g/cm^3. Also we find

$$\text{gravity at the Sun's surface } g_\odot = 2.74 \times 10^4 \text{ cm/sec}^2 \tag{12.6}$$

which is 27.9 times gravity at the surface of the Earth.

Disregarding for the moment radiation in far ultraviolet- and in radio-frequencies, the solar spectrum as observed using a tower-telescope is seen as a continuous spectrum crossed by many dark Fraunhofer lines (Fig. 12.2). The *Rowland Atlas* and the *Revision of Rowland's preliminary table of solar spectrum wavelengths* (1928) contains wavelengths λ and identifications of these lines. The Utrecht photometric atlas of the solar spectrum by M. Minnaert, G. F. W. Mulders and J. Houtgast (1940) contains the microphotometrically recorded intensity-distribution in the whole solar spectrum from 3332 Å to 8771 Å referred to the continuum as 100. These measurements refer to the centre of the solar disk. The solar disk shows a considerable centre-limb darkening; also the lines relative to the continuum at the same place exhibit slight centre-limb variations.

We denote the position on the solar disk to which an observation refers by giving the distance ϱ from the centre of the disk in terms of the radius as unity or, for theoretical purposes, the angle ϑ between the sightline and the normal to the Sun's surface (Fig. 11.3). Thus

$$\varrho = \sin\vartheta \quad \text{and} \quad \cos\vartheta = \sqrt{(1-\varrho^2)}. \tag{12.7}$$

In particular, the centre of the disk is $\varrho = \sin\vartheta = 0$ and the limb is $\varrho = \sin\vartheta = 1$, $\cos\vartheta = 0$. The radiation intensity at the Sun's surface (optical depth $\tau_0 = 0$, see below) at distance $\varrho = \sin\vartheta$ from the centre of the disk, and referred to the wavelength-scale, we denote by

$$I_\lambda(0, \vartheta) \; [\mathrm{erg/cm^2 \, sec \, ster}; \Delta\lambda = 1] \,. \tag{12.8}$$

One measures the centre-limb variation of the radiation-intensity

$$I_\lambda(0, \vartheta)/I_\lambda(0, 0) \tag{12.9}$$

by allowing the solar image to move across the slit of a spectrograph with, for instance, a photoelectric recording system. The essential difficulty in this and all other measurements of details of the solar disk is in the elimination of scattered light in the instrument and in the Earth's atmosphere. Fig. 12.3 shows a summary of recent results.

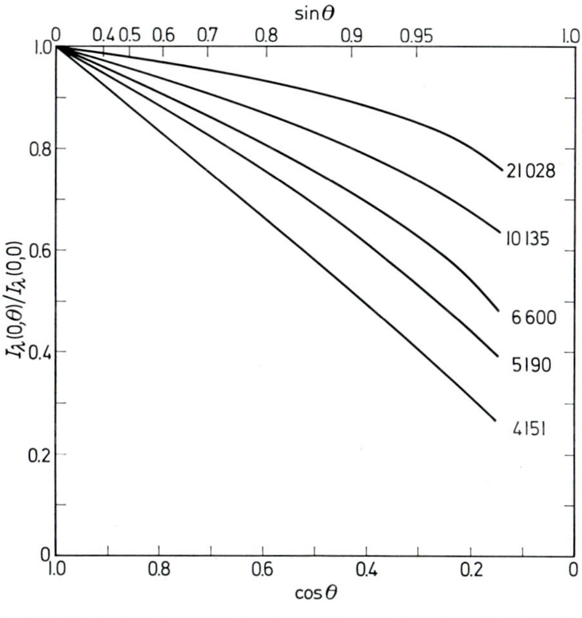

Fig. 12.3. *Limb-darkening of the Sun.* The ratio of radiation intensities $I_\lambda(0, \vartheta)/I_\lambda(0, 0)$ is plotted against $\cos\vartheta$ for several line-free wavelengths, ranging from the blue to the infrared, shown in angstrom units

The ratio of the mean intensity F_λ over the solar disk to the intensity $I_\lambda(0, 0)$ at the centre is given by

$$\frac{F_\lambda}{I_\lambda(0, 0)} = 2 \int_0^{\frac{1}{2}\pi} \frac{I_\lambda(0, \vartheta)}{I_\lambda(0, 0)} \cos\vartheta \sin\vartheta \, d\vartheta = \int_0^1 \frac{I_\lambda(0, \varrho)}{I_\lambda(0, 0)} \, d(\varrho^2) \,. \tag{12.10}$$

Nowadays one measures absolute values of the radiation intensity, say $I_\lambda(0, 0)$ for the centre of the disk, by comparison with a black body of known temperature. The latter in its turn is measured with reference to the highest exactly determined fixed point of the temperature-scale, namely

$$\text{melting point of gold } T_{\mathrm{Au}} = 1336.2\ °\mathrm{K} \text{ (or } 1063.0\ °\mathrm{C}).$$

Obviously the extinction in the Earth's atmosphere (Fig. 11.4) must be accurately determined for numerous wavelengths and allowed for in the measurements.

Since the solar spectrum (Fig. 12.2) is crossed by many Fraunhofer lines, it is best to make measurements for sharply defined wavelength-intervals, e. g. each of width $\Delta\lambda = 20$ Å, of the mean intensity in the spectrum including the lines $I_\lambda^L\,(0, 0)$. However, for the physics of the Sun the true continuum between the lines $I_\lambda\,(0, 0)$ is of greater interest. Its determination in the long wavelength region $\lambda > 4600$ Å presents no difficulty. On the other hand, in the blue and violet spectral region, about $\lambda < 4600$ Å, the lines are so closely packed that the determination of the true continuum $I_\lambda\,(0, 0)$ becomes more and more difficult until finally in the ultraviolet this is possible only in conjunction with a fully developed theory. The ratio

$$1 - \eta_{\bar{\lambda}} = I_\lambda^{\mathrm{L}}\,(0, 0)/I_\lambda\,(0, 0) \tag{12.11}$$

for a suitable restricted wavelength interval $\Delta\lambda$ of width, say, 20 to 100 Å centred on the wavelength $\bar{\lambda}$ is measured by using a planimeter to obtain the area under the microphotometer-tracing of a spectrum of high dispersion such as that in the Utrecht Atlas. In the ultraviolet (λ 3000 to 4000 Å) $\eta_{\bar{\lambda}}$ is of the order 25 to 45 percent; it falls to a few percent for $\lambda > 5000$ Å. Fig. 12.4 shows the results of the latest and probably the most accurate measurements by D. Labs and H. Neckel; earlier values by D. Chalonge, R. Peyturaux, A. K. Pierce and others are in good agreement with these.

By integrating over the solar disk according to equation (12.10) and integrating over all wavelengths, allowing by extrapolation for the parts of the spectrum cut off at the ends 3.9 percent in the ultraviolet ($\lambda < 3420$ Å) and 4.8 percent in the infrared ($\lambda > 23\,000$ Å), we obtain the total flux at the surface of the Sun

$$\pi F = \int_0^\infty \pi F_\lambda^L\, d\lambda = 6.32 \times 10^{10}\ \mathrm{erg/cm^2\ sec}. \tag{12.12}$$

Hence we easily derive the total emission of the Sun per second

$$\text{Luminosity } L_\odot = 4\,\pi R^2\,\pi F = 3.84 \times 10^{33}\ \mathrm{erg/sec}.$$

Further we obtain the flux S at the Earth's distance ($r=1$ astronomical unit) following equation (11.11) by multiplying the mean intensity F

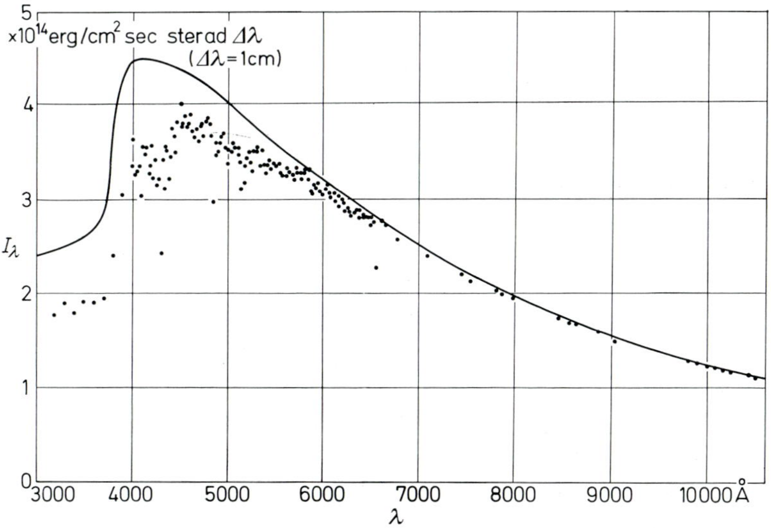

Fig. 12.4. *Intensity of the spectrum of the centre of the solar disk* in erg/cm² sec, $\Delta\lambda = 1$ cm, after D. Labs and H. Neckel. The dots give the *mean intensity* of the spectrum including spectral lines for sharply defined wavelength intervals each of width 20 Å. The continuous curve shows the *true continuum* I_λ (0, 0) between the spectral lines

the solar disk by the solid angle subtended at the Earth

$$\pi R^2/r^2 = 6.800 \times 10^{-5} \text{ steradian} \tag{12.13}$$

whence

$$S = 1.368 \times 10^6 \text{ erg/cm}^2 \text{ sec}\,. \tag{12.14}$$

After pioneering attempts by S. S. Pouillet (1837), this important quantity was first measured accurately by K. Ångström (c. 1893) and C. G. Abbot (from c. 1908) when they measured with a "black" receiver, the so-called pyrheliometer, the total solar radiation reaching the Earth's surface. However, this measurement has to be supplemented by absolute measurements of the radiation as distributed through the spectrum since only from these can one eliminate the atmospheric extinction in accordance with Fig. 11.4. The results of the pyrheliometer measurements are usually expressed in calories; including the ultraviolet and infrared corrections, one obtains the so-called

$$\text{solar constant } S = 1.961 \text{ calories/cm}^2 \text{ minute.} \tag{12.15}$$

Transforming to CGS units one recovers (12.14).

In view of the rapidly expanding energy-requirements of mankind it is interesting to express the energy-flux supplied by the Sun in commercial units. We derive

$$S = 1.37 \text{ kilowatt/m}^2 \text{ or } 1.54 \text{ British horse-power/sq.yard.} \tag{12.16}$$

This supply is available outside the Earth's atmosphere, but even after allowing for the extinction, a considerable energy-contribution does reach the ground, at anyrate in countries with a favourable climate. Nevertheless, direct practical use of solar energy has been made hitherto only for generating electric currents in satellites and to a small extent for cooking- and hot-water-installations in tropical countries.

We return once again to consider the total flux πF and the radiation intensity $I_\lambda(0, 0)$ at the Sun's surface, (12.12) and Fig. 12.4. As our first rather formal and provisional approach to the temperature in the solar atmosphere, we interpret the flux in the sense of the Stefan-Boltzmann radiation law and thence we define

$$\text{Sun's effective temperature:} \quad \pi F = \sigma T_{\text{eff}}^4 \, ; \; T_{\text{eff}} = 5780\ ^\circ\text{K} \tag{12.17}$$

with a mean error of $\pm 20^\circ$. Further, if we interpret the radiation-intensity $I_\lambda(0, 0)$ in the sense of Planck's radiation-law (11.23) we define the radiation-temperature T_λ for the centre of the disk as a function of wavelength.

Since the Sun does not radiate as a black body — otherwise $I_\lambda(0, \vartheta)$ would be independent of ϑ and the Sun would show no darkening towards the limb — we must not interpret T_{eff} and the T_λ quite literally. Nevertheless T_{eff} does to some extent correctly indicate the temperature of the layer of the solar atmosphere from which the total radiation comes to us, and T_λ that of the layer from which the radiation of wavelength λ comes. The effective temperature is thus an important parameter of the solar atmosphere (and in a corresponding way of stellar atmospheres) since in combination with the Stefan-Boltzmann radiation-law it represents the total radiation πF, i.e. the total energy-flux that comes out of the interior through 1 cm² of the Sun's surface.

The layers of the solar atmosphere from which the continuous radiation originates we call the *photosphere*. Formerly one distinguished between this and the reversing layer which was supposed to lie above it and to produce the dark Fraunhofer lines as in the well-known Kirchhoff-Bunsen experiment. Nowadays we know that at anyrate considerable parts of the lines arise in the same layers as the continuum, so that it is better not to use the term "reversing layer" any longer. At total eclipses of the Sun one observes at the solar limb the higher layers of the atmosphere which yield no appreciable continuum, but practically only *emission-lines* corresponding to the Fraunhofer lines. This part is called the *chromosphere.* As we now know, this makes only a small contribution to the intensity of the absorption-lines. Above it lies the *solar corona* which, as we shall see, merges into the *interplanetary medium.* Later on, we shall concern ourselves further with these extreme outer layers of the Sun as well as with the whole manifestation of solar activity (sunspots, prominences, eruptions, etc) i. e., with the "disturbed" Sun, as we now often call it.

13. Apparent Magnitudes and Colour Indices of Stars

Hipparchus and many of the earlier astronomers long ago catalogued the brightnesses of stars. We may speak of stellar photometry in the modern sense since N. Pogson in 1850 gave a clear definition of magnitudes and J. C. F. Zöllner in 1861 constructed his visual photometer with which one could accurately compare the brightness of a star with that of an artificial starlike image by employing two nicol prisms. (Two such prisms whose planes of polarisation are inclined at angle α to each other reduce the intensity of light passing through them both by a factor $\cos^2 \alpha$.)

If the ratio of the fluxes from two stars as measured with a photometer is S_1/S_2, then according to Pogson's definition, the difference of their apparent magnitudes is

$$m_1 - m_2 = -2.5 \log (S_1/S_2) \text{ magnitudes} \tag{13.1}$$

or conversely

$$S_1/S_2 = 10^{-0.4\,(m_1 - m_2)}. \tag{13.2}$$

The following are corresponding magnitude-differences and flux-ratios and the corresponding brightness-ratios:

$-\Delta m$	1	2.5	5	10	15	20	magnitudes
$\Delta \log S$	0.4	1	2	4	6	8	
Brightness ratio	2.512	10	100	10^4	10^6	10^8	

The zero-point of the magnitude scale is fixed by the international polar sequence, a set of stars in the vicinity of the north pole which have been very accurately measured and the constancy of their brightness established. Greater brightness corresponds to smaller and ultimately negative magnitudes and so it is recommended to say, for instance, "The star α Lyrae (Vega) of apparent (visual) magnitude $0^{\mathrm{m}}.14$ is 1.19 magnitudes brighter than α Cygni (Deneb) of magnitude $1^{\mathrm{m}}33$, or α Cyg is 1.19 magnitudes fainter than α Lyr".

The older photometric measurements were made visually. With K. Schwarzschild's *Göttinger Aktinometrie* photographic photometry in the years 1904 to 1908 was added, first using ordinary blue-sensitive plates. One soon learned to imitate the spectral sensitivity-distribution of the human eye using sensitized plates made sensitive in the yellow and with a yellow filter in front of them. Thus one had, in addition to the visual magnitudes m_{v}, the photographic m_{pg} and then the photovisual magnitudes m_{pv}. By combining suitable plates or photocells and photomultipliers with appropriate colour-filters, one can nowadays adjust the sensitivity-maximum of ones measuring devices, whether photographic or photoelectric, to be in any wave-length interval from the ultraviolet down to the infrared.

We now seek to clarify our ideas by asking, What precisely do our various magnitudes mean?

Suppose that a star of radius R emits at its surface the flux πF_λ or mean intensity F_λ at wavelength λ in the true continuum; suppose that a fraction η_λ of this is removed by Fraunhofer lines. Neglecting interstellar absorption for the time being, if the star is at distance r from the Earth then we receive outside the Earth's atmosphere a flux (see (12.13))

$$S_\lambda = \pi R^2 F_\lambda (1-\eta_\lambda)/r^2. \tag{13.3}$$

Now let the contribution to the stimulus of our measuring device made by a normal spectrum with $S_\lambda = 1$ be described by a sensitivity function E_λ. Then in the case of the star the stimulus is proportional to the integral

$$\int^{\infty} \frac{\pi R^2 F_\lambda (1-\eta_\lambda)}{r^2} E_\lambda \, d\lambda \tag{13.4}$$

and the apparent magnitude m becomes

$$m = -2.5 \log \int_0^{\infty} \frac{\pi R^2 F_\lambda (1-\eta_\lambda)}{r^2} E_\lambda \, d\lambda + \text{constant} \tag{13.5}$$

to within a normalizing constant which will have to be fixed by some convention. In (13.5) we have taken no account of extinction by the Earth's atmosphere. Since in practice E_λ has an appreciable value only in a small range of wavelength, the extinction can be determined following Fig. 11.4 as for monochromatic radiation at the centre of gravity of E_λ.

The *visual magnitude* m_v is defined by (13.5) using for the sensitivity function that of the human eye multiplied by the power of transmission of the instrument, of which the centre of gravity lies at the so-called isophotal wavelength of about 5400 Å in the green. In the same way, the centre of gravity for the *photographic magnitude* m_{pg} is about 4200 Å.

As a standard system of magnitudes and colours one nowadays mostly uses the *UBV*-system developed by H. L. Johnson and W. W. Morgan (1951) (U = ultraviolet, B = blue, V = visual); the corresponding magnitudes are written for short as

$$U = m_U \qquad B = m_B \qquad V = m_V\,. \tag{13.6}$$

The corresponding sensitivity functions E_λ, which can be realized either photographically or photoelectrically, are shown in Fig. 13.1.

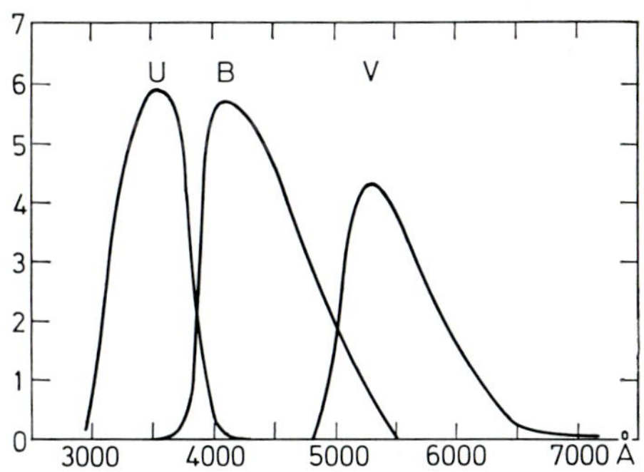

Fig. 13.1. *Relative response functions E_λ (referred to a light source with I_λ = const.)* of the UBV photometric system, after H. L. Johnson and W. W. Morgan

Their centres of gravity for average stellar colours are

$$\lambda_U \approx 3650\,\text{Å} \qquad \lambda_B \approx 4400\,\text{Å} \qquad \lambda_V \approx 5480\,\text{Å}. \tag{13.7}$$

For hot (blue) stars these are displaced towards shorter wavelengths; for cool (red) stars towards longer wavelengths.

By definition, the three magnitudes U, B, V are so related to each other, i. e. the constants in equation (13.5) are so chosen that for A0V-stars (for example α Lyr = Vega, see Chapter 15)

$$U = B = V. \tag{13.8}$$

As K. Schwarzschild first recognized, colour-indices such as, for example

$$U - B \quad \text{and} \quad B - V \tag{13.9}$$

provide a measure of the energy-distribution in the spectrum of a star. If as an approximation to the energy-distribution we take Planck's law for a black body or, so long as $h\nu/kT \gg 1$, Wien's law

$$F_\lambda \sim e^{-c_2/\lambda T} \quad \text{and} \quad \eta_\lambda = 0 \tag{13.10}$$

and if we concentrate the integration in (13.5) to the centre of gravity of the sensitivity function, we obtain

$$m_V = \frac{2 \cdot 5\, c_2 \log e}{\lambda_V T} + \text{const.}_V \tag{13.11}$$

or, with $c_2 = 1.4385$ cm deg and $\log e = M = 0.4343$,

$$V = \frac{1.562}{\lambda_V T} + \text{const.}_V. \tag{13.12}$$

Using (13.7) and the normalization (13.8) and taking for a A0V star $T = 15\,000°$, we find

$$B - V \approx 7000 \left(\frac{1}{T} - \frac{1}{15\,000} \right). \tag{13.13}$$

The values of the *colour temperatures* calculated, for example, from the colour indices $B - V$ have no profound significance because of the considerable departure of the stars from black bodies. To-day the significance of colour indices lies in a different direction: As the theory of stellar atmospheres shows, the colour-indices (or the colour-temperature) depend upon the fundamental parameters, in particular the effective temperature T_{eff} (total flux) and gravity g or absolute magnitude (see Chapter 18) of the stars. Since colour indices can be measured photoelectrically to an accuracy of $0\overset{\text{m}}{.}01$, we expect from (13.13) that near, say, 7000° K *temperature-differences* can be determined to an accuracy of about 1 percent, which cannot be attained by any other method. Naturally, the temperatures themselves are far less accurately determined.

In addition to the *UBV*-system, significance has been achieved by the *UGR*-system used by W. Becker with characteristic wavelengths

3660, 4630 and 6380 Å and by the six-colour system of J. Stebbins, A. E. Whitford and G. Kron which extends from the ultraviolet into the infrared ($\lambda_U = 3550$ Å ... $\lambda_I = 10\,300$ Å). The conversion of magnitudes and colour-indices from one photometric system to another with not greatly different characteristic wavelengths is performed mostly by using empirically established linear relations.

Besides the radiation in various wavelength intervals, we are interested in the total radiation of stars. Analogously to the solar constant, in the sense of equation (13.5) we therefore define apparent bolometric magnitude

$$m_{\text{bol}} = -2.5 \log \int_0^\infty \frac{\pi R^2 F_\lambda (1-\eta_\lambda)}{r^2}\, d\lambda + \text{constant}$$

$$= -2.5 \log (\pi R^2 F/r^2) + \text{constant} \tag{13.14}$$

where πF is again the total radiation flux at the surface of the star, including the effect of the spectral lines. Usually we define the constants so that, say at about the solar temperature, the

$$\text{bolometric correction BC} = m_{\text{v}} - m_{\text{bol}} \tag{13.15}$$

is equal to zero. Since the Earth's atmosphere completely absorbs considerable parts of the spectrum, we cannot measure the bolometric magnitude directly from ground-based observations; even the term bolometric magnitude in the strict sense may be misleading. Until observations from satellites are available, we can evaluate m_{bol} or BC only with the help of a theory as a function of other measurable parameters of stellar atmospheres.

14. Distances, Absolute Magnitudes and Radii of the Stars

As a result of the revolution of the Earth round the Sun, in the course of a year a nearer star must describe a small ellipse in the sky relative to the much more distant fainter stars (Fig. 14.1: to be distinguished from aberration, Fig. 5.7). Its semi-major axis, which is the angle that the radius of the Earth's orbit would subtend at the star, is called the heliocentric or annual *parallax* p of the star (παράλλαξιϛ = to and fro motion).

In the year 1838 F. W. Bessel in Königsberg succeeded, with the help of Fraunhofer's heliometer, in measuring directly (trigonometrically) the parallax of the star 61 Cygni $p = 0''.293$. We must mention also the simultaneous work of F. G. W. Struve in Dorpat and of T. Henderson at the Cape Observatory. The star α Centauri, with its companion Proxima Centauri, observed by the latter is our

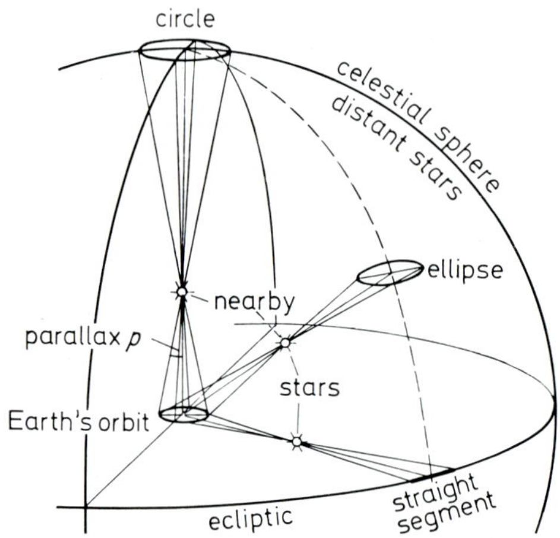

Fig. 14.1. *Stellar parallax p.* Relatively to the much more distant background stars, a nearby star in one year describes on the sky a circle of radius p if it is at the pole of the ecliptic, a straight segment $\pm p$ if it is in the ecliptic, and an ellipse if it is in between

nearest neighbour in space. According to recent measurements, their parallaxes are 0.″75 and 0.″76. It was a fundamental advance when in 1903 F. Schlesinger achieved the photographic measurement of parallaxes with an accuracy of about ±0.″01. The *General catalogue of trigonometric stellar parallaxes* and the *Catalogue of bright stars* (Yale 1952 and 1940) are amongst the most important aids to astronomers.

The first measurements of parallax signified not only the by then scarcely necessary vindication of the Copernican world-system, but above all the first quantitative advance into space. We next define the appropriate units of measurement:

A parallax $p = 1''$ corresponds to a distance $360 \times 60 \times 60/2\pi =$ 206 265 astronomical units or radii of the Earth's orbit. We call this distance 1 parsec (from *par*allax and *sec*ond) contracted to pc. Thus we have

$$1 \text{ parsec} = 3.086 \times 10^{13} \text{ km} = 3.086 \times 10^{18} \text{ cm or } 3.26 \text{ lightyears}$$
$$1 \text{ lightyear} = 0.946 \times 10^{18} \text{ cm}. \tag{14.1}$$

Thus light travelling at 300 000 km/sec takes 3.26 years to traverse

1 parsec. The distance $1/p$ parsec or $3.26/p$ lightyears corresponds to parallax p''. Corresponding to the accuracy with which trigonometric parallaxes can be measured, these take us out to distances of "only" some 15 pc or maybe 50 pc. We can reach out to about 2000 pc using "streaming"- or cluster-parallaxes (to be discussed later) for groups of stars that "stream" through the Milky Way system with a common velocity vector.

The observed brightness of a star (in any photometric system) depends according to equation (13.5) on its true brightness and its distance. We now define the *absolute magnitude* M of a star as the magnitude it would appear to have if it were transferred from its actual distance $r = 1/p$ parsec to a standard distance of 10 parsec. By the $1/r^2$ law of photometry (e. g. equation (13.4)), the brightness is then changed by the factor $(r/10)^2$; in magnitudes we have therefore

$$m - M = 5 \log (r/10) = -5\,(1 + \log p). \tag{14.2}$$

We call $m - M$ the *distance-modulus*. The following are corresponding values of the modulus and the distance:

(14.3)

$m - M =$	-5	0	$+5$	$+10$	$+25$ magnitudes
r $=$	1 pc	10 pc	100 pc	1000 pc = 1 kpc (1 kiloparsec)	10^6 pc = 1 Mpc (1 megaparsec)

In calculating in accordance with the $1/r^2$ law we have neglected interstellar absorption. We shall see that this can become significant for distances exceeding even 10 pc; then the relation (14.2) between distance modulus and distance (or parallax) is to be correspondingly modified.

Basically we can specify absolute magnitudes in every photometric system and we then attach the appropriate suffix to M; for example, visual absolute magnitude is written M_v. If no suffix is written then we always imply M_v.

We now collect the important properties of the Sun as a star; these serve also to make it a standard of comparison. We need scarcely emphasise the enormous technical difficulty of comparing the Sun photometrically with stars that are fainter by at least 10 powers of ten. The distance modulus of the Sun is determined by the definition of the parsec at the beginning of the Chapter directly as

$m - M = -31.57$. Thus we obtain:

SUN: Apparent magnitude			colour index	absolute magnitude
Ultraviolet	U	$= -26.06$	$U-B = +0.10$	$M_U = +5.51$
Blue	B	$= -26.16$	$B-V = +0.62$	$M_B = +5.41$
Visual	V	$= -26.78$	$\mathrm{BC} = +0.07$	$M_V = +4.79$
Bolometric	m_{bol}	$= -26.85$		$M_{\mathrm{bol}} = +4.72$

(14.4)

The absolute bolometric magnitude of a star M_{bol} is a measure of its total energy emission per second by radiation. We call this its *luminosity* L and we usually express it in terms of the Sun $L_\odot$ as unit. Thus

$$M_{\mathrm{bol}} - 4.72 = -2.5 \log L/L_\odot \text{ with } L_\odot = 3.84 \times 10^{33} \text{ erg sec}^{-1}. \quad (14.5)$$

Since the energy radiated away from the stars is generated by nuclear processes in their interiors, the luminosity is amongst the basic data for the investigation of the internal structure of the stars.

From equation (13.3) the flux S_λ that we receive from a star is $S_\lambda = F_\lambda (1 - \eta_\lambda) \pi R^2/r^2$. Here R is the radius and r the distance of the star. The ratio of these two quantities has a simple meaning: the very small angle α that the radius R subtends at distance r is $R/r = \alpha$ radian or $206\,265\, R/r = \alpha''$ arc-seconds.

Comparing with the Sun ($\alpha''_\odot = 959.6$) we have

$$m - m_\odot = -2.5 \log \{F_\lambda \alpha''^2 / F_{\lambda\odot} \alpha''^2_\odot\} \quad \text{or} \quad (\alpha''/\alpha''_\odot)^2 = (F_{\lambda\odot}/F_\lambda)\, 10^{-0.4\,(m - m_\odot)} \quad (14.6)$$

For the radii themselves we have the corresponding result

$$M - M_\odot = -2.5 \log \{F_\lambda R^2 / F_{\lambda\odot} R^2_\odot\} \quad \text{or} \quad (R/R_\odot)^2 = (F_{\lambda\odot}/F_\lambda)\, 10^{-0.4\,(M - M_\odot)} \quad (14.7)$$

As a crude estimate we may calculate the mean intensity F_λ according to the Planck radiation-law and in the Wien approximation ($c_2/\lambda T \gg 1$) we obtain

$$\frac{F_{\lambda\odot}}{F_\lambda} = \exp \frac{c_2}{\lambda} \left(\frac{1}{T} - \frac{1}{T_\odot} \right) \quad (14.8)$$

where λ is the isophotal wavelength and T, $T_\odot$ denote "the" temperatures of the star and the Sun.

If we wish to claim higher accuracy, however, we must calculate F_λ from the theory of stellar atmospheres and in terms of the parameters used in the theory. For certain bright red giant stars it has been possible to use the Michelson stellar interferometer to confirm the estimates got from (14.6)—(14.8). For example, it was found for α Orionis (Betelgeuse) with $m_v = 0.9$, $p = 0''.017$ and $T \approx 3200°K$

$$\alpha \approx 0''.024 \quad \text{or} \quad R \approx 300\, R_\odot \text{ (somewhat variable).} \qquad (14.9)$$

Thus the dimensions of this star correspond to about those of the orbit of Mars.

Table 15.1. *Classification of stellar spectra*

Spectral type	Temperature	Criteria for classification
O	50 000 °K	Lines of highly ionized atoms: He II, Si IV, N III . . .; hydrogen H relatively weak; occasionally emission lines.
B0	25 000 °K	He II absent; He I strong; Si III, O II; H stronger.
A0	11 000 °K	H I absent; H at maximum; Mg II, Si II strong; Fe II, Ti II weak; Ca II weak.
F0	7 600 °K	H weaker; Ca II strong; the ionized metals e. g. Fe II, Ti II reach maximum about A5; the neutral metals e. g. Fe I, Ca I now reach about the same strength.
G0	6 000 °K	Ca II very strong; neutral metals Fe I . . . strong.
K0	5 100 °K	H relatively weak, neutral atomic lines strong; molecular bands.
M0	3 600 °K	Neutral atomic lines, e. g. Ca I, very strong; TiO bands.
M5	3 000 °K	Ca I very strong; TiO bands stronger.
R, N (C)	3 000 °K	Strong CN, CH, C_2 bands; TiO absent; neutral metals as in K, M.
S	3 000 °K	Strong ZrO, YO, LaO bands; neutral atoms as in K, M.

(R, N currently designated C)

15. Classification of Stellar Spectra: Hertzsprung-Russell Diagram and Colour-Magnitude Diagram

When, following the discoveries of J. Fraunhofer, G. Kirchhoff and R. Bunsen, the observation of stellar spectra was begun, it soon appeared that in the main these could be arranged in a one-parameter sequence. The associated change of colour of the stars, or of their colour indices, showed the arrangement to be in order of decreasing temperature.

Starting from the work of Huggin's, Secchi, Vogel and others, in the 1880's E. C. Pickering and A. Cannon developed the Harvard classification of stellar spectra, which formed the basis for the *Henry Draper* catalogue. The sequence of spectral classes or "Harvard types"

$$\begin{array}{c} \qquad\qquad\qquad\quad \nearrow \mathrm{S} \\ \mathrm{O—B—A—F—G—K—M} \\ \qquad\qquad\qquad\quad \searrow R - N \\ \text{blue} \qquad \text{yellow} \qquad \text{red} \end{array} \tag{15.1}$$

where we have noted the colours of the stars, resulted after many modifications and simplifications. H. N. Russell's students in Princeton composed the well-known mnemonic O *B*e *A* *F*ine *G*irl *K*iss *M*e *R*ight *N*ow*.

Between two letters a finer subdivision is denoted by numbers 0 to 9 written after the first letter. For example, a B5 star comes between B0 and A0 and has about the same amount in common with both these types.

The Harvard sequence is established primarily by use of photographs of the spectra of certain standard stars (Fig. 15.1: the two spectra A0 I, A0 II should be disregarded for the moment). We describe it briefly on the preceding page 123, associating with it the lines used as classification criteria, the chemical elements to which they belong, and the ionization levels (I = neutral atom, arc-spectrum;

* Footnote for experts only: S is for Smack.

⟶

Fig. 15.1. *MK classification of stellar spectra,* from "An Atlas of Stellar Spectra" by W. W. Morgan, P. C. Keenan and E. Kellman (1942). The main sequence (luminosity class V) is shown from O9 to B9 (above), then A0V (centre) and F0—M2 (below). The *luminosity classes* I (supergiants) and II (luminous giants) have been added at A0 in order to show the effect of absolute magnitude (spectroscopic parallax determinations are based on this effect)

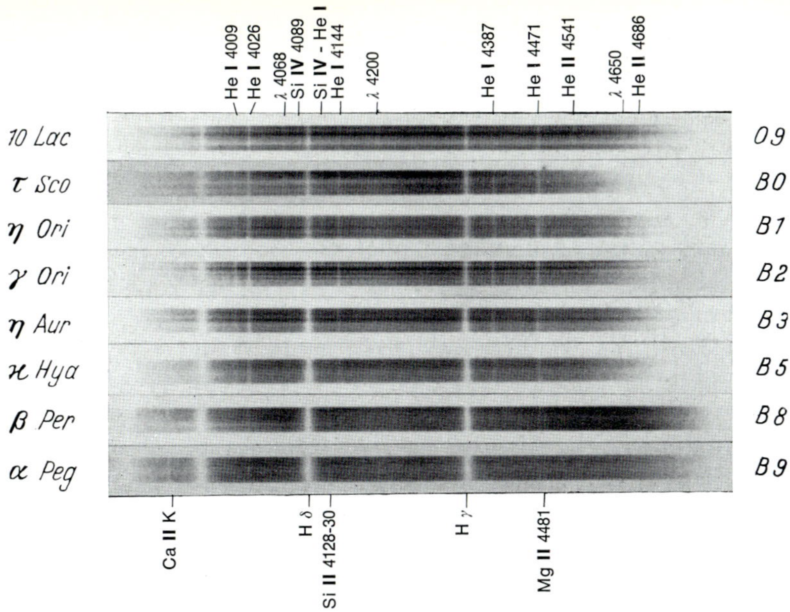
He I 4009
He I 4026
λ 4068
Si IV 4089
Si IV - He I
He I 4144
λ 4200
He I 4387
He I 4471
He II 4541
λ 4650
He II 4686
10 Lac
O9
τ Sco
B0
η Ori
B1
γ Ori
B2
η Aur
B3
κ Hya
B5
β Per
B8
α Peg
B9
Ca II K
H δ
Si II 4128-30
H γ
Mg II 4481

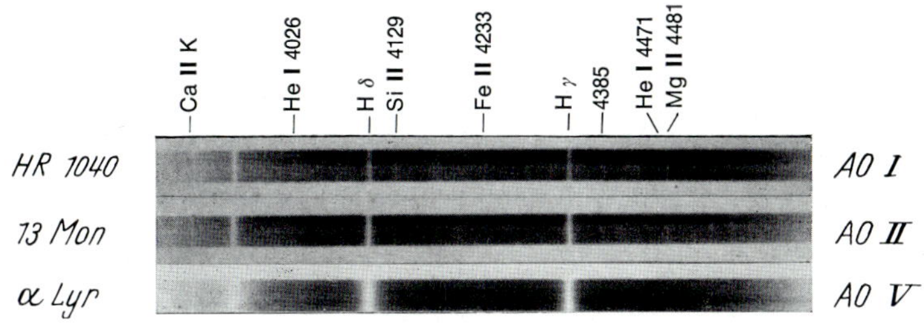
Ca II K
He I 4026
H δ
Si II 4129
Fe II 4233
H γ
4385
He I 4471
Mg II 4481
HR 1040
A0 I
13 Mon
A0 II
α Lyr
A0 V

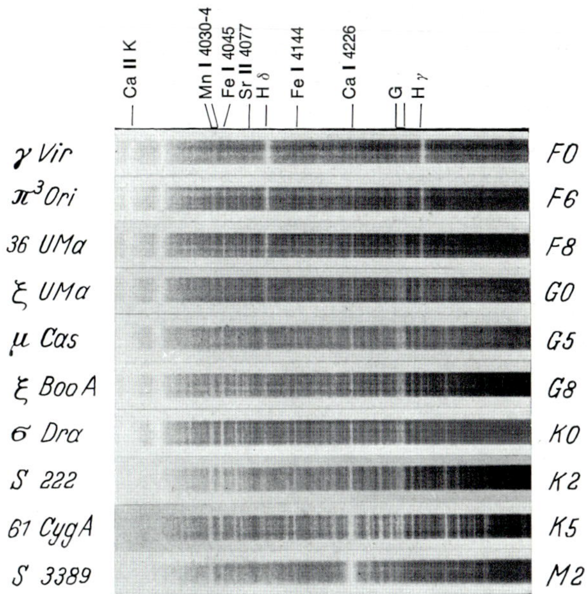
Ca II K
Mn I 4030-4
Fe I 4045
Sr II 4077
H δ
Fe I 4144
Ca I 4226
G
H γ
γ Vir
F0
π³ Ori
F6
36 UMa
F8
ξ UMa
G0
μ Cas
G5
ξ Boo A
G8
σ Dra
K0
S 222
K2
61 Cyg A
K5
S 3389
M2

II = singly ionized atom e.g. Si^{+}, spark spectrum; III = doubly ionized atom e.g. Si^{++} ...). The temperatures listed correspond somewhat to the colours of the stars and are meant to serve only for provisional orientation.

A series of standard stars for the establishment of spectral types is given in Fig. 15.1. Certain special features of many stellar spectra, which cannot be brought within the scope of a one-parameter classification, are denoted by the following symbols:

The prefix c signifies specially sharp lines, particularly of hydrogen (Miss Maury's c-stars, e. g. α Cyg cA2).

The suffix n (nebulous) signifies a specially diffuse appearance of the lines, and s (sharp) a specially sharp appearance without the other criteria of c-stars.

v indicates a variable spectrum and applies to most variable stars.

p (peculiar) indicates a peculiarity of any sort, e. g. anomalous strength of the lines of a particular element.

In the year 1913 H. N. Russell had the happy thought of studying the relationship between spectral type Sp and absolute magnitude M_v, and he plotted in a diagram with Sp as abscissa and M_v as ordinate all the stars for which the parallax was sufficiently well determined. Fig. 15.2 shows such a diagram which, using much improved observational material, Russell drew in 1927 for his textbook that served a whole generation as their "astronomical bible".

The majority of stars populate the narrow strip of the *main sequence* which runs diagonally from the (absolutely) bright blue-white B- and A-stars (e. g. the stars in the belt of Orion) through the yellow stars (e. g. Sun G2 and $M_v = +4.8$) to the faint red M-stars (e. g. Barnard's star M5 and $M_v = +13.2$).

In the right-hand upper part of the diagram there is the set of *giant stars;* by contrast, the stars that possess the same spectral type but much smaller luminosity are called *dwarf stars.* Since for about equal temperatures the difference in absolute magnitude can depend only upon a difference in stellar radius, this nomenclature seems very appropriate. The recognition and naming of giants and dwarfs goes back to the earlier work of E. Hertzsprung (1905), for which reason we now call the Sp, M_v-diagram the Hertzsprung-Russell diagram (HR diagram).

Instead of spectral type Sp one can also employ a colour index, say $B—V$, and one then obtains a colour-magnitude diagram which is equivalent to the HR diagram. Fig. 15.3 shows the ($B—V$, M_v)

colour-magnitude diagram, with now a very sharply defined main sequence, certain yellow giants (upper right) and white dwarfs (lower left), drawn for field stars and cluster stars in our vicinity in space that have well-determined parallaxes.

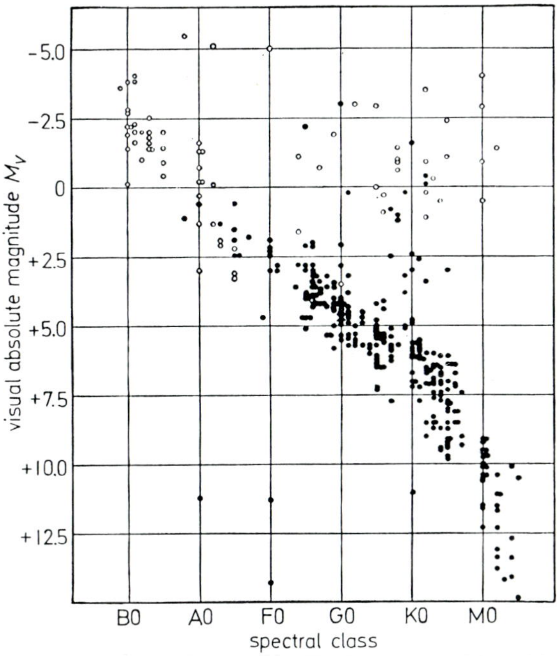

Fig. 15.2. *Hertzsprung-Russell diagram.* Absolute visual magnitude M_v plotted against spectral class. The Sun corresponds to $M_v = 4.8$ and G 2. The dots represent stars within 20 pc, which have reliable parallaxes. Open circles represent the less frequently occurring stars of greater absolute magnitude; for these the trigonometric parallaxes have been supplemented by spectroscopic and cluster parallaxes

Since colour indices can be measured with high accuracy even for faint stars, the colour-magnitude diagram has become one of the most important working tools of astronomy.

The extremely bright stars along the upper boundary of the HR diagram or colour-magnitude diagram are known as supergiants. For example α *Cygni* (Deneb cA2) has absolute magnitude $M_v = -7.2$; therefore it exceeds the brightness of the Sun ($M_v = +4.8$) by 12 magnitudes, i. e. a factor of about 63 000.

Another readily recognizable set are the *white dwarf* stars in the lower left of the diagram. Since in spite of relatively high temperature they have small luminosity, they must be very small; one can easily calculate the radius, which is scarcely larger than that of the Earth. For the companion of Sirius α CMa B and certain similar objects one

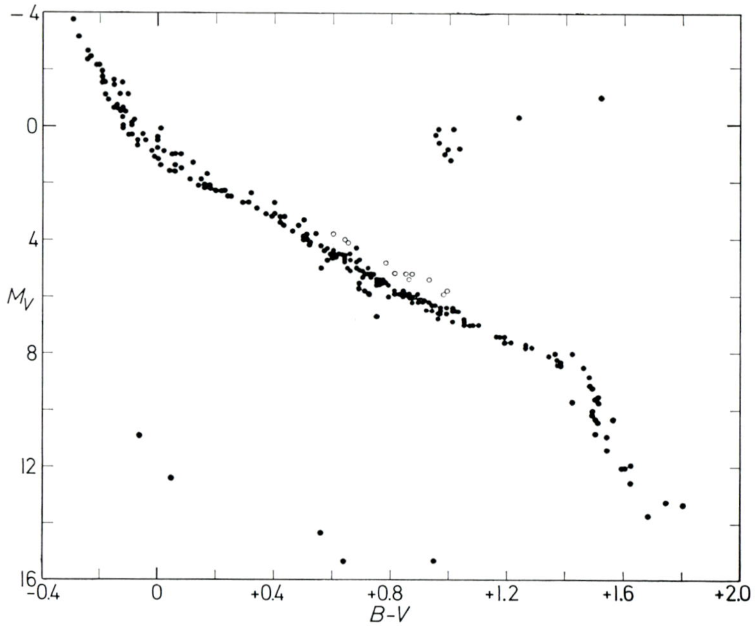

Fig. 15.3. *Colour-magnitude diagram* (M_V against $B-V$) after H. L. Johnson and W. W. Morgan, showing main sequence derived from stars with trigonometric parallaxes $p \geqq 0.''10$ and from stars taken from several galactic clusters with well determined parallaxes and corrected for interstellar absorption and reddening. Also included are five white dwarfs (lower left) and several yellow giants (upper right). The stars lying above the main sequence (from Praesepe) are probably binaries

also knows the mass and thence one derives mean densities of the order of 10^4 to 10^5 g/cm^3. The internal constitution of such stars must be quite different from the rest. In 1926 R. H. Fowler showed that in white dwarfs the material (more precisely, the electrons) is degenerate in the sense of Fermi statistics in the same way as shortly thereafter W. Pauli and A. Sommerfeld showed to be the case for electrons in metals. That is to say, almost all the quantum states are occupied in the sense of the Pauli principle, as are the inner shells of heavy atoms.

We shall mention further, mostly smaller and more special, sets of stars in the HR diagram in other contexts.

E. Hertzsprung remarked as long ago as 1905 that Miss Maury's c-stars with sharp hydrogen lines are distinguished by their great luminosity. Then in 1914 W. S. Adams and A. Kohlschütter showed that the spectra of stars of a given spectral type could be further subdivided according to luminosity on the basis of new spectroscopic criteria. In absolutely bright stars, for example, the lines of ionized atoms (spark lines) are enhanced relative to those of neutral atoms (arc lines); amongst the A stars, as has been said, the sharpness of the hydrogen lines serves as a luminosity-criterion.

If we calibrate such a criterion, which of course applies only to a limited range of spectral types, with the use of stars of known absolute magnitude, then with the help of the resulting calibration curve we can determine further absolute magnitudes by spectroscopic means. If we can neglect, or correct for, interstellar absorption (which had still not been dreamt of in 1914), then by combining these absolute magnitudes with the apparent magnitudes we derive *spectroscopic parallaxes* of the stars concerned (equation (14.2)). In part III we shall mention the significance of these for the investigation of the Milky Way system. Here we pursue the important insight afforded by the fact that we can classify the great majority of stars according to *two parameters*.

On the basis of the Harvard classification, etc., W. W. Morgan and P. C. Keenan developed the now generally used MK classification presented in *An atlas of stellar spectra with an outline of spectral classification* (1943 with E. Kellman and 1953). Its general principles hold good for any such classification:—

1. Only empirical criteria, i. e. directly observed absorption and emission phenomena are the basis of the classification.

2. The observational material is uniform. In order on the one hand to include sufficiently refined criteria and on the other hand to reach sufficiently far into the Galactic system, a uniform dispersion of 125 Å/mm at Hγ is employed, also for bright stars *.

3. The transferability of the system of classification from one instrument to another is validated by using a list of suitable standard stars, i. e. by direct application, and not by using specifications, that may very well be semi-theoretical.

* If we work with other spectrographs, in particular with higher dispersion, we must first photograph spectra of the standard stars (given in part in Fig. 15.1) and then classify the spectra of other stars by comparison with these.

4. The classification is according to

a) *Spectral type* Sp largely in agreement with the Harvard classification and with the same notation.

b) *Luminosity class* LC. Its calibration in terms of absolute magnitude will be considered in retrospect. An LC criterion over at anyrate a large range of spectral types is to depend in the first place as sensitively as possible upon the luminosity of the stars. Morgan's luminosity classes give at the same time the location of the stars in the HR diagram; they are

Ia–0, I	Supergiants
II	Bright giants
III	Giants
IV	Subgiants
V	Main sequence (dwarfs)
VI	Subdwarfs

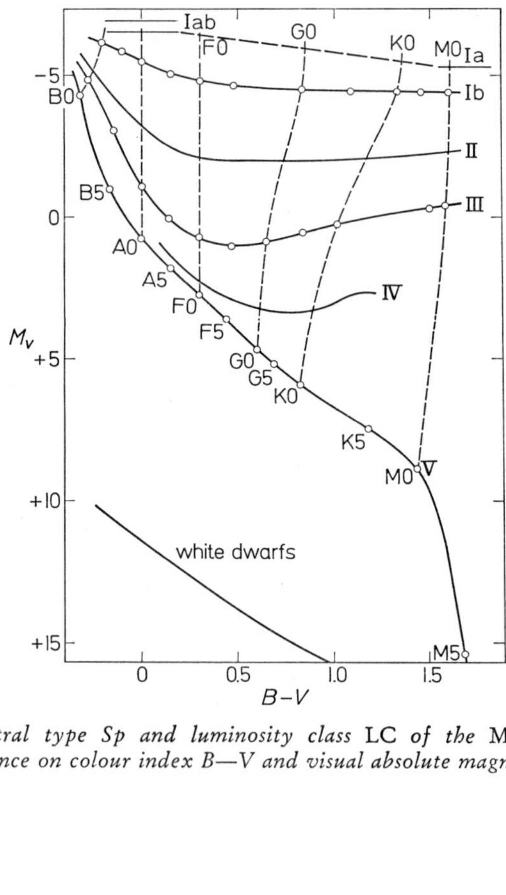

Fig. 15.4. *Spectral type Sp and luminosity class* LC *of the* MK *classification; dependence on colour index B—V and visual absolute magnitude* M_V

As needed, the luminosity classes I to V may be subdivided using suffixes a, ab and b. Fig. 15.1, extracted from the *Atlas of stellar spectra,* shows the important spectral types of the main sequence stars (LC = V) and, at least for spectral type A0, the division into luminosity classes I to V according to the width of the hydrogen lines.

About 90 percent of all stellar spectra can be dealt with by the MK classification; the rest are partly composite spectra of unresolved double stars and partly peculiar (p) spectra of pathological individuals.

The relationship between the parameter of the MK classification Sp and LC on the one hand, and the colour index $B-V$ and absolute magnitude M_V on the other, according to the best available calibration, is shown in Fig. 15.4.

16. Double Stars and the Masses of the Stars

In 1803 F. W. Herschel discovered that α Gem = Castor is a visual double star whose components move round each other under the influence of their mutual attraction. Thus the observation of double stars offers the possibility of determining the masses of stars or at anyrate of making quantitative assertions about them. Since the masses of stars are detectable only through their gravitational interaction, the investigation of double stars is even now of fundamental significance for the whole of astrophysics.

We first separate the optical pairs according to statistical criteria, also when possible by obtaining proper motions and radial velocities, from physical pairs or true binary stars. The apparent orbit of the fainter component, the "companion", around the brighter component is observed with a refractor provided with a filar micrometer (Fig. 9.1) in terms of the separation (in seconds of arc) and the position angle (N 0° — E 90° — S 180° — W 270°). If one plots this apparent orbit one obtains an ellipse. Were we to look at right angles to the plane of the orbit, we should see the bright component at the focus of the orbit. In general this does not occur because the plane of the orbit forms with the plane of the sky an angle of inclination i. Conversely, one can obviously determine the inclination i so that the orbit satisfies the Kepler laws. Let

a = semi-major axis of the true orbit in seconds of arc,

p = parallax of the binary star in seconds of arc
then a/p = semi-major axis of the true orbit in astronomical units.
If, further, P is the period of description of the orbit in years, then from Kepler's third law (6.33) we can easily write down the combined mass of the two stars $\mathfrak{M}_1 + \mathfrak{M}_2$ (in units of the solar mass) as

$$\mathfrak{M}_1 + \mathfrak{M}_2 = a^3/p^3 P^2. \tag{16.1}$$

If by use of a meridian circle or photographically (following E. Hertzsprung) one measures the absolute motion of the two components, (i. e. relative to the background stars, after allowing for parallactic motion and proper motion), one derives the semi-major axes a_1 and a_2 of their true paths around the mass-centre and one has from (6.31)

$$a_1 : a_2 = \mathfrak{M}_2 : \mathfrak{M}_1 \quad \text{and} \quad a = a_1 + a_2 \tag{16.2}$$

so that one can now calculate the individual masses $\mathfrak{M}_1$, $\mathfrak{M}_2$.

If the fainter component of a binary star is not visible its presence can nevertheless be inferred from the motion of the brighter component about the mass-centre, measured in an absolute fashion. If a_1 is the semi-major axis of this orbit, again in arc-seconds, since $a_1/a = \mathfrak{M}_2/(\mathfrak{M}_1 + \mathfrak{M}_2)$ we obtain

$$(\mathfrak{M}_1 + \mathfrak{M}_2)\left(\frac{\mathfrak{M}_2}{\mathfrak{M}_1 + \mathfrak{M}_2}\right)^3 = \frac{a_1^3}{p^3 P^2}\,. \tag{16.3}$$

Using meridian observations, F. W. Bessel in 1844 found in this way that Sirius must have a dark companion. A. Clark in 1862 actually discovered Sirius B as about $10^{\mathrm{m}}.14$ fainter than Sirius itself. Sirius B has absolute magnitude only $M_{\mathrm{v}} = +11.54$ although its mass is 0.96 solar mass. Since the surface temperature is quite normal, as already remarked, it must be very small. In 1923 F. Bottlinger concluded that "here we have to do with something entirely new", namely a *white dwarf* star.

In recent times K. A. Strand and P. van de Kamp have found dark companions of certain nearer stars that represent an intermediate stage between stars and planets (i. e. between bodies with and without their own energy-sources). In particular, our second-nearest neighbour, Barnard's star, of spectral type M5 V and of about 0.15 solar mass, possesses a companion of only about 0.0015 solar mass or about 1.6 times the mass of Jupiter. Thus we observe a second planetary system within a distance of 1.84 pc.

In 1889 E. C. Pickering observed that in the spectrum of *Mizar* = ζ U Ma the lines twice become double in time intervals of $P = 20^{\mathrm{d}}.54$. Thus Mizar shows itself to be a spectroscopic binary. In this particular system two similar A2 stars revolve around each other; their angular separation is too small for telescopic resolution. In other systems only one component is recognizable in the spectrum; the other is obviously considerably fainter. If we plot the radial velocity as derived from the doppler effect for either or both components as a function of time, we get the *velocity curve*. After removing the mean velocity or the velocity of the mass-centre, we can read off the sight-line component of the orbital velocity. Without going into the details here, we can calculate thence, not the semi-major axis itself, but the quantity $a_1 \sin i$ if only component 1 is detectable in the spectrum, and also $a_2 \sin i$ if component 2 is detectable as well, where i is the (unknown) inclination of the orbit.

If only one spectrum is visible, we derive from Kepler's third law and from the mass-centre formula (16.2)

$$\frac{(a_1 \sin i)^3}{P^2} = (\mathfrak{M}_1 + \mathfrak{M}_2)\left(\frac{\mathfrak{M}_2}{\mathfrak{M}_1 + \mathfrak{M}_2}\right)^3 \sin^3 i = \frac{\mathfrak{M}_2^3 \sin^3 i}{(\mathfrak{M}_1 + \mathfrak{M}_2)^2}\,. \qquad (16.4)$$

The last quantity is called the *mass-function*. For statistical purposes we can make use of the fact that the mean value of $\sin^3 i$ for all possible inclinations is 0.59, or, taking account of the probability of discovery, about 2/3. Since $\mathfrak{M}_2 < \mathfrak{M}_1$, in all cases the factor $(\mathfrak{M}_2/(\mathfrak{M}_1 + \mathfrak{M}_2))^3 < 1/8$.

If both spectra are visible then we obtain $\mathfrak{M}_1 \sin^3 i$ and $\mathfrak{M}_2 \sin^3 i$ and thence also the mass ratio $\mathfrak{M}_1 : \mathfrak{M}_2$.

If the inclination of the orbital plane of a spectroscopic binary is near to 90° then eclipses occur and we observe an eclipsing variable or *eclipsing binary*. The classic example is β Persei = Algol, with period $P = 2^{\mathrm{d}}\, 20^{\mathrm{h}}\, 49^{\mathrm{m}}$ which J. Goodricke interpreted as such in 1782. From the magnitudes of an eclipsing variable measured (preferably photoelectrically) over a long enough time, we obtain first the period P and then the light-curve. Hence we derive the radii of both stars in terms of the radius of the relative orbit as unit and also the inclination of the orbit i (Fig. 16.1). If we can determine in addition the velocity curve for one or both components, we derive also the absolute dimensions of the system as well as the masses and, thence the mean densities, of both stars. In favourable cases we can derive even the ellipticity (flattening) and the centre-limb darkening of the component stars. Thanks to the methodology for the determination of the elements of eclipsing variables developed to the greatest completeness by H. N. Russell and H. Shapley these are at the present time amongst

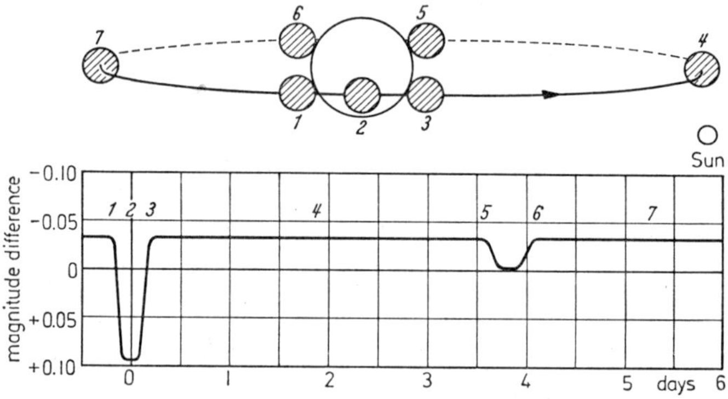

Fig. 16.1. *Apparent relative orbit and light curve of the eclipsing binary* IH *Cassiopeiae.* Corresponding points of the orbit and light curve are indicated by numbers. In this case the primary eclipse of the brighter by the fainter and smaller component is annular

the most accurately known stars. As a result of penetrating analyses of the spectra, O. Struve later showed that in close pairs the two components are in a state of material interaction with each other. Common gaseous envelopes and streams of gas flowing from one component to the other sometimes give direct insight into the *evolution* of such systems.

We now seek to give a sketch in broad terms; on account of unavoidable selection effects a detailed statistical discussion would be of doubtful significance:

The various sorts of visual, spectroscopic and eclipsing binaries, which are distinguished only by the way in which they are observed, join together continuously and with a certain degree of overlapping. The periods range from a few hours to many thousands of years. The binaries with short periods mostly have circular orbits; long-period systems favour larger eccentricities. Besides double stars, multiple systems also frequently occur, which mostly include one or more close pairs. For instance, the "double star" α Gem = Castor first discovered by F. W. Herschel was later shown to consist of three pairs A, B, C with periods 9.22, 2.93, 0.814 days; A, B move round each other in several hundred years, Castor C moves round A and B in several thousand years. According to P. van de Kamp (1953) our neighbourhood within a distance of 5 parsec contains 31 single stars (with maybe some invisible companions), 9 double stars and 2 triple systems. Of the total of 55 stars almost half are therefore members of double or multiple systems.

In the spectra of double stars and eclipsing variables of short period, the components of which revolve around each other in close proximity, the Fraunhofer lines are mostly extraordinarily broad and diffuse. This depends on the fact that the two components rotate together like a rigid body, as a result of tidal friction. The periods of revolution and rotation are equal. If the component of the equatorial velocity along the sight-line is $v \sin i$, then at wavelength λ there is a doppler displacement $\Delta\lambda = \pm \lambda (v/c) \sin i$. Were the spectral line sharp for a stationary star, it would now appear as spread out into a band of width $2\Delta\lambda$, the profile of which would indicate the brightness-distribution across the stellar disk, e. g. in the case of no limb-darkening we get a line profile of elliptic shape. For instance, if a B star of radius $5 R_\odot$ rotates with period 1 day and if $i = 90°$, then its projected equatorial velocity $v \sin i = 250$ km/sec and half the resulting width of the line MGII 4481 is $\Delta\lambda = \pm 3.73$ Å.

O. Struve and his colloborators then discovered that there are also single stars in whose spectra all the lines are strongly widened in this way and that therefore rotate with equatorial velocities of up to about 300 km/sec. Like the rapidly rotating binaries, the rapidly rotating single stars belong predominantly to spectral types O, B and A in the upper part of the main sequence.

We shall later return to consider the significance of rotation of single and binary stars and therewith to the rôle of angular momentum in problems of stellar evolution.

To conclude this Chapter we take a look at the *masses* of the stars. Their values as determined from binaries of all kinds range from about $0.15\,\mathfrak{M}_\odot$ to $50\,\mathfrak{M}_\odot$. Their relation to other stellar parameters long remained obscure until A. S. Eddington in 1924 in connexion with his theory of the internal constitution of the stars discovered the *mass-luminosity relation.* We can understand this in principle from a present-day standpoint as follows: The stars of the main sequence are evidently in analogous states of development (their energy requirements being supplied by the conversion of hydrogen into helium); in the main, they are therefore built to the same pattern. To a given mass, therefore, there belong energy-sources of a well-determined amount, and these in turn determine the luminosity of the star. We therefore expect there to be a relation between the mass $\mathfrak{M}$ and the luminosity L or absolute bolometric magnitude M_{bol}. As the analysis of all the observational material shows (Fig. 16.2) there is

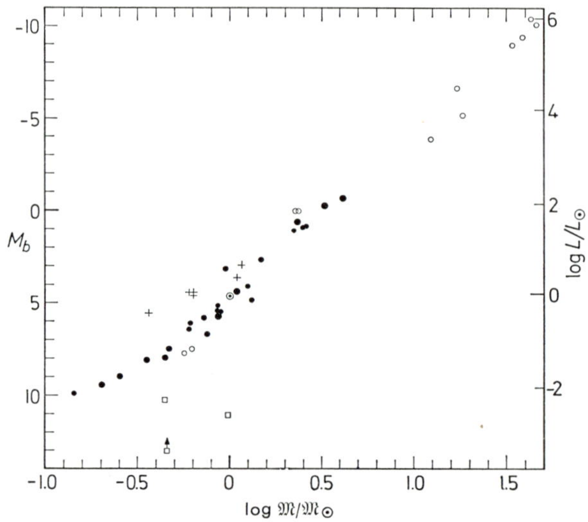

Fig. 16.2. *Empirical mass-luminosity relation.* The absolute bolometric magnitude M_b or the luminosity L plotted as a function of the mass $\mathfrak{M}$ (after G. P. Kuiper). ● visual binaries, ○ spectroscopic binaries, + Hyades binaries, □ white dwarfs

actually such a relation for main-sequence stars. As expected, the white dwarfs do not conform to this. Also "the" mass-luminosity relation cannot without further discussion be applied to the red giant stars for which no reliable empirical masses are yet available.

If we regard the radii R of the stars as known (calculated from the absolute magnitude and, roughly speaking, the temperature) we can evaluate the *surface gravity* $g = G\mathfrak{M}/R^2$ cm/sec², which is important for the theory of the spectra. As we shall verify more exactly, we find that for main sequence stars of spectral types from B0 V to M3 V the value is constant to within about a factor 2 with

$$g \approx 20\,000 \text{ cm/sec}^2 \quad \text{or} \quad \log g \approx 4.3\,. \tag{16.5}$$

For giants and supergiants the value is considerably less (down to $\log g \approx 0.5$), and for white dwarfs considerably greater ($\log g \approx 8 \pm 0.5$). In Chapter 19 we shall see how the value of surface gravity g can influence the spectrum of a star.

17. Spectra and Atoms: Thermal Excitation and Ionization

After important pioneering work by N. Lockyer, in the year 1920 M. N. Saha's *Theory of thermal excitation and ionization* led to the interpretation of stellar spectra and their classification. It rests essentially upon the quantum theory of atoms and their spectra developed by N. Bohr, A. Sommerfeld and others from 1913 onwards. Here we briefly recall the fundamentals without their full derivation:

We represent the possible energy-levels of an atom graphically in an energy-level- or Grotrian-diagram (Fig. 17.1). We distinguish:

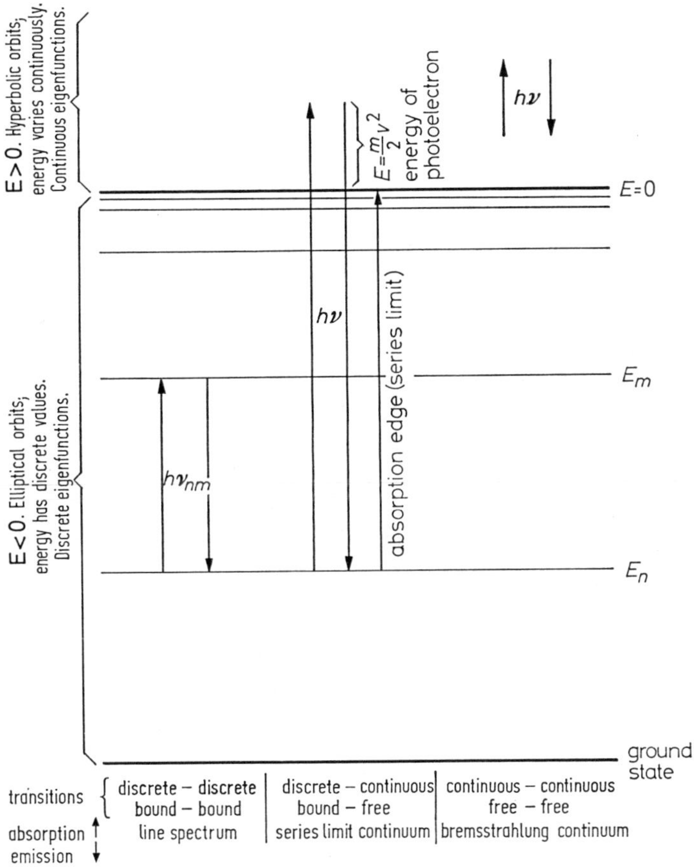

Fig. 17.1. *Energy-level or Grotrian diagram of an atom (schematic only) showing various transitions*

a) *Discrete negative energy-levels* $E<0$ corresponding to the bound or elliptic orbits of the electrons in the Bohr atom model or the discrete eigenfunctions (standing de Broglie waves) of quantum mechanics. Each energy-level is characterized by a set of several integral or half-integral quantum numbers which we represent by a *single* symbol n, m, s or the like.

b) *Continuous positive energy values* $E>0$ corresponding to the free or hyperbolic orbits of electrons in the Bohr model or to "continuous" eigenfunctions (progressive de Broglie waves) in quantum mechanics. At a great distance from the atom such an electron has only kinetic energy $E_{\text{kin}}=\frac{1}{2}m v^2$, where m is its mass and v its velocity.

In a transition between two energy levels E_m, E_n a photon of energy

$$h\nu=|E_m-E_n| \tag{17.1}$$

is absorbed (↑) or emitted (↓), where $h=6.62\times10^{-27}$ erg sec is Planck's quantum of action. The frequency ν sec^{-1} or Hz (Hertz) corresponds to a wave-number (number of light-waves per cm in the vacuum) $\bar{\nu}=(\nu/c)$ cm^{-1} or Kayser and a wavelength $\lambda=1/\bar{\nu}=c/\nu$ cm. Besides the centimetre we use as unit 10^{-8} cm $=1$ Å (one angstrom). We often reckon the energy-values, not from $E=0$ but from the ground state of the atom. As unit we use generally not 1 erg but 1 cm^{-1} or 1 Kayser and then we speak of the terms and the term scheme of the atom, or else we use 1 eV (one electron volt) or the energy that an electron gains on traversing a potential increase of 1 volt. In thermal equilibrium we have always to deal with energies of the order of magnitude kT ($k=1.38\times10^{-16}$ erg/deg). In this context we specify the temperature in degrees Kelvin corresponding to an energy E. There correspond thus

$$\begin{aligned}1\ \text{eV} &\triangleq 1.602\times10^{-12}\ \text{erg}\triangleq 8066\ \text{cm}^{-1}\triangleq 12\,398\ \text{Å}\\ &\triangleq 11\,605\ {}^\circ\text{K}.\end{aligned} \tag{17.2}$$

We naturally divide the transitions undergone by an atom as a result of the absorption or emission of a single photon into the following classes:

(a a) $E_m<0$, $E_n<0$; elliptic-elliptic, discrete-discrete or bound-bound transitions under absorption or emission of a spectral line whose wave-number $\bar{\nu}$ cm^{-1} we obtain as the difference of the term-values in Ch. E. Moore's *Atomic energy levels.*

(a b) $E>0$, $E_n<0$; hyperbolic-elliptic, continuous-discrete or free-bound transitions. At the series limit or absorption edge $h\nu_n=E_n$ the continuous-discrete absorption $\nu>\nu_n$ follows on with the ejection of a photoelectron having kinetic energy $\frac{1}{2}\, m\, v^2 = h\, \nu - |E_n|$ which is Einstein's photo-electric formula. Thereby the atom becomes ionized or goes into the next higher state of ionization. We denote the spectra of neutral, singly-ionized, doubly-ionized ... atoms, e. g. calcium, by CaI, CaII, CaIII and so on.

The inverse process is the capture of a free electron of energy $\frac{1}{2}\, m\, v^2$ with the emission of a photon

$$h\, \nu = \tfrac{1}{2} m\, v^2 + |E_n|$$

the so-called *two-body recombination.*

(b b) $E' > 0$, $E'' > 0$ gives hyperbolic-hyperbolic, continuous-continuous or free-free transitions. A photon $h\, \nu = |E' - E''|$ is absorbed or emitted; the free electron gains or loses the corresponding amount of kinetic energy in its encounter with the atom.

Bound-free and free-free absorption and emission were first found in the X-ray region and they were explained theoretically by N. Bohr and H. A. Kramers.

From now on* we shall be concerned with the *discrete terms* ($E_n < 0$) of atoms and ions.

We describe a particular energy level of an atom or ion having a single emitting or valence-electron (i. e. the remaining electrons undergo no transition) by *four quantum numbers:*

n the principal quantum number. In the case of hydrogen-like (coulomb-field) orbits, in the language of Bohr's theory, $n^2\, a_0/Z$ is the semi-major axis of the orbit, $-\frac{e^2\, Z^2}{2\, a_0}\frac{1}{n^2}$ erg is the corresponding energy, and $R_\infty\, Z^2/n^2$ cm^{-1} the term-value. Here $a_0 = 0.529$ Å is the radius of the first Bohr orbit for hydrogen, $R_\infty = 109\,737.30$ cm^{-1} is the Rydberg constant, and Z the effective nuclear charge-number ($Z = 1$ for a neutral atom, arc spectrum; $Z = 2$ for a singly-ionized atom, the first spark spectrum, and so on).

l is the orbital angular momentum of the electron measured in the quantum unit $\hbar = h/2\,\pi$, and l can assume integral values 0, 1, 2, ..., $n - 1$.

---

* The newcomer to the subject need not work through all the details of the following short introduction to the quantum theory and classification of atomic spectra (about the next two pages). It is important that he should understand the meaning of the energy-level- or term-scheme in Figures 17.1, 17.3, 17.4.

$l = 0 \; 1 \; 2 \; 3 \; 4 \; 5$

gives respectively a

s p d f g h - electron*.

s is the spin angular momentum (Goudsmit-Uhlenbeck) in the same units. For a single electron $s = \pm\frac{1}{2}$

j is the total angular momentum, again in units of $\hbar$. j is given by vector addition of l and s and can only have one of the values $l \pm \frac{1}{2}$

As an example, one electron having $n = 2, l = 1$ and $j = {}^3/_2$ is denoted as a $2\,p_{1^1/_2}$-electron.

In atoms or ions with several electrons, as H. N. Russell and F. A. Saunders found in 1925 in the case of the alkaline earths, the angular momentum vectors are normally coupled as follows (Russell-Saunders or *LS coupling):*

The orbital angular momenta $\vec{l}$ are added vectorially to give the resultant orbital angular momentum $L = \sum \vec{l}$; similarly the spin angular momenta $\vec{s}$ give the resultant $S = \sum \vec{s}$. Then L and S combine,

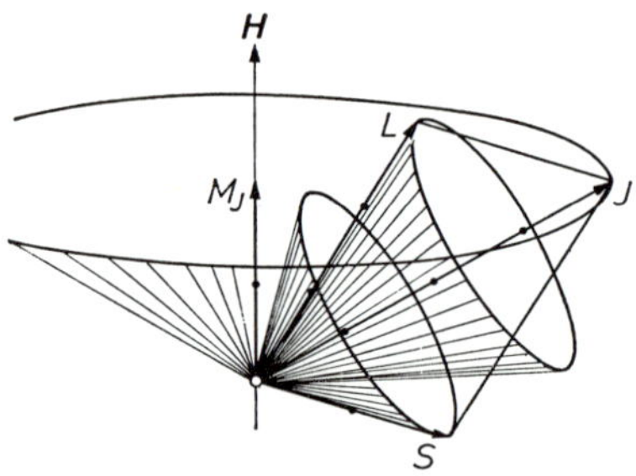

Fig. 17.2. *Vector model for an atom with Russell-Saunders coupling.* The total orbital angular momentum vector L and the spin angular momentum vector S combine to give the total angular momentum vector J (all being measured in units of $\hbar = h/2\pi$) M_J denotes the component of J along the direction of an external field H. The vectors L and S precess about J, while itself precesses about H. The diagram corresponds to the particular energy level $L=3$, $S=2$, $J=3$ i.e. to the $^5F_3$ level

again vectorially, to give the total angular momentum J (Fig. 17.2) so that

$$|L - S| \leqq J \leqq L + S. \qquad (17.3)$$

L is always integral; S and J are half-integral/integral for atoms with an odd/even number of electrons.

* The notation referred originally to the upper term of the series: s = sharp secondary series; p = principal series; d = diffuse secondary series; f = fundamental series.

A particular pair of values of S and L constitutes, as we say, a *term*. By analogy with the single-electron system, the orbital angular momentum quantum number L corresponds to a term designated as follows:

$$\begin{array}{llllll} L=0 & 1 & 2 & 3 & 4 & 5 \\ \mathrm{S} & \mathrm{P} & \mathrm{D} & \mathrm{F} & \mathrm{G} & \mathrm{H}. \end{array}$$

So long as $L \geqq S$, the term splits into $r=2S+1$ energy-levels with different J. Then, and also if $L<S$, the number r is called the multiplicity of the term and it is written to the upper left in the term-symbol; J is written to the lower right as a subscript, in order to designate the particular levels in the term. Table 17.1 gives a synopsis of possible terms of various multiplicities, their energy levels and the usual designation.

Table 17.1. *Terms and J-values of their levels for different quantum numbers L and S for Russell-Saunders coupling*

		$S=0$	$\frac{1}{2}$	1	$\frac{3}{2}$
		$r=2S+1=1$	2	3	4
		singlet	doublet	triplet	quartet
$L=0$	S-term	$J=0$	$J=\frac{1}{2}$	$J=1$	$J=\frac{3}{2}$
1	P-term	1	$\frac{1}{2}\ \frac{3}{2}$	0 1 2	$\frac{1}{2}\ \frac{3}{2}\ \frac{5}{2}$
2	D-term	2	$\frac{3}{2}\ \frac{5}{2}$	1 2 3	$\frac{1}{2}\ \frac{3}{2}\ \frac{5}{2}\ \frac{7}{2}$
3	F-term	3	$\frac{5}{2}\ \frac{7}{2}$	2 3 4	$\frac{3}{2}\ \frac{5}{2}\ \frac{7}{2}\ \frac{9}{2}$

Example [dotted box] Quartet P-term with energy-levels $^4\mathrm{P}_{\frac{1}{2}}$, $^4\mathrm{P}_{1\frac{1}{2}}$, $^4\mathrm{P}_{2\frac{1}{2}}$.
Statistical weight of the term $g\,(^4\mathrm{P}) = 4 \times 3 = 2+4+6$.

In the presence of an external field (e. g. a magnetic field) the vector of total angular momentum J sets itself so that its component M_J in the direction of the direction of the field is half-integral or integral. M_J can thus assume the values $J, J-1, \ldots, -J$; thus the directional quantization of J gives $2J+1$ possible configurations (Fig. 17.2). When the external field vanishes, these $2J+1$ energy-levels fall together; we say that the J level is $(2J+1)$-ply *degenerate*. Furthermore we divide the terms according to their parity into two groups, the *even* and *odd* terms according as the arithmetic sum of the l-values of the participating electrons is even or odd. Odd terms are distinguished by an affix ^{o} written to the upper right.

A *line* arises from a transition between two energy levels; the possible transitions between all the levels in two *terms* produce a group of neighbouring lines called a *multiplet*. The possibilities for transitions (in the emission or absorption of electric-dipole radiation, analogous to that of the familiar Hertz dipole) are restricted by the following selection rules:

1. Transitions occur only between odd and even levels.

2. J changes only so that $\Delta J = 0, \pm 1$. The transition $0 \rightarrow 0$ is forbidden.

For LS-coupling we have further:

3. $\Delta L = 0, \pm 1$.

4. $\Delta S = 0$, i. e. there are no intercombinations (e. g. singlet-triplet).

Russell-Saunders or LS-coupling is recognized by the fact that multiplet splitting, arising from the magnetic interaction between orbital momenta and spin-momenta, is small compared with the separation between neighbouring terms or multiplets.

If the selection rules 1, 2 are *not* satisfied, forbidden transitions can occur with much smaller transition-probability through electric quadrupole radiation or magnetic dipole radiation (the analogue of the frame antenna).

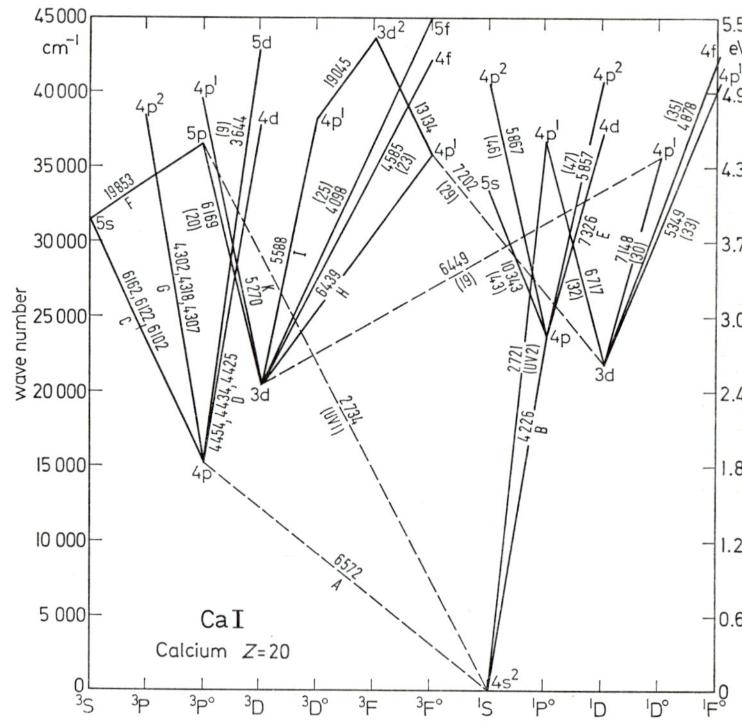

Fig. 17.3. *Term scheme or Grotrian diagram for the (arc) spectrum of neutral calcium* Ca I

As an illustration, Fig. 17.3 shows the term-scheme of neutral calcium CaI. The most important multiplets are designated by their numbers in *A multiplet table of astrophysical interest* or *An ultraviolet multiplet table* by Ch. E. Moore (Washington) and the wavelength in Å of the strongest lines is shown.

The theory of atomic spectra sketched here made it possible in the first place to classify the wavelengths λ or wave numbers $\bar{\nu}$ of most of the elements in their various stages of ionization (I = arc spectrum, II = first spark spectrum, ...) as measured in the laboratory. This means that for each spectral line we can give the lower and upper terms (usually measured in cm^{-1} from the ground-state) and their classification. For example, Fraunhofer's K-line, the strongest line in the solar spectrum λ 3933.664 Å is CaII $4\,^2S_{1/2} - 4\,^2P^0_{1\,1/2}$.

Using the expression first in a qualitative sense and not yet in a precisely defined sense, the *intensity* of a line must depend upon what fraction of the atoms of the element concerned are in the appropriate state of ionization and what fraction of these are in the appropriate state of excitation in which they can absorb the line. Saha's theory supplies the answer to this question in so far as we may assume that the gas is in a state of thermodynamic equilibrium, i. e. one that corresponds sufficiently closely to the conditions inside a cavity at temperature T (degrees Kelvin).

We first consider an ideal gas at temperature T that is composed of neutral atoms (Fig. 17.4). In unit volume ($1\ cm^3$) let

N = total number of atoms of a particular element
N_0 = number of atoms in the ground-state 0
N_s = number of atoms in excited state s with excitation energy χ_s.

(We shall shortly extend this provisional notation by a suffix r inserted before suffix s to distinguish neutral, single, doubly ... ionized particles by $r = 0, 1, 2, \ldots$)

If all the quantum states are simple (i. e. if each occupies a volume h^3, one quantum cell, in phase-space) then according to the fundamental principles of statistical thermodynamics developed by Boltzmann

$$N_s/N_0 = e^{-\chi_s/kT} \tag{17.4}$$

where k again denotes the Boltzmann constant.

[We can think of (17.4) as a generalization of the barometric formula given by (7.4), (7.5) according to which the density distribution in an isothermal atmosphere at temperature T is given as a function of height y by

$$N(y)/N_0 = \exp\left\{-\frac{\mu g y}{RT}\right\} \equiv \exp\left\{-\frac{m g y}{kT}\right\}.$$

Here μ means the molecular weight and m the mass of a molecule, R is the usual gas-constant and k is the Boltzmann constant, g is the value of gravity. Thus $m g y$ is the potential energy of a molecule at height y above the ground; in (17.4) the excitation energy χ_s corresponds to this. While in classical statistics the potential energy can be varied continuously, in quantum statistics there are quantized states and all simple quantum states have statistical weight 1.]

If the energy-level is g_s-ply degenerate, i. e. if it would split into g_s simple levels under the influence of a suitable magnetic field, or equivalently if it occupies volume $g_s h^3$ in phase-space, we must assign it multiplicity or statistical weight g_s. Correspondingly, let the ground-state have weight g_0. Then the general Boltzmann-formula applies

$$\frac{N_s}{N_0} = \frac{g_s}{g_0} e^{-\chi_s/kT}. \tag{17.5}$$

The content of this formula is illustrated graphically on the right in Fig. 17.4. If we wish to relate N_s to the total number of atoms $N = \Sigma_s N_s$, instead of the number in the ground-state N_0, we have

$$\frac{N_s}{N} = \frac{g_s e^{-\chi_s/kT}}{\Sigma_s g_s e^{-\chi_s/kT}} \tag{17.6}$$

As denominator we have the important

$$\text{sum over states} \quad u = \Sigma_s g_s e^{-\chi_s/kT} \tag{17.7}$$

or *partition function.*

We obtain the statistical weights g_s from the theory of spectra: A level with angular momentum quantum number J shows, for example, in a magnetic field $2J+1$ different M_J (Fig. 17.2) and so has weight

$$g_J = 2J+1. \tag{17.8}$$

If we treat the levels of a multiplet term with quantum numbers S and L together, then this term has weight

$$g_{S,L} = (2S+1)(2L+1). \tag{17.9}$$

The summation of the corresponding g_J in Table 17.1 obviously leads to the same result.

The thermal excitation of the atoms into quantum states of higher and higher energy χ_s, described by the Boltzmann formula (17.5) or

(17.6), passes over continuously into the excitation of states of positive energy $E > 0$. The atom then receives the ionization energy χ (Fig. 17.4), which suffices to remove an electron from the atom, plus the kinetic energy $E = \frac{1}{2} m v^2$ with which the electron is ejected.

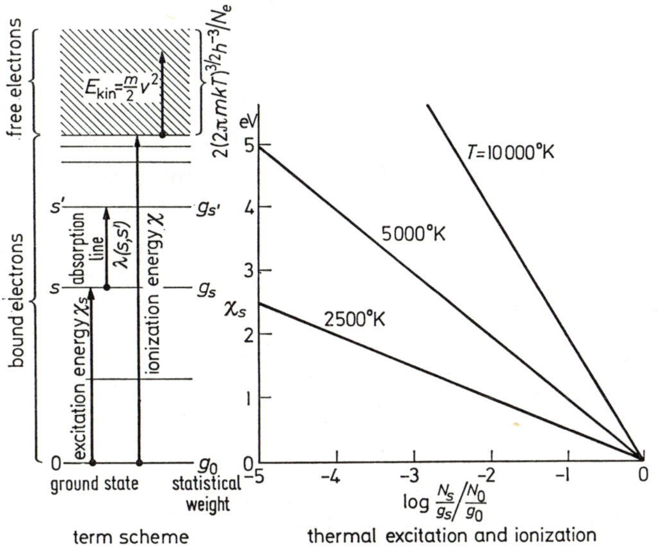

Fig. 17.4. *Thermal excitation and ionization of neutral atoms.* The term scheme (schematic only) on the left illustrates the basic concepts. The diagram on the right shows the relative number of atoms in a single (i. e. statistical weight 1) quantum state as a function of its excitation energy χ_s (χ_s in eV is plotted as ordinate). The ionization potential $\chi = 5.14$ eV would correspond to NaI

We now denote the number of (simply) ionized atoms with a first suffix, thus N_1, which corresponds to the number of neutral atoms which we now write as N_0 and so on for later stages, reckoning per cm³ throughout:

Ionization stage	Neutral	Singly ionized	Doubly ionized		r-ply ionized
Free electrons per atom	0	1	2	...	r
Ionization energy	χ_0	χ_1	$\chi_2 \cdots$	χ_{r-1}	
Spectra e. g. iron Fe	FeI	FeII	FeIII		Fe$(r+1)$
All atoms in the ionization stage	N_0	N_1	N_2		N_r
Atoms in the ground-state of the ionization stage	$N_{0,0}$	$N_{1,0}$	$N_{2,0}$		$N_{r,0}$
Atoms in level $s, s', \ldots$	$N_{0,s}$	$N_{1,s'}$	$N_{2,s''}$		$N_{r,s^{(r)}}$

We employ a corresponding notation for statistical weights. How now do we calculate $N_{1,0}/N_{0,0}$ that is the ratio of the numbers of singly-ionized and neutral atoms in the relevant ground states having statistical weights $g_{1,0}$ and $g_{0,0}$? (In the first place, higher stages of ionization will be supposed to play no part.)

Clearly the problem amounts to that of calculating the statistical weight of the ionized atom in its ground-state plus its free electron. The atom has statistical weight $g_{1,0}$; the electron alone, corresponding to the two possible orientations of its spin in an applied field, has statistical weight equal to 2. In addition, we must reckon the weight i. e. the number of quantum cells h^3, corresponding to the motion of the one free-electron. Statistical thermodynamics shows that the electron with mass m, takes up a volume $(2\pi m k T)^{3/2}$ in momentum space *; in ordinary space, if we have N_e electrons per cm³, we have $1/N_e$ cm³ per electron. Thus we have statistical weight

$$\left.\begin{array}{l}\text{Ionized atom in ground state}\\ +1\text{ free electron}\end{array}\right\}\quad g=g_{1,0}\,2\,\frac{(2\pi m k T)^{3/2}}{h^3 N_e}\,. \tag{17.10}$$

Substituting in the Boltzmann formula (17.5) we obtain, referred to the ground states of the atom and ion

Saha formula
$$\frac{N_{1,0}}{N_{0,0}}\,N_e=\frac{g_{1,0}}{g_{0,0}}\,2\,\frac{(2\pi m k T)^{3/2}}{h^3}\,e^{-\chi_0/kT}. \tag{17.11}$$

For the total number of ionized or neutral atoms we obtain from (17.6) the corresponding ionization formula

$$\frac{N_1}{N_0}\,N_e=\frac{u_1}{u_0}\,2\,\frac{(2\pi m k T)^{3/2}}{h^3}\,e^{-\chi_0/kT}. \tag{17.12}$$

In a completely analagous way for the passage from the rth to the $(r+1)^{\text{th}}$ stage of ionization, where the $(r+1)^{\text{th}}$ electron is removed with ionization-energy χ_r, we have quite independently of other ionization processes

$$\frac{N_{r+1}}{N_r}\,N_e=\frac{u_{r+1}}{u_r}\,2\,\frac{(2\pi m k T)^{3/2}}{h^3}\,e^{-\chi_r/kT} \tag{17.13}$$

and so on. Instead of the number of free electrons per cm³ we equally

---

* The a priori probability of a state with momentum $p=mv$ (mass × velocity) or of kinetic energy $\frac{1}{2}mv^2=p^2/2m$ is $\exp(-p^2/2mkT)$. Integrating over momentum space this gives

$$\int_0^\infty \exp(-p^2/2mkT)\,4\pi p^2\,dp=(2\pi m k T)^{3/2}.$$

well introduce the electron pressure P_e, i. e. the partial pressure of free electrons

$$P_e = N_e k T. \tag{17.14}$$

If we take logarithms in the Saha formula, insert the numeral constants, reckon χ_r in eV and P_e in dyn/cm$^2 \approx 10^{-6}$ atm, we obtain

$$\log \frac{N_{r+1}}{N_r} P_e = -\chi_r \frac{5040}{T} + \frac{5}{2} \log T - 0.48 + \log \frac{2\, u_{r+1}}{u_r}. \tag{17.15}$$

The important temperature parameter $5040/T$ we denote following H. N. Russell by

$$\Theta = 5040/T. \tag{17.16}$$

The quantity $\log(2\, u_{r+1}/u_r)$ is in general small. Originally, M. N. Saha derived his formula without this last term, using thermodynamic calculations applying Nernst's heat theorem and the chemical constant of the electron. One considers the process of ionization of an atom A and the recombination of the ion A^+ with a free electron e as a chemical reaction in which, in a state of chemical (thermodynamic) equilibrium, the reaction proceeds with equal frequency in both directions

$$A \rightleftarrows A^+ + e. \tag{17.17}$$

If we restrict ourselves to sufficiently low pressures, the number of recombination processes (per cm^3 per sec) $\leftarrow$ will be proportional to the number of encounters between ions and electrons i. e. $\sim N_1 N_e$. The number of ionizations by radiation $\rightarrow$ will be proportional to the density of neutral atoms N_0. The factor of proportionality (here left undetermined) depends only on the temperature. Thus one understands the form of the ionization equation (17.12), as well as its generalizations, as an application of the Guldberg-Waage law of mass-action

$$N_1 N_e / N_0 = \text{function of } T. \tag{17.18}$$

[At higher pressure, instead of ionization by radiation, ionization by electron collisions, and instead of two-body recombination of ion + electron, three-body encounters involving a further electron (in order to satisfy the energy and momentum balance) play the essential parts. Thus both processes in (17.17) acquire an additional factor N_e and once again we arrive at the law of mass-action (17.18).]

In Fig. 17.5 we show in logarithmic measure the fraction of atoms of H, He, Mg, Ca in certain states of thermal ionization and excitation, according to a combination of the formulae (17.6) and

(17.12) or (17.13), for an electron pressure $P_e = 100$ dyn/cm² and for temperatures from 3000° to 50 000° K. This value of P_e can be regarded as about the mean value for the atmospheres of main sequence

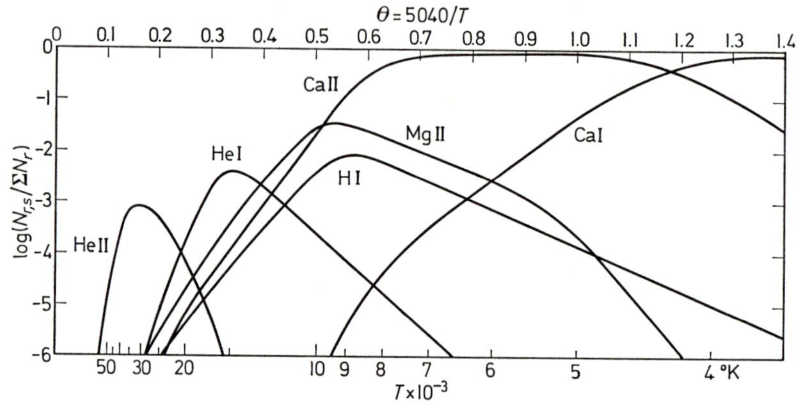

Fig. 17.5. *Thermal ionization (equation 17.13) and excitation (equation 17.6) as a function of temperature T or Θ (where Θ=5040/T) for an electron pressure P_e = 100 dyn/cm² (approx. mean value for stellar atmospheres).* The temperature scale covers the whole range from O stars (on left) to M stars (on right). The Sun (G2) would fit in at about T=5700 °K. The curves demonstrate M. N. Saha's interpretation (1920) of the Harvard sequence of spectral types (see Chapter 15). For example, hydrogen (H I) is predominantly neutral up to $T \approx 9000$ °K. The visible Balmer lines are produced by absorption from the second quantum state, the excitation of which increases with T. Above 9000 °K, however, the hydrogen is rapidly removed by ionization. We therefore see that the hydrogen lines will be at their maximum intensity in the A0 stars, for which $T \approx 9000$ °K.

Spectrum	Ionization potential χ_0 eV	Excited state and excitation potential	$\chi_{r,s}$ eV
H I	13.59	n = 2;	10.15
He I	24.58	2^3 P⁰;	20.87
He II	54.40	n = 3;	48.16
Mg I	7.64	—	—
Mg II	15.03	3^2 D;	8.83
Ca I	6.11	4^1 S;	0.00
Ca II	11.87	4^2 S;	0.00

stars. The absorption lines of these atoms and ions play a part in the Harvard and MK-classification of stellar spectra.

The maxima of the curves, which correspond approximately to the greatest strengths of the spectral lines concerned, result from the fact that at a particular stage of ionization the degree of excitation first increases with increasing T. But as T increases far enough, this stage itself gets "ionized away", so that the fraction of effective atoms

Table 17.2. *Elements contributing significantly to the electron-pressure $P_e = N_e kT$ in stellar atmospheres ($P_e \approx 100$ dyn/cm²). Table shows charge number Z, atomic weight μ, solar abundance ε by numbers of atoms referred to hydrogen=100; ionization potential and statistical weight of first three ionization-levels. The three sets grouped according to the ionization potential of the neutral atom χ_0 are effective in different temperature-ranges.*

Z	Element		Atomic weight μ	Abundance ε (H=100)	Neutral atom χ_0	g_0	Singly-ionized χ_1	g_1	Doubly ionized χ_2	g_2	Temperature-range of effectiveness for $P_e \approx 100$ dyn/cm²
1	H	Hydrogen	1.008	100.	13.59	2	—	—	—	—	$T > 5700$ °K
2	He	Helium	4.003	16.	24.58	1	50.40	2	—	—	
12	Mg	Magnesium	24.32	0.0025	7.64	1	15.03	2	80.12	1	$6000° > T > 4500°$
14	Si	Silicon	28.06	0.0040	8.15	9	16.34	6	33.46	1	
26	Fe	Iron	55.85	0.0004	7.90	25	16.18	30	30.64	25	
11	Na	Sodium	23.00	0.00020	5.14	2	47.29	1	71.65	6	$T < 4700°$
19	K	Potassium	39.10	0.000005	4.34	2	31.81	1	46	6	
20	Ca	Calcium	40.08	0.00014	6.11	1	11.87	2	51.21	1	

decreases again. Taking the temperatures for the maxima as known (e. g. the maximum of the Balmer lines of hydrogen occur in spectral type A0 V at about 9000° K) R. H. Fowler and E. A. Milne in 1923 were first able to estimate the electron pressure in stellar atmospheres.

The Saha theory was able also to account qualitatively for the increase in the intensity-ratio of spark- to arc-lines in passing from main sequence stars to giant stars as a consequence of the increase of the degree of ionization, i. e. of N_1/N_0, resulting from the lower *pressure*. On the other hand, the well-known differences between the spectra of sunspots and of the normal solar atmosphere can be accounted for by the lower *temperature* in the spots. In the laboratory, the Saha theory plays an important part in application to the King furnace, to the electric arc, to very high-temperature plasmas for nuclear-fusion experiments, and so on.

In dealing with a mixture of several elements, we can calculate the degree of ionization most simply by treating the temperature T, or else $\Theta = 5040/T$, and the electron pressure P_e or else $\log P_e$, as independent parameters, and then applying the Saha equation (17.12) to each element and to its various stages of ionization. We can afterwards calculate the gas pressure P_g as kT times the sum of all the particles, including electrons, per cm³. Then we can write down also the mean molecular weight μ. For example, fully ionized hydrogen has mean molecular weight $\mu = 0.5$ since as a result of the ionization unit mass appears in the form of one proton and one electron.

With regard to the theory of stellar atmospheres and of stellar interiors we give in Table 17.2, in percentage numbers of atoms, the

Table 17.3. *Gas pressure P_g as function of electron pressure P_e and of temperature T, or of $\Theta = 5040/T$, for stellar material (Table 17.2): pressures expressed as logarithms*

$\Theta = \frac{5040}{T}$	T	$\log P_e = -1.0$	0.0	1.0	2.0	3.0	4.0	5.0
0.10	50 400	−0.73	0.27	1.27	2.27	3.27	4.27	5.28
0.20	25 200	−0.72	0.29	1.30	2.30	3.30	4.30	5.30
0.30	16 800	−0.70	0.30	1.30	2.30	3.31	4.33	5.36
0.50	10 080	−0.67	0.33	1.34	2.38	3.63	5.36	7.29
0.70	7 200	−0.62	0.68	2.45	4.40	6.34	8.18	
0.90	5 600	+1.39	3.32	5.00	6.41	7.86	9.56	
1.10	4 582	3.19	4.40	5.78	7.21	8.70		
1.30	3 877	3.66	5.13	6.42	7.84			
1.50	3 360	4.35	5.55	6.98	8.82			$\log P_g$

composition of stellar material of which, as we shall see, the Sun and most of the stars consist. Table 17.3 gives the values of P_g for this mixture (the formation of hydrogen molecules being taken into account at the lower temperatures). For $\log P_e \approx 2$ (about the mean value for stellar atmospheres) the stellar material is almost wholly ionized for temperatures $T > 10\,000°$ and so $P_g/P_e \approx 2$ or $\log P_g \approx \log P_e + 0.3$. At the solar temperature $T \approx 5600°$ the metals (Mg, Si, Fe) are essentially ionized; corresponding to their relative abundance $\approx 7 \times 10^{-5}$, we have therefore $\log P_g \approx \log P_e + 4.15$. At still lower temperatures electrons come only from the most easily ionized group, Na, K, Ca.

18. Stellar Atmospheres: Continuous Spectra of the Stars

Looking forward to the quantitative interpretation of the continuous spectra and later of the Fraunhofer lines of the Sun and stars we turn to the physics of stellar atmospheres. We can characterize the atmosphere, that is those layers of a star that transmit radiation to us directly, by the following parameters:

1. The effective temperature T_{eff} which is so defined that, corresponding to the Stefan-Boltzmann radiation law, the radiation flux per cm² of the surface of the star is

$$\pi F = \sigma T_{\text{eff}}^4 . \tag{18.1}$$

For the Sun we obtained in (12.17) directly from the solar constant $T_{\text{eff}} = 5780°$ K.

2. The acceleration due to gravity g [cm/sec²] at the surface.

For the Sun we found in (12.6) $g_{\odot} = 2.74 \times 10^4$ cm/sec²; according to (16.5) gravity for other main sequence stars is not very different.

3. The chemical composition of the atmosphere i. e. the abundances of the elements.

In Table 17.2 we have anticipated some values.

Other possible parameters like rotation or pulsation, or stellar magnetic fields, will not be considered for the moment.

We easily convince ourselves that for given T_{eff}, g and chemical composition the structure, i. e. the distribution of temperature and pressure in a static atmosphere can be completely calculated. For this purpose we need two equations:

(a) The first describes the transfer of energy, by radiation, convection, conduction, mechanical or magnetic energy, and thus determines the *temperature-distribution.* (b) The second is the hydrostatic equation, or, generally speaking, the basic equations of hydrodynamics or magnetohydrodynamics, which determine the *pressure-distribution.* The state of ionization, the equation of state, and all physical constants can all be got from atomic physics if we know only the chemical composition. Then the theory of stellar atmospheres makes it possible, starting from the data 1, 2, 3, to compute a model stellar atmosphere and thence to discover how certain *measurable quantities,* e. g. the intensity-distribution in the continuum or the colour indices, or the intensity of the Fraunhofer lines of a certain element in this or that state of ionization or excitation, depend upon the parameters 1, 2, 3.

If we have solved this problem of theoretical physics, then we can — and this is the decisive consideration — reverse the procedure and by a process of successive approximation determine what T_{eff}, g and what abundance of the chemical elements the atmosphere of a particular star possesses, in the spectrum of which we have measured the energy distribution in the continuum (or colour indices), and/or the intensities of various Fraunhofer lines? In this way we should have a procedure for the quantitative analysis of the spectra of the Sun and stars.

We have remarked that the temperature distribution in a stellar atmosphere is determined by the mechanism of energy-transport. As K. Schwarzschild appreciated in 1905, in stellar atmospheres energy is transported predominantly by *radiation;* we speak then of radiative transfer and of radiative equilibrium.

In order to describe the radiation field, we consider (Fig. 18.1) at depth t, measured from an arbitrary fixed zero-level, a surface element of 1 cm² whose normal makes angle ϑ with the normal to the surface ($0 \leqq \vartheta \leqq \pi$). Then the energy flowing through this surface element within an element of solid angle $d\omega$ about the normal to the element in the frequency interval ν to $\nu + d\nu$ per second is $I_\nu(t, \vartheta)\, d\nu\, d\omega$ (see Chapter 11). According to equation (11.14), in a path-element $ds = -dt \sec\vartheta$ the radiation intensity I_ν suffers by absorption a diminution of amount $-I_\nu(t, \vartheta)\, \varkappa\, ds$, where $\varkappa = \varkappa(\nu)$ is the continuous absorption coefficient per cm at frequency ν. On the

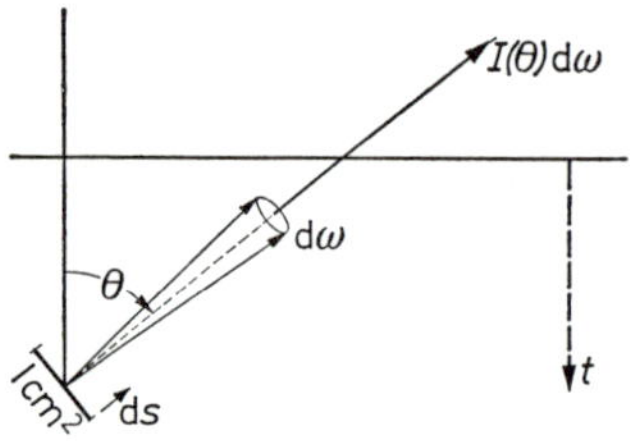

Fig. 18.1. *Radiative equilibrium*

other hand, by the application of Kirchhoff's law to each volume element of the atmosphere assuming local thermodynamic equilibrium (LTE), the intensity I_ν is enhanced by emission by an amount which according to equations (11.19), (11.20) can be written as $+\varkappa B_\nu(T)\,ds$ where $B_\nu(T)$ is the Kirchhoff-Planck function (11.23) for the local temperature T at depth t and frequency ν. Combining these we have therefore

$$dI_\nu(t,\vartheta) = I_\nu(t,\vartheta)\,\varkappa\,dt\sec\vartheta - B_\nu(T(t))\,\varkappa\,dt\sec\vartheta. \tag{18.2}$$

In order to describe the various layers of the atmosphere, instead of geometrical depth t it is better to use the optical depth τ for ν-radiation given by (equation 11.8)

$$\tau = \int_{-\infty}^{t} \varkappa\,dt \quad \text{or} \quad d\tau = \varkappa\,dt. \tag{18.3}$$

So we derive from (18.2) the equation of transfer of radiation

$$\cos\vartheta\,dI_\nu(t,\vartheta)/d\tau = I_\nu(t,\vartheta) - B_\nu(T(t)). \tag{18.4}$$

If there is radiative equilibrium, i. e. if all energy transfer is by radiation, then here we may bring in the energy-equation which states that the total flux must be independent of the depth t. According to (11.8) it follows that

$$\pi F = \pi\int_0^\infty F_\nu(t)\,d\nu = \int_0^\infty\int_0^\pi I_\nu(t,\vartheta)\cos\vartheta\cdot 2\pi\sin d\vartheta\,d\nu = \sigma T_{\text{eff}}^4. \tag{18.5}$$

The solution of the system of equations (18.4), (18.5), with the boundary conditions that at the surface the incident radiation $I_\nu(0,\vartheta)$ for $(0 \leqq \vartheta \leqq \frac{1}{2}\pi)$ must vanish and that at large depths the radiation field must approach black-body radiation, is relatively simple if we postulate that $\varkappa$ is independent of the frequency ν, or if in the transfer equation (18.4) instead of $\varkappa(\nu)$ we use a harmonic mean taken over all frequencies with suitable weight-factors. This is the so-called Rosseland absorption coefficient $\bar{\varkappa}$ and we use the corresponding optical depth

$$\bar{\tau} = \int_{-\infty}^{t} \bar{\varkappa}\,dt.$$

For such a "gray" atmosphere, following E. A. Milne and others we obtain as a good approximate solution of (18.4), (18.5) the temperature distribution

$$T^4(\bar{\tau}) = \tfrac{3}{4}\,T_{\text{eff}}^4(\bar{\tau} + \tfrac{2}{3}). \tag{18.6}$$

According to this the temperature is equal to the effective temperature at optical depth $\bar{\tau}=\frac{2}{3}$. At the stellar surface $\tau=0$, T approaches a finite boundary temperature for which (18.6) gives the value $T_0=2^{-1/4}\,T_{\text{eff}}=0.84\,T_{\text{eff}}$.

For the actual analysis of stellar spectra the "gray" approximation (18.6) is too inexact. Before all else, therefore, in regard to the theory of stellar continuous and line spectra we must first calculate the continuous absorption coefficient $\varkappa$. Several atomic processes contribute to $\varkappa$. From Chapter 17 and Fig. 17.1 we already know:

1. The bound-free transitions of hydrogen. From the limit of the Lyman series at 912 Å, of the Balmer series at 3647 Å, of the Paschen series at 8206 Å, and so on, there extends on the short wavelength sides a series-continuum for which $\varkappa$ falls off approximately like $1/\nu^3$. At low frequencies such continua crowd together and pass over continuously into the continuum produced by the free-free transitions of hydrogen.

2. As R. Wildt pointed out in 1938, the bound-free and free-free transitions of the *negative hydrogen* ion H^- play an important part in stellar atmospheres. This ion can arise by the attachment of a second electron to the neutral hydrogen atom with a binding, or ionization, energy of 0.75 eV. Corresponding to this small ionization energy, the long-wave end of the bound-free continuum lies in the infrared at 16 550 Å. Free-free absorption increases beyond that point towards longer wavelengths, as in the case of the neutral atom.

The atomic coefficients for absorption from a particular energy level have been calculated with great accuracy on a quantum-mechanical basis. If we want then to obtain the absorption coefficient for stellar material of specified composition as a function of frequency, temperature T and electron pressure P_e or gas pressure P_g, we must calculate for the various energy-levels the degree of ionization and excitation, using the formulae of Saha and Boltzmann. In addition we obtain the relation between gas pressure P_g, electron pressure P_e and temperature T, as we have presented it for a particular example in Table 17.3.

We find thus that in the hot stars (say, $T>7000°$) the continuous absorption by hydrogen atoms predominates, and in cooler stars that by H^--ions. Besides these two most important processes, in more exact calculations account must be taken of:—

Bound-free and free-free absorption by HeI and HeII (in hot stars) and by metals (in cooler stars, see Table 17.2); further, scattering of light by free electrons (Thomson scattering; in hot stars) also by neutral hydrogen (Rayleigh scattering; in cooler stars). We still know very little about the continua of molecules in cool

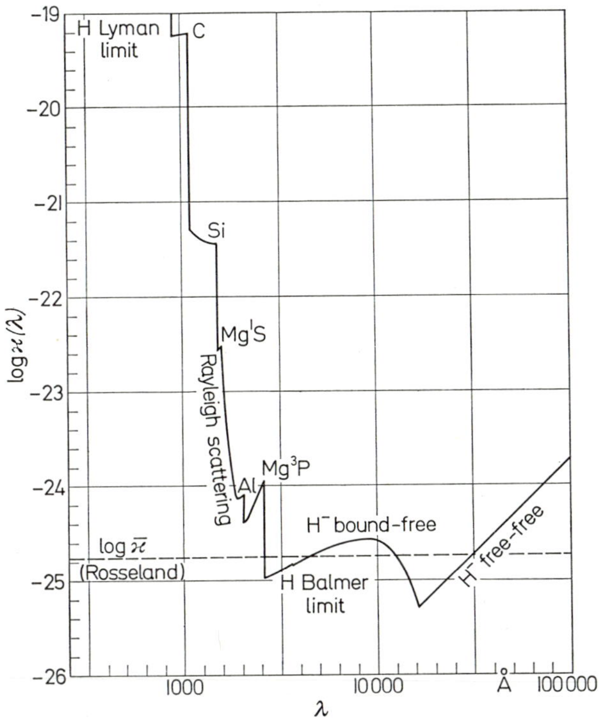

Fig. 18.2. *Continuous absorption coefficient $\varkappa(\lambda)$ in the solar atmosphere.* (G 2 V) at $\tau_0 = 0.1$ (τ_0 at λ 5000 Å). Here we have $T = 5040$ °K ($\Theta = 1$) and $P_e = 3.2$ dyn/cm^2 ($P_g = 5.8 \times 10^4$ dyn/cm^2)

stars. From the extensive calculations of G. Bode we reproduce in Figs. 18.2, 18.3 the absorption coefficient $\varkappa$ as a function of wavelength λ for mean conditions (see below) in the atmospheres of the Sun and of the B0 V-star τ Scorpii.

With the help of the Rosseland opacity coefficient $\bar{\varkappa}$ shown in these diagrams, and the "gray" approximation (18.6) we first calculate for a given effective temperature T_{eff}, the temperature T (in the sense of local thermodynamic equilibrium) as a function of optical depth $\bar{\tau}$ corresponding to opacity $\bar{\varkappa}$. The electron pressure P_e on which most features do not depend very sensitively, has first to be estimated. Sometimes it is advantageous to use, instead of $\bar{\tau}$, an optical depth τ_0

that corresponds to the absorption coefficient $\varkappa_0$ referring to a particular wavelength, usually $\lambda_0 = 5000$ Å. The relation between $\bar{\tau}$ and τ_0 (similarly also to that between optical depths for two different

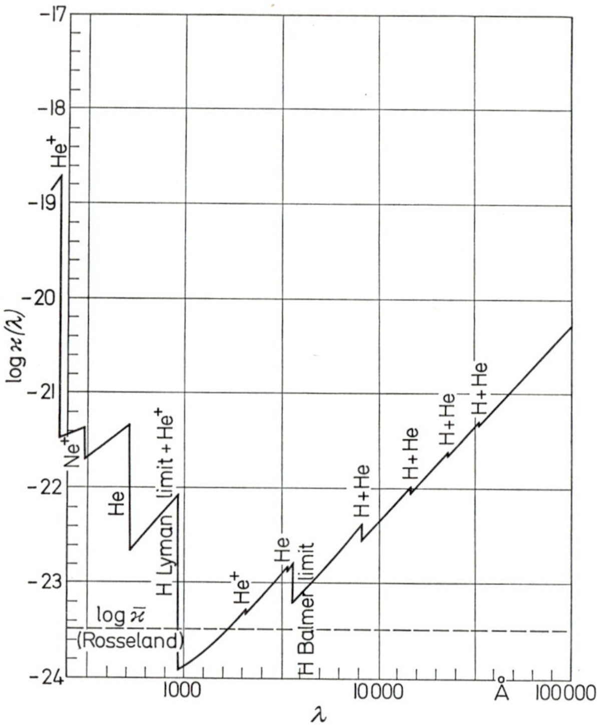

Fig. 18.3. Continuous absorption coefficient $\varkappa(\lambda)$ in the atmosphere of τ Scorpii (B0 V) at $\bar{\tau} \approx 0.1$. Here we have $T = 28\,300$ °K ($\Theta = 0.18$) and $P_e = 3.2 \times 10^3$ dyn/cm² ($P_g = 6.4 \times 10^3$ dyn/cm²)

wavelengths) is easily got from

$$\left.\begin{aligned} d\bar{\tau} &= \bar{\varkappa}\, dt \\ d\tau_0 &= \varkappa_0\, dt \end{aligned}\right\} \quad \frac{d\bar{\tau}}{d\tau_0} = \frac{\bar{\varkappa}}{\varkappa_0} \text{ for a definite depth.} \tag{18.7}$$

For higher demands on the accuracy of model atmospheres the frequency-dependence of the continuous absorption coefficient and ultimately also the very strongly frequency-dependent line-absorption must be taken into account. In radiative equilibrium again the total radiation flux πF must be the same at all depths t. The theory of radiative equilibrium of such "non-gray" atmospheres offers considerable mathematical difficulties, which hitherto have been overcome only by processes of successive approximation.

If we know the temperature-distribution $T(\bar{\tau})$ or $T(\tau_0)$ in a stellar atmosphere, the calculation of the pressure-distribution offers no further difficulty. The increase of gas pressure P_g with depth t in a static atmosphere is determined by the hydrostatic equation

$$dP_g/dt = g\,\varrho \tag{18.8}$$

where the density ϱ is related to P_g and T by the equation of state of a perfect gas. Again g means the value of gravity. If we divide both sides by $\varkappa_0$, since $\varkappa_0\, dt = d\tau_0$ we get

$$dP_g/d\tau_0 = g\,\varrho/\varkappa_0\,. \tag{18.9}$$

Since the right-hand side of this equation is known as a function of P_g and T, together with T as a function of τ_0, we can now integrate equation (18.9) numerically.

In hot stars, radiation pressure as well as gas pressure must be taken into account. If there are currents in the atmosphere whose speed v is not small compared with local sound-speed (for example, 12 km/sec in atomic hydrogen at 10 000 °K), then the dynamical pressure $\frac{1}{2}\,\varrho\, v^2$ also contributes. In sunspots and in Ap stars with their magnetic fields of several thousand gauss, magnetic forces have to be reckoned with as well.

Important aids to the calculation of model stellar atmospheres are extensive tables that give for various mixtures of elements (e. g. Table 17.2) the *gas pressure* P_g (e. g. Table 17.3), the *mean molecular weight* μ (from Table 17.2 for neutral stellar material $\mu = 1.50$ and for fully ionized material $\mu = 0.70$), the *continuous absorption coeffi-*

Tables 18.1 and 18.2. *Model stellar atmospheres.*
The optical depth τ_0 refers to $\varkappa$ (5000 Å); $\bar{\tau}$ is calculated for the Rosseland mean $\bar{\varkappa}$; temperature T is in °K; gas pressure P_g and electron pressure P_e are in dyn cm^{-2}, logarithms being tabulated

1. Sun (G2 V). Effective temperature $T_{eff} = 5780$ °K; surface gravity $\log g = 4.44$

τ_0	T	$\log P_g$	$\log P_e$
0.001	3875	3.74	−0.92
0.01	4295	4.26	−0.28
0.1	5030	4.79	+0.45
0.2	5280	4.96	+0.69
0.5	5805	5.14	+1.20
1.0	6400	5.23	+1.79
2.0	7180	5.28	+2.44

2. τ Scorpii (B0 V). Effective temperature $T_{eff} = 32\,800$ °K; surface gravity $\log g = 4.45$

$\bar{\tau}$	T	$\log P_g$	$\log P_e$
0.01	23 200	3.05	2.75
0.1	28 300	3.80	3.50
0.2	30 600	4.03	3.73
0.5	34 100	4.32	4.02
1.0	38 000	4.53	4.23
2.0	42 700	4.74	4.47
5.0	50 100	5.03	4.76

cient $\varkappa$ for numerous wavelengths and its Rosseland mean $\bar{\varkappa}$ all as functions of temperature T and electron pressure P_e.

The procedures for computation outlined above can be refined by successive approximations. In recent times large electronic computers have become an indispensable aid to the theory of stellar atmospheres.

In Tables 18.1, 18.2 we summarize the models of the atmospheres of the Sun and of the B0 V star τ Scorpii according to the present state of the art. Here we have anticipated the step by step adjustment of theory and observation. We shall return to this later but now we apply ourselves to the calculation of the *continuous spectrum*.

The radiation that a star sends out at its surface originates in layers at various depths in its atmosphere, in such a way that the radiation coming from the deeper layers is naturally more weakened by absorption than that coming from higher up.

The "ergiebigkeit" or source function, defined as the emission coefficient ε divided by the absorption coefficient $\varkappa$, at optical depth τ in the case of local thermodynamic equilibrium is equal to the Kirchhoff-Planck function $B_\nu(T(\tau))$ for the local temperature (equation (18.2) or (18.4)). From (11.19), the ν-radiation in a direction inclined at angle ϑ to the normal that comes from this depth, is weakened by an absorption-factor $\exp\{-\tau \sec\vartheta\}$ before it escapes at the stellar surface. Therefore we obtain for the radiation intensity at $\tau = 0$

$$I_\nu(0, \vartheta) = \int_0^\infty B_\nu(T(\tau))\, e^{-\tau \sec\vartheta}\, d\tau \sec\vartheta\,. \tag{18.10}$$

Since on the one hand according to the theory of radiative equilibrium (e. g. equation (18.6)) we can calculate the temperature T as a function of optical depth τ, and on the other hand, with the help of the theory of the continuous absorption coefficient $\varkappa$, we can formulate the connection between the various optical depths $\bar{\tau}, \tau, \tau_0 \ldots$ as in (18.7), the whole theory of the continuous spectrum of a star is contained in (18.10). In the case of the Sun, this includes the theory of limb-darkening ($\vartheta = 0$ corresponds to the centre of the disk and $\vartheta = 90°$ to the limb).

Without going into details, with the model solar atmosphere in Table 18.1 the full calculations yield the continuous spectrum $I_\nu(0, 0)$ for the centre of the solar disk plotted in Fig. 12.4 and the centre-limb variation in Fig. 12.3. With sufficient accuracy for many purposes, we can simplify the calculation of $I_\nu(0, \vartheta)$ if we expand the "ergiebigkeit" $B_\nu(\tau)$ in a series starting from that at a particular optical depth τ^*, which remains to be selected,

$$B_\nu(\tau) = B_\nu(\tau^*) + (\tau - \tau^*)\,(\mathrm{d}B_\nu/\mathrm{d}\tau)_{\tau^*} + \cdots. \tag{18.11}$$

Substituting in (18.10) we obtain

$$I_\nu(0, \vartheta) = B_\nu(T(\tau^*)) + \{\cos\vartheta - \tau^*\}\,(\mathrm{d}B_\nu/\mathrm{d}\tau)_{\tau^*} + \cdots. \tag{18.12}$$

If we now choose $\tau^* = \cos\vartheta$, the second term on the right becomes zero and we have the so-called Eddington-Barbier approximation

$$I_\nu(0, \vartheta) \approx B_\nu(T(\tau = \cos\vartheta)). \tag{18.13}$$

This means that the intensity radiated at the surface of, say, the Sun corresponds in the sense of Planck's law to the temperature at optical depth $\tau = \cos\vartheta$ measured perpendicular to the surface which means $\tau \sec\vartheta = 1$ measured along the sight-line. This is instructive; with some reservation we can say that the radiation we see originates at unit optical depth along our line of sight.

In the case of the stars we can deal with the mean intensity over the disk, i. e. apart from a factor π, the flux of radiation at the surface. From (11.7) we first have

$$F_\nu(0) = 2\int_0^{\frac{1}{2}\pi} I_\nu(0, \vartheta)\cos\vartheta\sin\vartheta\,\mathrm{d}\vartheta. \tag{18.14}$$

With the use of (18.12) we easily obtain the approximation

$$F_\nu(0) \approx B_\nu(T(\tau = \tfrac{2}{3})). \tag{18.15}$$

Thus the radiation of a star at frequency ν corresponds to the local temperature T at an optical depth $\tau = \frac{2}{3}$ in this frequency, corresponding to $\cos\vartheta = 2/3$ giving a mean emission angle of 54°44′.

As further illustrations we show in Fig. 18.4 the energy distribution in the continuum $F_\nu(0)$ calculated for the model of the B0 V star τ Scorpii in Table 18.2, and immediately below the corresponding result for the A0 V star α Lyrae = Vega. Here the similarity to the spectrum of a black body has disappeared. The F_ν-curve looks much more like a mirror image of the wave-length dependence of the continuous absorption coefficient $\varkappa$ (shown in Fig. 18.3 for τ Sco). Since $\mathrm{d}\tau/\mathrm{d}\bar{\tau} = \varkappa/\bar{\varkappa}$, for large $\varkappa$ the radiation comes from layers near the surface having small $\bar{\tau}$; these are relatively cool and the radiation curve F_ν drops down. In this way we understand how at each absorption-edge of hydrogen (λ 912 Å Lyman-edge, λ 3647 Å Balmer-edge ...) F_ν drops abruptly on the short wavelength side. The hotter of these two stars τ Sco radiates mostly between the Balmer and Lyman limits, the cooler α Lyr mostly between the Paschen and Balmer limits. We indicate the small range |—| of the visible spectrum and also the

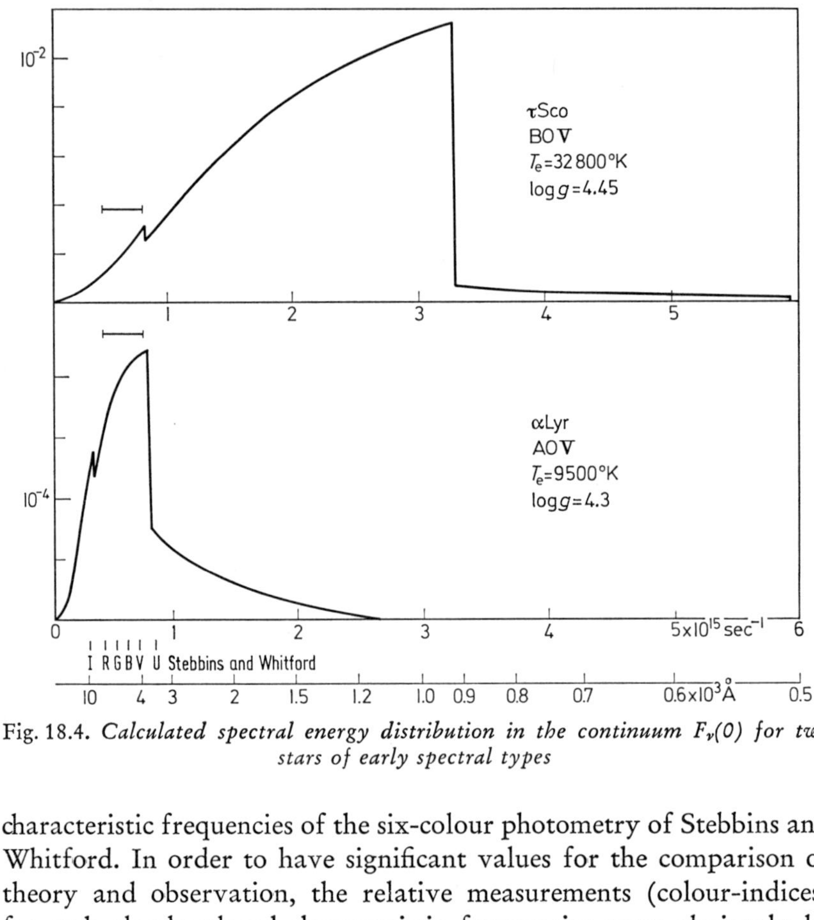

Fig. 18.4. *Calculated spectral energy distribution in the continuum $F_\nu(0)$ for two stars of early spectral types*

characteristic frequencies of the six-colour photometry of Stebbins and Whitford. In order to have significant values for the comparison of theory and observation, the relative measurements (colour-indices) for such closely-placed characteristic frequencies must obviously be made with high accuracy.

Finally, in comparing theory and observation we must remember that the theory relates to the true continuum F_ν or F_λ while stellar photometry and spectra of low dispersion measure the continuum with smeared-out lines $F_\lambda^{\mathrm{L}} = (1 - \eta_\lambda)\, F_\lambda$, where η_λ is the fraction of the true continuum obstructed by the Fraunhofer lines. In Fig. 18.5 we have assembled measured values of η_λ from coudé spectra of large dispersion for a number of stars. In hot stars the broad Balmer lines of hydrogen predominate. In cooler stars, from about F 5, the numerous metal-lines obstruct a considerable part of the spectrum, especially in the blue and ultraviolet. Finally, in K and M stars absorption in molecular bands also play a part.

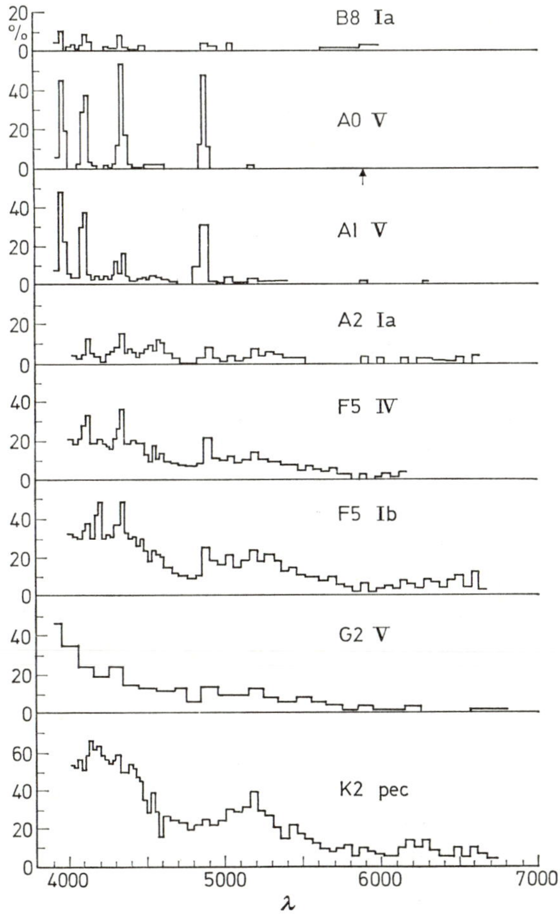

Fig. 18.5. *Percentages* η_λ *of energy absorbed by the Fraunhofer lines in stars of various spectral types*

The integration of $(1-\eta_\lambda) F_\lambda$ over the whole spectrum yields the total flux and thence the bolometric correction BC (see (13.14) (13.15)) of the star concerned. Here we cannot go into the performance of this difficult calculation. However, it need scarcely be remarked that great demands are made upon the theory when, in the case of hot stars for example, we have to infer the area under the whole curve from measurements in the tiny part of the spectrum that penetrates to the surface of the Earth (shown by $\vdash\!\!\dashv$ in Fig. 18.4). Direct measurements of the total flux from hot stars, by far the greater part of which is in the far ultraviolet, can be made only with the help of rockets and space-vehicles.

19. Theory of Fraunhofer Lines: Chemical Composition of Stellar Atmospheres

From the theory of the continuous spectra of stars we now turn to the quantitative study of the Fraunhofer lines.

Referring to observations of the solar spectrum we mentioned in Chapter 12 the classical work of Rowland; with modern diffraction gratings resolving powers up to $\lambda/\Delta\lambda \approx 10^6$ are attained. With the coudé spectrographs of large telescopes, the spectra of at anyrate the brighter stars from about 3200 Å to 6800 Å can be got with dispersion of a few Å/mm.

Besides reference to the Rowland tables and monographs on individual stars (Table 19.1), the tabulations of Charlotte E. Moore are indispensable for the identification and classification of lines (see also Chapter 17):

1. *A multiplet table of astrophysical interest* — revised ed. (Nat. Bur. of Standards Washington, Techn. Note No. 36, 1959).

2. *An ultraviolet multiplet table (ibid.* Circular 488, Sect. 1—5, 1950—1961).

3. *Atomic energy levels* — several volumes *(ibid.* Circular 467, 1949—).

Also, including numerous term-schemes, there is Paul W. Merrill's *Lines of the chemical elements in astronomical spectra* (Carnegie Inst. Washington; Publ. 610, 1956).

By automatically taking account of the blackening-curve with the help of intensity-marks, the microphotometer nowadays directly yields the intensity distribution in the spectrum (Fig. 19.1). Besides such photographic photometry, the direct recording of spectra by means of photocells or photo-multipliers gains ever-increasing significance. In practice the next step of fixing the true continuum is very difficult; its intensity is then normalized to unity, or 100 percent, from which the depression in the lines $R_\nu = 1 - I_\nu$ is measured. Aside from weak disturbing lines, socalled "blends", we then have the intensity distribution in the line given by I_ν or R_ν as a function of ν or of λ, that is the *line-profile*. The spectrograph itself would reproduce an infinitely sharp line as a line of finite width, the so-called instrumental profile. So we must note that the width and structure of weak Fraunhofer lines always include an instrumental contribution. This one can determine by measuring the profiles of sharp lines in a laboratory source (the lines of the iron arc being sufficient in reference to stellar spectra). The energy absorbed in the line is obviously independent of instrumental distortion of the line-profile. As M. Minnaert has remarked, it is therefore useful to measure the equivalent width W_λ; this is the width in angstroms or milli-angstroms of a rectangular strip of the spectrum having the same area as that of the line-profile (Fig. 19.2). The revision of the Rowland tables by

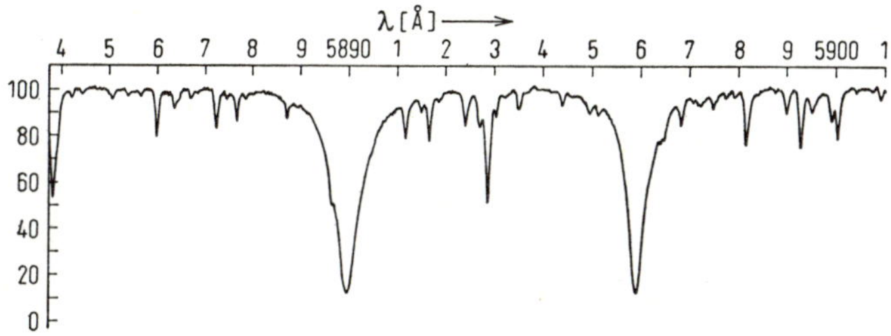

Fig. 19.1. *Microphotometer tracing; the NaD lines of the solar spectrum*

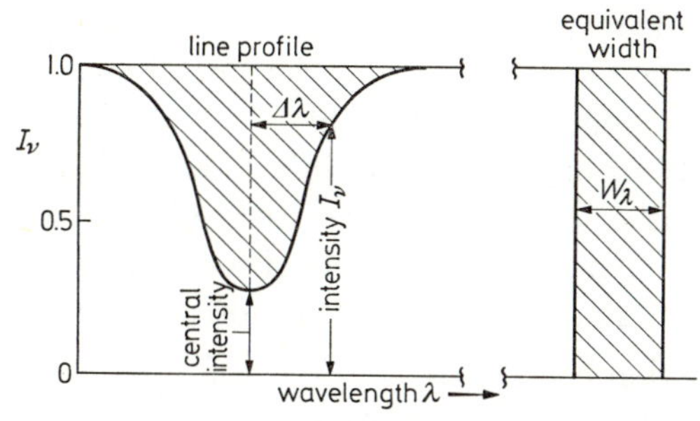

Fig. 19.2. *Profile and equivalent width of a Fraunhofer line*

Ch. E. Moore and M. Minnaert contains equivalent widths in the spectrum of the centre of the solar disk; some references for stellar spectra are given in Table 19.1.

From spectro-photometric measurements of lines and the continuum we must then derive the parameters that we have repeatedly mentioned as characterizing the stellar atmosphere, that is to say the effective temperature T_{eff}, the gravitational acceleration g and the relative abundances of the elements.

The complexity of the problem suggests the use of successive approximations. We first ask the question: For an atmosphere of given T_{eff}, g and assigned chemical composition, what values would the theory predict for the equivalent widths W_λ of various lines? In Chapter 18 we have already studied the corresponding problem for the continuous spectrum. Here again we use model atmospheres. Then we investigate the way in which measurable quantities, i. e. in addition to the colour-indices already considered, principally the W_λ for particular elements in particular states of ionization and excitation, depend upon the parameters that were chosen at the outset. Finally, we improve these parameters until we obtain the best possible agreement between the computed spectrum of the model atmosphere and the measured spectrum of the actual star under investigation.

As already indicated in Chapter 10, the theory of Fraunhofer lines consists of two different parts:

1. The theory of *radiative transfer* in the lines which gives the dependence of the profiles and equivalent widths for a given model atmosphere upon the line-absorption coefficient $\varkappa_\nu$.

2. The atomic theory of the *line-absorption coefficient* $\varkappa_\nu$ itself.

We turn to the first problem:

In addition to the continuous absorption coefficient $\varkappa$, which varies only slowly with wavelength and may be treated as constant in the vicinity of a line, we now consider the line-absorption coefficient $\varkappa_\nu$ which, as a function of the interval $\Delta\lambda$ or $\Delta\nu$ from the line-centre (Fig. 19.3), falls off more or less steeply from a high maximum to the value zero on either side. Thus we distinguish between the following absorption coefficients (which in their different ways depend upon the quantities T and P_e) and the associated optical depths:

$$\begin{aligned} &\varkappa_\nu = \text{line-absorption coefficient} && \tau_\nu = \int \varkappa_\nu \, dt \\ &\varkappa = \text{continuous absorption coefficient} && \tau = \int \varkappa \, dt \end{aligned} \Bigg\} \; x_\nu = \int (\varkappa_\nu + \varkappa)\, dt$$

$$\bar{\varkappa} = \text{Rosseland opacity coefficient} \qquad \bar{\tau} = \int \bar{\varkappa} \, dt\,. \tag{19.1}$$

All the integrals are to be taken from $-\infty$ to t.

If we again make the assumption of local thermodynamic equilibrium (LTE) in regard to the radiative exchange (which should be quite good in the case of atoms or ions having rather complicated term-schemes)*, then for the calculation of the line profile we may take over the whole formulation and computation in Chapter 18, simply replacing $\varkappa$ by $\varkappa + \varkappa_\nu$ and the optical depth τ by x_ν.

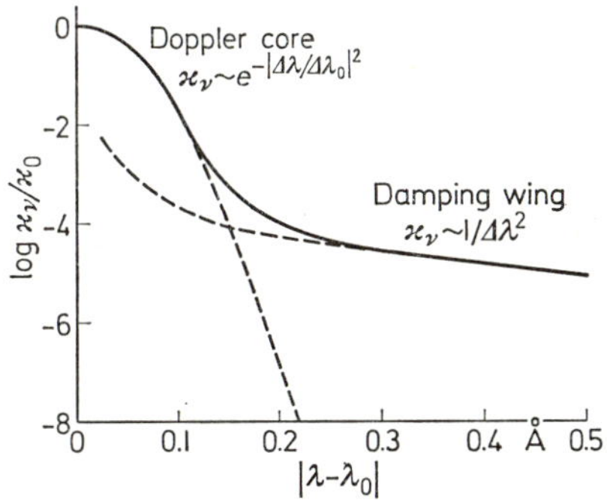

Fig. 19.3. *Coefficient of line absorption $\varkappa_\nu$ (compared with $\varkappa_0$ for the line centre).* The Doppler core and damping wings of the NaD lines have been calculated from equations 19.9 and 19.10 respectively for $T = 5700$ °K and pure radiation damping. Their superposition (full line) gives the so-called Voigt profile

Thus the intensity of the emergent radiation that makes an angle ϑ with the normal to the surface of the atmosphere ($\vartheta = 0$ again corresponding to the centre of the solar disk and $\vartheta = \frac{1}{2}\pi$ to the limb) becomes

Line (frequency ν) $$I_\nu(0, \vartheta) = \int_0^\infty B_\nu(T(x_\nu))\, e^{-x_\nu \sec\vartheta}\, dx_\nu \sec\vartheta$$

Neighbouring continuum (suffix 0)

$$I_0(0, \vartheta) = \int_0^\infty B_0(T(\tau))\, e^{-\tau \sec\vartheta}\, d\tau \sec\vartheta, \qquad (19.2)$$

where the relation between the two optical depths is expressed by

$$\frac{dx_\nu}{d\tau} = \frac{\varkappa_\nu + \varkappa}{\varkappa} \quad \text{or} \quad x_\nu = \int \frac{\varkappa_\nu + \varkappa}{\varkappa}\, d\tau. \qquad (19.3)$$

The depression in the line is then

$$r_\nu(0, \vartheta) = \frac{I_0(0, \vartheta) - I_\nu(0, \vartheta)}{I_0(0, \vartheta)} \qquad (19.4)$$

* We cannot here deal with the special features of the scattering of light in the continuum (Thomson or Rayleigh scattering) and in lines (resonance fluorescence at low pressures). In the outcome the departures from LTE mostly remain within moderate bounds.

and the equivalent width defined in accordance with Fig. 19.2 is

$$W_\lambda = \int r_\lambda(0, \vartheta)\, d\lambda. \tag{19.5}$$

We can again make use of the approximation (18.13) and so obtain

$$r_\nu(0, \vartheta) \approx \frac{B_\nu(T(\tau = \cos\vartheta)) - B_\nu(T(x_\nu = \cos\vartheta))}{B_\nu(T(\tau = \cos\vartheta))}. \tag{19.6}$$

In calculating the Kirchhoff-Planck function $B_\nu(T)$ the small frequency-differences in the vicinity of a line are naturally of no significance; an essential point is that the radiation in the line comes from a higher layer $x_\nu = \cos\vartheta$, with a correspondingly lower temperature, than that in the neighbouring continuum that comes from $\tau = \cos\vartheta$. If the absorption coefficient in the line-centre is very much greater than in the continuum ($\varkappa_\nu \gg \varkappa$), the central intensity of the line is simply the Kirchhoff-Planck function for the boundary temperature $T(\tau = 0)$ of the atmosphere. The line then reaches the common maximum depression $r_c(0, \vartheta)$ of all sufficiently strong lines in the same region of the spectrum.

For application to *stars*, in which we cannot observe the centre-to-limb variation, we again need the corresponding expressions for the flux, or the mean intensity F_ν in the line and F_0 in the neighbouring continuum. The depression in the line then becomes

$$R_\nu(0) = \frac{F_0(0) - F_\nu(0)}{F_0(0)} \approx \frac{B_\nu(T(\tau = 2/3)) - B_\nu(T(x_\nu = 2/3))}{B_\nu(T(\tau = 2/3))} \tag{19.7}$$

and the corresponding equivalent width in the stellar spectrum

$$W_\lambda = \int R_\lambda(0)\, d\lambda. \tag{19.8}$$

[We see from equation (19.7) that in a stellar spectrum the continuum and also the wings of the lines (where $\varkappa_\nu \ll \varkappa$) come mainly from those layers of the atmosphere whose optical depth is $\tau = \frac{2}{3}$ in the continuum in the region of the spectrum concerned. The lines of neutral metals (FeI, TiI, ...) in the spectra of the Sun and similar fairly cool stars are exceptions. On account of increasing ionization, the concentration of the atoms decreases so rapidly with increasing depth in the atmosphere that the "centre of gravity" for forming the lines lies no deeper than $\tau \approx 0.05$ to 0.1. The centres of strong lines, where $\varkappa_\nu \gg \varkappa$, come from correspondingly higher layers where $\tau \approx \frac{2}{3}\varkappa/(\varkappa + \varkappa_\nu)$.]

In order to be able actually to apply the theory of radiative transfer to the lines *, we now turn to the second point of our program, the calculation of the line-absorption coefficient as a function of temperature T, electron pressure P_e or gas pressure P_g and distance from the line-centre $\Delta\nu$ in frequency or $\Delta\lambda$ in wavelength. The following mechanisms affect $\varkappa_\nu$:

1. *Doppler effect* of thermal velocities and maybe of turbulent motions. For instance, the thermal velocities of Fe atoms in the solar atmosphere for $T \approx 5700°$ K are about 1.3 km/sec which gives for, say, the line at λ 3860 Å a doppler width $\Delta\lambda_D = 0.017$ Å. Corresponding to the Maxwell velocity-distribution of the atoms, the doppler-distribution of the line absorption coefficient is

$$\varkappa_\nu \sim \exp\{-(\Delta\lambda/\Delta\lambda_D)^2\}. \tag{19.9}$$

Turbulent motions in stellar atmospheres (see Chapter 20) often produce similar velocities and make a corresponding contribution to $\Delta\lambda_D$.

2. *Damping*. In classical optics a wave-train of limited duration with a damping constant γ sec^{-1}, or a characteristic time γ^{-1} sec, corresponds in accordance with a well-known theorem in Fourier analysis to a spectral line in which the absorption coefficient $\varkappa_\nu$ exhibits the typical damping-distribution

$$\varkappa_\nu \sim \frac{\gamma}{(2\pi\,\Delta\nu)^2 + (\gamma/2)^2}\,. \tag{19.10}$$

Thus γ is the half-width $2 \cdot 2\pi\,\Delta\nu_{1/2}$ of the absorption coefficient in units of angular frequency.

According as the temporal limitation of the radiative process is determined by the emission of the atom itself or by its collisions with other particles, following H. A. Lorentz we speak of *radiation-damping* or *collision-damping*.

a) Radiation damping. According to quantum theory, the radiation-damping constant γ_{rad} is equal to the sum of the decay constants (reciprocal lives) of the two energy-levels between which the transition takes place. Since γ is thus of the order 10^7 to 10^9 sec^{-1}, we expect from this mechanism (half) half-widths of the absorption coefficient, say at λ 4000 Å, of about 4×10^{-6} to 4×10^{-4} Å.

---

* In passing, we remark that in calculating the temperature distribution in a "non-gray" model atmosphere account must be taken of line-absorption as well as continuous absorption. Since the outward flow of radiation is held back in the lines, the radiation from the higher layers is thereby enhanced, thus causing a steeper temperature drop in the high layers.

b) Collision-damping. The collision-damping constant is $\gamma_{\text{coll}} = 2 \times$ number of effective collisions per second. According to W. Lenz, V. Weisskopf and others, we count as such those encounters of the radiating particle with perturbing particles in which the phase of the radiative oscillation is displaced by more than about one-tenth of an oscillation.

In fairly cool stellar atmospheres like that of the Sun in which hydrogen is mostly neutral collision-damping by neutral hydrogen atoms preponderates. The hydrogen atoms affect the radiating atom by van der Waals forces (interaction energy $\sim$ (distance)$^{-6}$). The damping constant γ_{coll} is then proportional to the gas-pressure P_g.

In the case of spectral lines that show a large quadratic Stark effect, and in mainly ionized atmospheres, collision damping by free electrons may be predominant. Then the interaction-energy varies as the square of the field-strength that the electron produces at the position of the radiating particle; so it is proportional to the (distance)$^{-4}$. The damping constant is now proportional to the electron-pressure P_e.

For about 10^9 effective collisions per second we expect the (half) half-width of the line-absorption coefficient to be of the order 10^{-3} Å. It makes no difference whether the collisions, as in the solar atmosphere, are mostly with hydrogen atoms or, as in hot stars, are mostly with free electrons.

3. Combination of doppler effect and damping. The ratio of the half damping constant $\frac{1}{2}\gamma$ to the doppler width $\Delta\omega_D = c\,\Delta\lambda_D/\lambda^2$ (measured in circular frequency) is

$$\alpha = \tfrac{1}{2}\gamma/\Delta\omega_D \tag{19.11}$$

and in stellar atmospheres almost without exception $\alpha < 0.1$. So one might suppose that collision broadening can be neglected in comparison with doppler broadening. However, this is not correct because the doppler distribution (19.9) falls off exponentially away from the line centre, while the collision distribution (19.10) falls off only like $1/\Delta\lambda^2$. Each moving atom produces a damping distribution with a sharp central peak and broad wings, which is doppler-shifted as a whole corresponding to the velocity. Thus we obtain (Fig. 19.3) a distribution of the line-absorption coefficient $\varkappa_\nu$ having a fairly sharply bounded doppler core in accordance with equation (19.9) to which are joined almost directly the damping wings where in accordance with (19.10) $\varkappa_\nu$ varies as $1/\Delta\lambda^2$.

The absolute value of the absorption coefficient $\varkappa_\nu$ (cm^{-1}) is always normalized by the quantum-theoretical relation, the integral being taken over the whole line

$$\int \varkappa_\nu \, d\nu = \frac{\pi e^2}{m c} N f. \tag{19.12}$$

Here e, m again denote the charge and mass of the electron; c is light-speed; N is the number per cm^3 of absorbing atoms in the energy level from which the absorption occurs. Within the framework of classical electron theory, which sought to ascribe the spectral lines to harmonic electron-oscillators of the corresponding frequencies, the formula would be complete without f (that is, with $f=1$). In agreement with laboratory measurements (see below), quantum theory requires that the formula be extended to include the so-called *oscillator-strength* f. After some simple manipulation, we finally obtain from (19.9), (19.10) in combination with (19.12) the absorption coefficient in cm^{-1} at interval $\Delta\lambda$ from the centre of a line of wavelength λ_0

Doppler core $$\varkappa_\nu = \sqrt{\pi}\,\frac{e^2}{m c^2}\,\frac{\lambda_0^2 N f}{\Delta\lambda_D}\,\exp\{-(\Delta\lambda/\Delta\lambda_D)^2\} \tag{19.13}$$

Damping wings $$\varkappa_\nu = \frac{1}{4\pi}\,\frac{e^2}{m c^2}\,\frac{\lambda_0^4}{c}\,\frac{N f \gamma}{\Delta\lambda^2}\,. \tag{19.14}$$

The quantity $c^2/m c^2$ is the so-called classical radius of the electron $r_0 = 2.817 \times 10^{-13}$ cm.

Since the number of atoms N is proportional to the statistical weight g of the corresponding state, instead of f we usually state the value of $g f$. The intensity of an emission or absorption line produced by an optically thin layer is directly proportional to this quantity.

One can calculate the relative $g f$-values within a multiplet (Chapter 17) using quantum-theoretical formulae found by A. Sommerfeld and H. Hönl, H. N. Russell and others in connexion with the Utrecht measurements made by H. C. Burger and H. B. Dorgelo (1924). The $g f$-values for a doublet such as the sodium D-lines $3\,^2S_{\frac{1}{2}} - 3\,^2P_{\frac{1}{2},\frac{3}{2}}$ are in the ratio 1 : 2, for a triplet such as the calcium lines $4\,^3P_{2,1,0} - 5\,^3S_1$ (λ 6162, 6122, 6103 Å) they are in the ratios 5 : 3 : 1, and so on.

There exist corresponding formulae for aggregates of higher order, the so-called super-multiplets and transition arrays.

A general idea of the absolute values of the oscillator strengths is given by the sum-rule of W. Kuhn and W. Thomas: Let absorption transitions $n \rightarrow m$ be possible with oscillator-strengths f_{nm} from a particular energy level n of an atom or ion with z electrons (more precisely, we always restrict ourselves to consideration of the radiating electrons that take part in the relevant transitions) and let transitions $m \rightarrow n$ to n from lower levels occur with oscillator-strengths f_{mn}. Then we have

$$\sum_m f_{nm} - \sum_m \frac{g_m}{g_n} f_{mn} = z\,, \tag{19.15}$$

where g_n and g_m are the statistical weights (Chapter 17) of the corresponding energy-levels. If there is effectively a single strong transition from the ground state of an atom or ion to the next-higher term then for the multiplet as a whole we can put approximately $f \approx z$. For instance, we have about $f = \frac{1}{3} + \frac{2}{3} = 1$ for the D-lines of sodium taken together.

For hydrogen and ionized helium one can calculate the f-values exactly from quantum theory.

For certain hydrogen-like spectra (in particular systems with 1, 2 or 3 radiating electrons) D. R. Bates and A. Damgaard have developed a very practical quantum theoretical approximation procedure.

For the so-called complex spectra of atoms and ions with several outer electrons (such as the astrophysically very important spectra of the metals Fe I, Ti I, ..., Fe II, Ti II ...) one can measure *relative* f-values in emission (in the arc or King's furnace) or in absorption (King's furnace). The main difficulty lies in measuring absolute f-values, even for a few selected lines of the atom or ion, since here the number of absorbing or emitting particles has somehow to be measured directly. One can, for instance, introduce into an electric furnace a quartz absorption vessel in which there is established a certain vapour pressure, depending upon the temperature, of the metal under investigation. In recent times the decay-constants or transition-probabilities of individual atomic states, equivalent to the f-values, have been successfully measured electronically. This method, in which a determination of the particle number becomes unnecessary, appears to be very promising, especially for ionized atoms.

B. M. Glennon and W. L. Wiese (Nat. Bur. of Standards Washington, Monograph 50, 1962) have published a bibliography of f-values and atomic transition probabilities arranged according to elements.

We now consider how the profile and the equivalent width of an absorption line grows if the concentration N of the atoms producing the line (or the product of N with the oscillator strength f) increases. The absorption coefficient $\varkappa_\nu$ in the central part of the line is determined by the doppler effect according to (19.13); this part is connected to the damping wings in the outer parts, the damping constant being determined by radiation-damping and collision-damping.

The connexion between the depression in the line R_ν and the absorption coefficient $\varkappa_\nu$ (cm^{-1}) for an absorption tube of length H in the laboratory, without re-emission, would be from equation (11.7)

$$R_\nu = 1 - e^{-\varkappa_\nu \mathrm{H}}. \tag{19.16}$$

For a stellar atmosphere, it is given by the formulae (19.2) to (19.7). With not too great concern for accuracy, we can often replace this somewhat complicated calculation by the approximate or interpolation-formula

$$R_\nu = \left(\frac{1}{\varkappa_\nu H} + \frac{1}{R_c}\right)^{-1}. \tag{19.17}$$

Here H denotes an effective height, or NH an effective number of absorbing atoms above 1 cm² of the stellar surface. For $\varkappa_\nu H \ll 1$ (absorption in a thin layer) $R_\nu \approx \varkappa_\nu H$, and for $\varkappa_\nu H \gg 1$ (optically thick layer) R_ν tends to the limit R_c for very strong lines which has already been mentioned. One can calculate H or NH by comparing (19.17) with the formulae (19.2) to (19.7); in a not too large wavelength-interval (often in practice some hundreds of angstroms) one may treat the effective thickness of an atmosphere as constant.

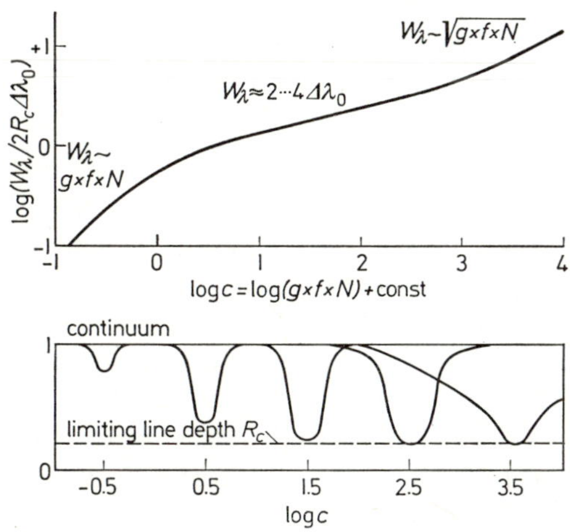

Fig 19.4. *Curve of growth (upper diagram).* The equivalent width W_λ is plotted as a function of the concentration of absorbing atoms, log (gfN)+const. Here W_λ is expressed in terms of the area of a strip of dimensions twice the doppler width $\Delta\lambda_D$ by the limiting depth R_c. The line profiles (lower diagram) illustrate how the curve of growth arises

We now exhibit the results of these calculations: In the lower part of Fig. 19.4 we have drawn the line-profile for various values of the quantity* log $N H f$ + constant. The depression in the weak line ($C \ll 1$) with $R_\nu \approx \varkappa_\nu H$ reflects simply the doppler distribution of the absorption coefficient in the core of the line. With increasing $N H f$ or with $C \gg 1$, the line centre approaches the maximum depth R_c, since here we receive only radiation from the uppermost layer at the boundary temperature $T(\tau = 0)$. At first, the line becomes only a little wider, because the absorption coefficient falls off steeply with $\Delta\lambda$. This situation changes only if with increasing $N H f$ the effective optical depth becomes appreciable also in the damping wings (Fig. 19.3). Since now $\varkappa_\nu H \sim N H f \gamma/\Delta\lambda^2$, the line acquires wide damping wings and, other things being equal, its width at a given depth R_ν is proportional to $(N H f \gamma)^{1/2}$.

By integration over the line-profile we easily obtain the equivalent width of the line. We show it as a multiple of $2 R_c \Delta\lambda_D$ that is of a strip of depth equal to the maximum depth and width equal to twice the doppler width. Thus we obtain the *curve of growth* showing $\log (W_\lambda/2 R_c \Delta\lambda_D)$ as a function of $\log C = \log N H f$ + constant, which is important for the evaluation of stellar spectra (Fig. 19.4, upper part).

As one can see from our discussion, for weak lines (left) $W_\lambda \sim N H f$ and we are in the linear part of the curve of growth. Then follows the flat, or doppler, interval in which the equivalent width is about 2 to 4 times the doppler width $\Delta\lambda_D$. For strong lines (right), corresponding to the growth in the width of the profile, $W_\lambda \sim (N H f \gamma)^{1/2}$ we get into the damping, or square-root region of the curve of growth. Quite generally, the damping constant γ is here important. In Fig. 19.4 we have given the ratio of damping to doppler width $\alpha = \gamma/2 \Delta\omega_D$ the numerical value 1/30, corresponding to a mean value for metal lines in the solar spectrum. The strong D-lines of Na I as well as the H- and K-lines of Ca II in the solar spectrum, for example, lie on the damping part of the curve of growth.

The hydrogen lines claim special consideration. Since in an electric field these show specially large linear Stark-effect splitting, their

---

* More precisely, we are concerned with the quantity $C = \varkappa_0 H/R_c$, where from (19.13) $\varkappa_0$ is the absorption coefficient at the centre of the doppler core. $C = \frac{1}{R_c} \frac{\sqrt{\pi} e^2}{m c^2} \frac{\lambda_0^2 N H f}{\Delta\lambda_D}$ is the effective optical depth of the atmosphere for the centre of the line.

widening in a partially ionized gas results in the first place from a quasi-static Stark-effect of the statistically distributed electric field produced by the slowly moving ions. Working from the consideration that in the distance interval r to $r+\mathrm{d}r$ from a hydrogen atom a perturbing ion is present with probability proportional to $4\pi r^2\,\mathrm{d}r$ and that it then produces a field proportional to $1/r^2$ (to which also the line splitting is proportional), we can show that in the wings of the lines the absorption coefficient $\varkappa_\nu \sim 1/\Delta\lambda^{5/2}$. The theory was developed originally by Holtsmark and has recently been refined by Griem, Kolb and others by taking account of non-adiabatic effects, of collision damping by electrons and of an improved calculation of the microfield in the plasma.

Now we turn to our main problem, *the quantitative analysis of stellar spectra.* For a first orientation in the field, we usually begin with an approximation procedure, the so-called coarse analysis, in which we work with constant mean values of temperature, electron pressure, effective thickness H (in some considerable part of the spectrum). Then we can apply the universal curve of growth calculated from the interpolation formula (19.17) in order to obtain the number of absorbing atoms $N\,H$ above 1 cm² of the stellar surface for any particular species of atom or ion in the energy level that produces the line, using the measured equivalent width W_λ of the line and the known values of f and γ.

By comparing the $N\,H$-values for different excitation energies χ_s and for different stages of ionization (for example Ca I and Ca II) for one and the same element, from the formulae of Boltzmann and of Saha (Chapter 17) one can now calculate the temperature T and the electron-*pressure* P_e. Knowledge of these then leads from the numbers of atoms in particular energy-levels to the total number of all particles of the element concerned (for all stages of excitation and ionization) and thus to the *abundance distribution of the elements.* If we know this and the degree of ionization of the various elements fairly completely, then we can go over from the electron-pressure P_e to the gas-pressure P_g and calculate the *gravitational acceleration* g from the equation of hydrostatic equilibrium. Indeed, the gas pressure is nothing other than the weight, that is the mass times the gravitational acceleration g, of all the particles above 1 cm² of the selected reference-surface, or, as we usually say, the surface of the star.

Upon the basis of such a coarse analysis we can build the more exact, but also much more laborious, fine analysis of stellar spectra.

Table 19.1. *Relative abundances of the elements in stellar atmospheres: log N and giants, high-velocity stars, subdwarfs and horizontal branch (Population II). gravity g determined from the spectrum (except for the Sun where they are stars with small abundances of the heavy elements, values (logarithms) of N times to facilitate comparison with*

Spectraltype	O9 V	B0 V		B 2.5 V	A0 V	G2 V		B1 Ib
Star	10 Lac	τ Sco		γ Peg	α Lyr	Sun		ζ Per
		a)	b)					
T_{eff}	37 450	32 800	35 000	24 000	9 500	5 780		27 000
log g	4.45	4.45	4.30	~4.0	4.5	4.44		3.6
1 H	12.00	12.00	12.00	12.00	12.0	12.00	12.00	12.00
2 He	11.23	11.23	11.04	11.17	11.4·			11.31
6 C	8.37	8.37	7.7	8.58			8.72	8.26
7 N	8.37	8.57	8.26	8.01	8.8		7.98	8.31
8 O	8.77	9.12	8.63	8.63	9.3		8.96	9.03
10 Ne	8.72	8.72	8.86	8.73				8.61
11 Na					7.3	6.12	6.30	
12 Mg	8.22	7.73	8.3	7.95	7.7	7.27	7.36	7.77
13 Al	7.07·	6.58	6.4	5.76	5.7	6.12	6.20	6.78
14 Si	7.75	7.95	7.63	7.03	8.2		7.45	7.97
16 S				7.80			7.30	7.48
20 Ca					6.3	6.16	6.15	
21 Sc					3.4		2.82	
22 Ti					4.8		4.68	
23 V					4.0		3.70	
24 Cr					5.6		5.36	
25 Mn					5.3		4.90	
26 Fe					6.5	6.47	6.47	
27 Co							4.64	
28 Ni					7.0		5.91	
38 Sr					2.8		2.60	
39 Y					2.1		2.25	
40 Zr					2.9		2.23	
56 Ba							2.10	

10 Lac — Traving, G.: Z. Astrophys, **41**, 215 (1957).

τ Sco — a) —Z. Astrophys. **36**, 1 (1955); **44**, 142 (1958).
b) Aller, L. H., G. Elste, and J. Jugaku: Astrophys. J. Suppl. **3**, 1 (1956); Jugaku, J.: Publ Astron. Soc. Japan **11**, 161 (1959).

γ Peg — Aller, L. H., and J. Jugaku: Astrophys. J. Suppl. **4**, 109 (1959).

α Lyr — Hunger, K.: Z. Astrophys. **49**, 129 (1960) and earlier work.

Sun: — Weidemann, V.: Z. Astrophys. **36**, 101 (1955).
Goldberg, L., E. A. Müller, and L. H. Aller: Astrophys. J. Suppl. **5**, 1 (1960).
Corrections due to Zwaan, C.: Bull. Astr. Inst. Netherlands **16**, 225 (1962) have been taken into account.
In the solar spectrum one can determine, or at least estimate, abundances for many rare elements; cf. work by Goldberg, Müller, and Aller as well as the classical investigation of Russell, H. N.: Astrophys. J. **70**, 11 (1929).

ζ Per — Cayrel, R.: Suppl. Ann. d'Astrophys. No 6 (1958).

referred to log N=12 for hydrogen. Normal main sequence, normal supergiants For each star is shown its MK-classification, and the values of T_{eff} and surface determined directly). Uncertain values of log N are indicated by·. In the case of a common reduction-factor are shown on the right-hand side of the column in order stars of "normal" composition

B3 Ia 55 Cyg	A2 Ia α Cyg	G8 III ε Vir	F6 IV – V γ Ser		Subdwarf * HD 140 283		Horizontal branch * HD 161 817	
?	9 170	4 940	6 350		5 940		7 630	
2.90	1.13	2.7	4.0		4.6		3.0	
12.00	12.00	12.00	12.00	(log N +0.24)	12.00	(log N +2.32)	12.00	(log N +1.11)
11.18	11.63·							
8.41	8.19	8.60	8.43	8.67	6.4·	8.7·	7.50·	8.61·
8.63	9.40							
8.98	9.36		9.07	9.31				
		6.60	6.06	6.30	3.50	5.82	5.01	6.12
	7.81	7.40	7.50	7.74	4.96	7.28	6.47	7.58
	6.59	6.34	6.14	6.38	3.48	5.80	4.83	5.94
7.46	7.88	7.58	7.36	7.60	5.18	7.50	6.20	7.31
		7.39	7.17·	7.41				
	6.47	6.25	5.90	6.14	3.88	6.20	5.13	6.24
	3.19	2.76	2.45	2.69	1.41·	3.73·	1.72	2.83
	5.13	4.67	4.28	4.52	2.51	4.83	3.82	4.93
	3.88	3.67	3.23	3.47			2.17	3.28
	5.67	5.36	4.92	5.16	2.98	5.30	4.14	5.25
	5.57·	4.97	4.72	4.96	2.57	4.89	3.20	4.31
	7.62	6.47	6.17	6.41	4.34	6.66	5.36	6.47
	3.73	4.61	4.21	4.45	2.44	4.76	3.40·	4.51·
	4.82·	5.94	6.41	6.65	4.23·	6.55·	5.55	6.66
	3.11	2.62	2.43	2.67	0.11·	2.43·	1.73	2.84
		2.08					0.95·	2.06·
		2.08					0.98·	2.09·
		2.01	2.09	2.33			0.94·	2.05·

* Halo population II

55 Cyg — Aller, L. H.: Astrophys. J. **123**, 133 (1955).

α Cyg — Groth, H. G.: Z. Astrophys. **51**, 206 and 231 (1961) (model III b).

ε Vir — Cayrel, G., and R. Cayrel: Astrophys. J. **137**, 431 (1963). The abundances relative to the Sun given in the paper are here combined with figures given by Goldberg, Müller and Aller.

γ Ser — Kegel, W. H.: Z. Astrophys. **48**, 95 (1959); **56**, 207 (1962).

HD 140 283 — Baschek, B.: Z. Astrophys **48**, 95 (1959); **56**, 207 (1962). (C obtained from CH-bands compared with the Sun.) See also Aller, L. H., and J. L. Greenstein: Astrophys. J. Suppl. **5**, 139 (1960).

HD 161 817 — Kodaira, K.: Z. Astrophys. **59**, 139 (1964).

As in Chapter 18, we construct a model of the stellar atmosphere under investigation, selecting the most plausible T_{eff}, g and chemical composition. Invoking the whole theory of radiative transport, of continuous and line-absorption coefficients (including their dependence on depth) and so on, we calculate the equivalent widths W_λ of the lines for this model. We then compare the results of these calculations for the model with the measurements for such elements as are represented by several different ionization and excitation levels. Hydrogen lines also play an important part since we know that hydrogen is the most abundant element so that its abundance does not enter here. Further, we can appeal to the energy-distribution in the continuum or to the colour indices (Chapter 18). From the preceding considerations one can estimate which criteria depend more strongly on T_{eff}, or on g, or on the abundance of a particular element. Thence we can improve the starting values of these quantities using step by step approximation-procedures.

The newcomer is often alarmed by the great number of lines in a spectrum. Yet in carrying out an analysis it is usually found that the available measurements scarcely suffice to determine all the parameters of interest! The situation gets worse if, for instance, we treat turbulence in the stellar atmosphere as a function of depth and then require that this function be extracted from the measurements.

Having thus sought briefly to present the methodology of a quantitative analysis of stellar spectra, we turn to the results obtained.

In Table 19.1 we collect the results of careful spectral analysis first for stars of the main sequence, then for certain supergiants and giants. In the case of these "normal" stars we give the spectral type Sp and the luminosity class LC (cf. Fig. 15.4) according to Morgan and Keenan. The table contains the logarithm of the number of atoms $\log N$, referred as usual to $\log N = 12$ for hydrogen. In the case of hot stars, helium and the lighter elements up to Si or S are represented by highly excited lines, for which f-values can be calculated fairly exactly. Here the heavier elements are expected only in states of high ionization that have no lines in the accessible part of the spectrum. Below about 10 000° the lines of ionized, and then of neutral, metals make their appearance in ever greater numbers. In the case of the Sun we have incomparably better observational material than for any other star, and also T_{eff} and g are already known. In Table 19.1 we give the latest analysis of the solar spectrum by Goldberg, Müller and Aller; for certain elements, Weidemann had earlier attained about the same accuracy (possible errors in the absolute f-values naturally do not appear in the comparison). For K and M stars there still exist no satisfactory analyses.

Naturally, the accuracy with which the abundance of a particular element can be determined is different for stars of different temperatures. It depends upon the number of lines by which it is represented, the part of the curve of growth in which these occur and, above all, the accuracy with which the oscillator strengths f are known. It is also to be remembered that the abundances can be determined only along with T_{eff} and g; for instance, in middle spectral types an underestimate of T_{eff} mimics an underestimate of metal abundances. Generally speaking, on account of their high ionization potentials, helium and the lighter elements are best determined (relatively to hydrogen) in hot stars, while the more easily ionized metals can be studied better in cooler stars. At the present time, the accuracy of a careful determination would correspond to about $\Delta \log N = \pm 0.3$.

If we compare the relative abundances of the elements in the "normal" stars we have been discussing up to now, i. e. those which conform to the MK-classification, we can say that they agree amongst themselves within the limits of uncertainty. We shall see later that these agree also with the abundances in interstellar gas and in galactic nebulae such as the Orion nebula.

The situation is different if we turn our attention to high velocity stars and to subdwarfs as shown on the right in Table 19.1. In the case of the F6 IV–V star γ Serpentis, which has a space-velocity of 80 km/sec relative to our surroundings, an exact analysis carried to the limit of the accuracy of the observations shows that all the heavy elements referred to $\log N = 12$ for hydrogen are less abundant by a factor 1.7 than in the stars we considered first. In the case of the considerably faster-moving subdwarf HD 140 283 (with magnitude and temperature roughly matching the Sun, but for which the MK-classification of "normal" stars loses its significance) relative to hydrogen all the detectable heavy elements C, Na, Mg, Al, Si, Ca, Sc, Ti, Cr, Mn, Fe, Co, Ni, Sr are reduced by a factor of 200 as compared with the Sun. Differences between the factors for these various elements are not detectable within the limits of error. Certain other red giants (HD 122 563, 165 195 and 221 170) have similar low metal-abundances. While γ Ser and HD 140 283 lie in the vicinity of the main sequence in the HR diagram, the star HD 161 817 on the extreme right in Table 19.1 belongs to the horizontal branch in the HR diagram (see Chapter 26). As Albitzky's star it is also known to be the high-velocity star having the greatest measured radial velocity of −363.4 km/sec; its space-velocity relative to the neighbourhood of the Sun is only a little greater. In this interesting star the abundances of all the heavy elements are reduced relative to the Sun by a factor of 13. In order to simplify the comparison of high velocity stars, etc., with "normal" stars, in the last three columns of Table 19.1 we have shown the abundances of the heavy elements also multiplied by the mean "reduction factor". Then we recognize that the *relative* abundances of the heavy elements (unfortunately we do not yet know about helium) are again the same as in main-sequence stars, giants and supergiants, within the limits of error.

The relative abundances of the elements obviously depend upon the history of their production and upon their transformations by nuclear processes in the course of stellar evolution. We shall see (Chapter 26) that the high velocity stars which we describe for short, but not wholly accurately, as *metal-poor,* belong to the so-called Population II stars of the galactic halo. These are the oldest stars in our Milky Way; in particular we should have for the extreme case of HD 140 283 an age of 10 to 15×10^9 years. In the younger stars of the galactic disk population (e. g. the Sun) and in the quite young stars of the spiral-arm population (10 Lac, τ Sco, etc.) the abundances of the heavy elements have grown to an appreciably higher value. This has obviously scarcely changed since the formation of the galactic disk.

Here we have anticipated some ideas, which we have still to establish by more exact investigation, on the one hand of abnormal stars (p = peculiar stars) and on the other hand of the Milky Way itself.

20. Motions and Magnetic Fields in the Solar Atmosphere: Solar Cycle

When looked at carefully, the solar surface is seen to be mottled in appearance. The *granulation* is composed of bright granules the temperature of which exceeds that of the darker interstices by 100° to 200° K. The diameters of the granulation elements range from about 5″ down to the limit of telescopic resolving power at about 1″ corresponding to 725 km on the Sun. From sequences of photographs, the lifetime of a granule is found to be about 8 minutes.

Then we can easily observe the dark sunspots that were discovered by Galileo and his contemporaries. They appear predominantly in two zones of equal northern and southern heliographic latitude. A typical sunspot has the following structure and dimensions as regards order of magnitude:

	Diameter	Area
Umbra (dark centre)	18 000 km	80 millionths of solar
Penumbra (somewhat brighter fringe)	37 000 km	350 hemisphere

(20.1)

The reduced brightness in the spot results from a lowering of temperature. In the largest spots the effective temperature drops from 5780° K for the normal solar surface to about 3700° K. Correspondingly, the

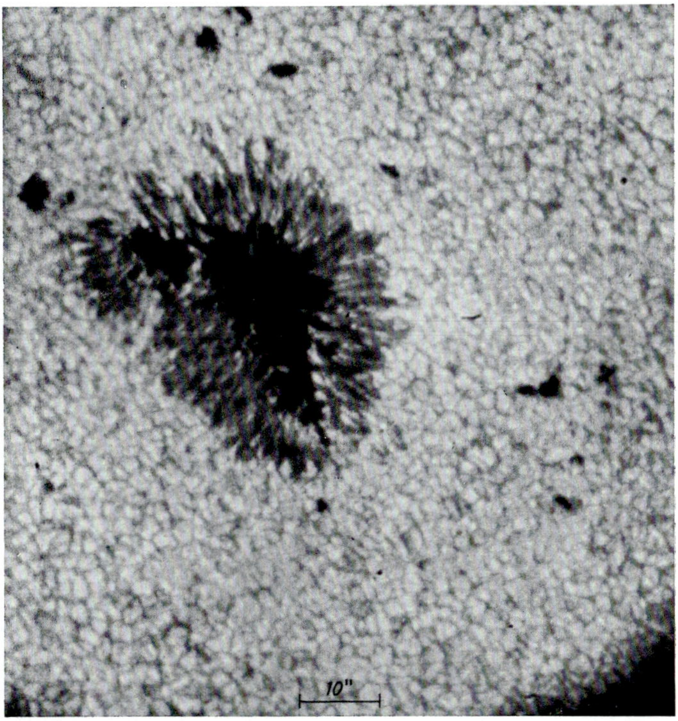

Fig. 20.1. *Granulation and sunspot* (Mount Wilson sunspot number 14 357). Several dark pores, with diameters of a few seconds of arc, can be seen in the vicinity of the spot. Photograph; 12-inch (30 cm) stratospheric telescope (M. Schwarzschild) at a height of 15 miles (24 km), 1959 August 17. Exposure time 0.0015 secs and passband 5470 ± 370 Å

spectrum of a large spot is much like that of a K-star; we anticipated the explanation by means of Saha's theory of ionization in Chapter 17.

Sunspots make their appearance on the solar surface mostly in *groups*. Such a group (Fig. 20.1) is surrounded by bright *faculae*. There exist also the so-called *polar faculae*, independently of sunspots. The brightness of the faculae exceeds that of the normal solar surface by a few percent only near the solar limb. Using equation (18.13), we infer that in the faculae only the outermost layers of the Sun (about $\tau \leqq 0.2$) are a few hundred degrees hotter than elsewhere.

Making use of spots and — in higher heliographic latitudes (see below) — of faculae, etc., we can study the *rotation of the Sun*. We find that the Sun does not rotate as a rigid body but that the regions of high latitude lag behind the equatorial region.

Heliographic latitude	0° (equator)	20°	40°	70°	
Mean sidereal rotation	14°.5	14°.2	13°.5	~11°.7	per day
Sidereal period of rotation	24.8	25.4	26.7	~31	days.

(20.2)

Within their limits of accuracy, measurements of the doppler effect at the solar limb (equatorial speed 2 km/sec, approx.) confirm this picture. The synodic period of rotation — seen from the Earth — is correspondingly longer; for the sunspot zones we find a value in round figures of 27 days, which agrees with quasi-periodical recurrence of a number of geophysical phenomena.

As the apothecary Schwabe was able to show about 1843, the spottedness of the Sun varies with a period on the average of 11.2 years. The other manifestations of solar activity, which we shall consider again later on, also follow the sunspot cycle so that nowadays we often speak of the 11.2 year *solar cycle* or cycle of solar activity. As a measure of solar activity we use the quantity R introduced by R. Wolf in Zürich.

$$\text{Relative sunspot-number } R = k\,(10 \times \text{number of visible spot-groups} + \text{number of all spots}) \qquad (20.3)$$

where k denotes a constant depending upon the telescope employed. Or else we use the areas of the umbrae of all spots and of the faculae as measured photographically at Greenwich since the time of Carrington (a) directly in projection, in units of one-millionth of the solar disk, (b) corrected for foreshortening, in millionths of the solar hemisphere.

A more exact insight into the higher layers of the solar atmosphere (which are obviously strongly affected by the solar cycle) is afforded us by observation in the light of their spectral lines. According to equation (19.6), the latter comes from an optical depth for continuous absorption plus line-absorption given by

$$x_\nu = \int (\varkappa + \varkappa_\nu)\, dt \approx \cos\vartheta\,. \tag{20.4}$$

In this way we can separately observe layers having an optical depth in the continuum of only $\tau \approx 10^{-3}$ or less, that do not appear at all in ordinary photographs of the Sun. The instruments we use are:—

a) The *spectroheliograph* (G. E. Hale and H. Deslandres 1891), a large grating-monochromator with which the Sun is photographed piece by piece in a sharply bounded wavelength-interval of about 0.03 Å to 0.1 Å inside a Fraunhofer line. We obtain the most interesting pictures using the very strong K-line of Ca II (λ 3933 Å) or the hydrogen line Hα (λ 6563 Å). The calcium or hydrogen spectroheliograms, which relate to fairly high layers of the solar atmosphere (the chromosphere), form one of the most important aids to the study of solar activity.

b) The *Lyot polarization filter* (B. Lyot 1933/38) which makes it possible in a single short exposure to photograph the entire solar image, for example in the red hydrogen line Hα 6563 Å in a, to some extent adjustable, wavelength band of only about 0.5 to 2 Å. Consequently the Hα-"filtergram" attains somewhat better definition than the corresponding spectroheliogram. Lyot constructed similar filters also for the coronal lines (see below).

The observation of the highest layers at the solar limb is strongly encroached upon by scattered light. This comes partly from contamination of the Earth's atmosphere and of the optics of the instrument, but also to a considerable degree from the diffraction of light at the entrance aperture. The *Lyot coronograph* to a great extent eliminates instrumental scattered light; its principle is shown in Fig. 20.2. Further, *total solar eclipses* (Chapter 4) still form an indispensable aid to the investigation of the outermost layers of the Sun. In future, observation from space vehicles may offer serious competition.

If we observe the solar limb in the continuum, its brightness falls away very rapidly towards the outside as soon as the optical thickness along the sight-line becomes less than unity. The corresponding optical thickness in a strong Fraunhofer line remains greater than unity for hundreds, or even thousands, of kilometres further out. This means that the Fraunhofer spectrum with its absorption lines goes over into an emission spectrum of the highest layers of the solar atmosphere

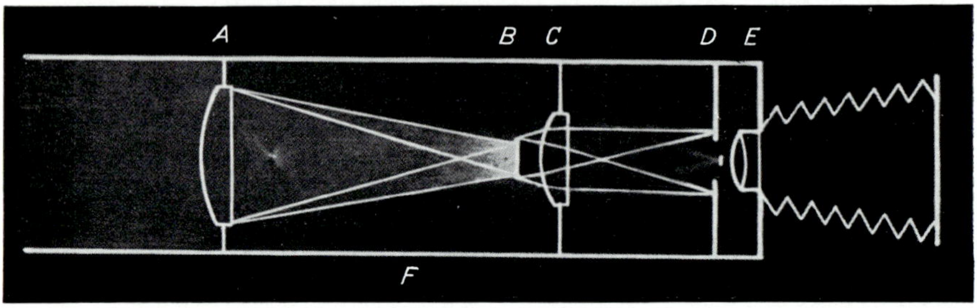

Fig. 20.2. *Coronograph* (after B. Lyot, about 1930). Light passing through the objective *A*, an optically pure simple plano-convex lens, forms a solar image on the disk *B*. This disk is introduced in order to cut out the light coming from the solar disk, and projects 10 to 20 seconds of arc beyond the solar limb. Light which was originally diffracted at the edge of the objective must next be removed. This is effected by inserting an annular stop at *D*, where an image of the aperture stop has been formed by the field lens *C*. The final image is formed by the objective *E* upon either a photographic plate or a spectrograph slit, where the corona, prominences and other features may be studied

called the *chromosphere*. This was first noticed by J. Janssen and N. Lockyer at the total eclipse of 1868. When the Moon covered the Sun except for its extreme limb, the emission spectrum of the chromosphere flashed out for a few seconds and consequently it became known as the *flash spectrum* (Fig. 20.3). Recent spectrographic ob-

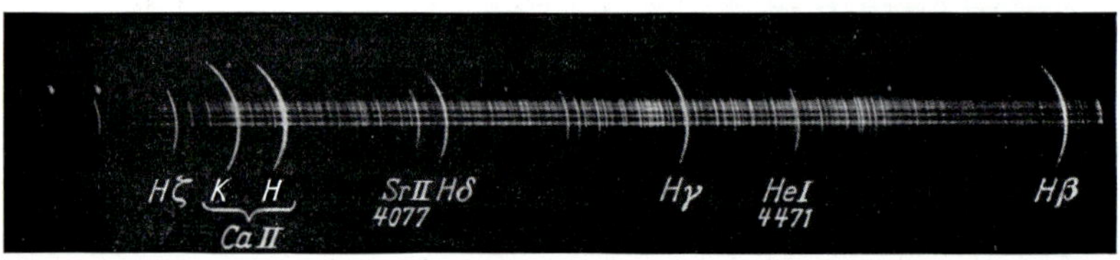

Fig. 20.3. *Flash spectrum* (i. e. emission spectrum of the solar chromosphere). Photograph taken at total solar eclipse, Khartoum, 1952 February 25 by J. Houtgast. Objective-prism camera 9 foot (270 cm) focal length and 1½ inches (4 cm) aperture. Exposure time 0.2 to 0.9 secs after second contact

servations using a ciné camera (with time-resolution about 1/20 sec), in combination with the motion of the Moon relative to the Sun calculated from the ephemerides, give remarkable information about the stratification of the solar chromosphere: its scale-height (corresponding to a decrease in density by a factor $e = 2.72$) is greater than that expected for an isothermal atmosphere at the boundary temperature $T_0 \approx 4000°$ and it even increases outwards.

During totality in a solar eclipse, or in the clearest possible air on a high mountain such as the Pic du Midi (2870 m) using a Lyot corono-

Fig. 20.4. *Solar corona, near sunspot minimum.* Photograph taken at eclipse, Khartoum, 1952 February 25, $9^h 10^m$ U.T. by G. van Biesbroeck. Camera 20 foot (6.1 metre) focal length and 6 inch (15 cm) aperture. Kodak 103a-E emulsion and yellow filter. Exposure time 1.5 minutes. At sunspot minimum the corona displays extended steamers in the neighbourhood of the sunspot zones, and the finer "polar tufts" over the polar caps. At sunspot maximum the corona has a more rounded form

graph, the solar *corona* is observed extending out to several solar radii (Fig. 20.4). Its form (flattening, ray-structure, etc.) and brightness, especially for the inner parts, depend on the phase in the solar cycle. Spectroscopic analysis distinguishes the following phenomena which we attempt now to interpret:—

1. The *inner corona* ($r \approx 1$ to 3 solar radii) exhibits a completely *continuous spectrum*, the energy-distribution in which otherwise corresponds to that of normal sunlight. Following W. Grotrian, we call this the K-corona. Its light is partly linearly polarized. In agreement with K. Schwarzschild and others, we ascribe it to Thomson scattering of photospheric light by the free electrons of the obviously completely ionized gas (plasma) of the corona. Thanks to the doppler effect resulting from the high velocities of the electrons, the Fraunhofer lines are blurred out.

We can calculate the mean (i. e. disregarding fluctuations) electron density N_e in the K-corona, as a function of distance r from the centre of the Sun, from its brightness distribution *. We find, for instance, at about coronal maximum:

r	$=1$	1.1	1.5	2.0	2.5	3.0	solar radii (7×10^{10} cm)
N_e	$=4.1\times10^8$	1.6×10^8	1.6×10^7	2.8×10^6	8.3×10^5	3.2×10^5	electrons cm^{-3}

2. In the *inner corona* we observe in emission the so-called *coronal lines* whose identification remained a great puzzle in astrophysics until in 1941 B. Edlén succeeded interpreting them as forbidden transitions of highly ionized atoms. The strongest and most important are those in Table 20.1.

It was possible with confidence to identify further corresponding lines of the elements A, K, Ca, V, Cr, Mn, Fe, Co. The ionization potentials of several hundred electron volts signified emphatically an electron temperature of some millions of degrees. And then it was realized that in fact a whole series of other phenomena indicate a coronal temperature of 1 to $3\times10^{6\circ}$ K:—

a) The great *extent of the corona* in combination with the hydrostatic equation, or the barometric formula, demands a mean temperature $\geqq 1.7\times10^6$ °K.

b) If we ascribe the *widths of the coronal lines* (in the sense of (19.9)) measured by B. Lyot, D. E. Billings and others to the thermal doppler effect, we obtain for various lines 1.7 to $3.7\times10^{6\circ}$ K.

c) Since a radiation field corresponding to 10^6 degrees is certainly not present, we cannot calculate the *ionization* in the corona from

* We know the distribution of light-intensity and the Thomson scattering coefficient per free electron $\sigma_{el}=\frac{8\pi}{3}\left(\frac{e^2}{m c^2}\right)^2=0.665\times10^{-24}$ cm^2, which to within the factor 8/3 is equal to the classical cross-section of the electron.

Saha's formula (LTE). The latest calculation of individual processes of ionization and recombination yields temperatures of 0.5 to $2.5 \times 10^{6\circ}$ K for the maxima of the lines shown in Table 20.1.

Table 20.1. Coronal lines

	Red coronal line 6374.51	Green coronal line 5302.86	Yellow coronal line 5694.42 Å
Spectrum	[Fe X]	[Fe XIV]	[Ca XV]
Preceding ionization potential	235	355	820 eV
Transition	$3s^2\,3p^5\,{}^2P_{\frac{3}{2}} - {}^2P_{\frac{1}{2}}$	$3s^2\,3p\,{}^2P_{\frac{1}{2}} - {}^2P_{\frac{3}{2}}$	$2s^2\,2p^2\,{}^3P_0 - {}^3P_1$
Electron temperature T_e (ionization)	0.5×10^6	1.1×10^6	2.5×10^6 °K
Doppler temperature T_D (line width)	1.7×10^6	2.5×10^6	3.7×10^6 °K

[It is not known whether the excess of the temperatures in (b) over those in (c) depend upon the existence of systematic flow.]

d) We can account for the *thermal radio emission* of the Sun as free-free radiation from the solar corona. We observe it in millimetre, centimetre and decimetre wavelengths while non-thermal components mostly predominate at longer wavelengths. The changes in the thermal radiation intensity in the course of days and months admit its separation into the always-present radiation of the quiet Sun and the slowly varying radiation originating in coronal condensations. At the solar limb and in condensations in the corona an optical thickness greater than unity is attained in places, so that we can measure the black-body radiation directly. Its temperature, which again corresponds to the electron temperature, also has values from 1 to $3 \times 10^{6\circ}$ K.

Recent measurements from space-vehicles (Mariner II, etc.) show that at large distances from the Sun, say in the vicinity of the Earth but still outside the magnetosphere, the plasma of the K-corona streams out into interplanetary space in a way that is highly variable in space and time, so that near the Earth we have a stream of about 5 electrons/cm^3 (with the protons required to give electrical neutrality) moving at about 400 km/sec. According to L. Biermann and others this *solar wind* blows the tails of comets away from the Sun. However, we must return to the corona and consider:

3. *The outer corona.* At some distance outside the solar limb and increasing steeply (relative to the K-corona) outwards from the Sun, we observe scattered photospheric radiation showing the *Fraunhofer lines unchanged.* Following W. Grotrian, who designated this component the F-Corona (Fraunhofer corona), C. W. Allen and H. C. van de Hulst (1946—47) showed that it arises from "Tyndall scattering", i. e. predominantly forward scattering by small particles somewhat larger than the wavelength of the radiation. These particles are far enough from the Sun not to be heated to the point of vaporization. Measurements of the distribution of brightness and of polarization in the outer corona, especially Blackwell's observations from an aircraft in the stratosphere, have shown that the F-corona is simply the innermost part of the *zodiacal light.* Grotrian had already conjectured this. Thus the F- or dust-corona, or the zodiacal light, do not belong to the Sun and are relatively little influenced by it.

Here we shall not concern ourselves further with interplanetary matter, but pursue our study of the K-corona, the true *solar corona.* Since the temperature must somehow increase from the 4000-degrees of the lower chromosphere to the 1 to 3 million degrees of the corona, it is no longer astonishing that lines of high ionization and excitation potential make their appearance with great intensity even in the higher atmosphere, including amongst others

	Excitation potential eV	Ionization potential eV
Hydrogen Balmer lines	10.15	13.54
HeI lines, e. g. D_3 λ 5876 Å	20.87	24.48
HeII lines, e. g. $4-3$; λ 4686 Å	48.16	50.80

Incidentally, P. J. C. Janssen first observed the D_3-line of the "solar element" in the Sun in 1868; it was not until 1895 that W. Ramsay succeeded in isolating helium from terrestrial minerals.

The spectrum of the upper chromosphere (about 2000 to 10 000 km above the solar limb) is generally like that of prominences. These "clouds in the corona" radiate chiefly in the light of the Hα line λ 6563 and the other hydrogen lines, the calcium H and K lines λ 3933, 3968 and the helium D_3 line λ 5876. Quiescent prominences (Fig. 20.5) retain their form with little change (flow at about 10 km/sec) often for weeks. From time to time, however, without warning, they may accelerate more or less jerkily to velocities of 100 km/sec up to, on occasions, 600 km/sec. Such eruptive prominences can then escape into interplanetary space. All prominences have a characteristic filamentary structure. Also the upper part of the chromosphere under

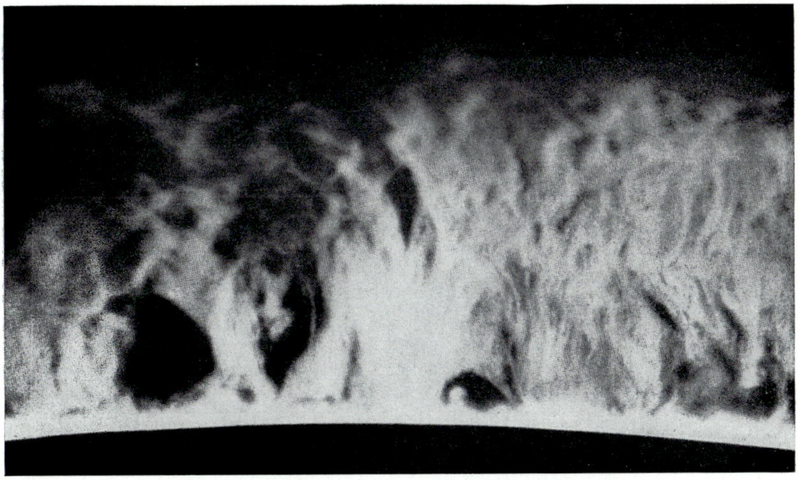

Fig. 20.5. *Quiescent prominences* (filaments) usually have the form of a thin sheet, standing almost perpendicular to the solar surface on several "feet". The dimensions of this sheet are: thickness ~ 6600 km (4000—15 000 km), height ~ 42 000 km (15 000—120 000 km) and length ~ 200,000 km (up to 1.1×10^6 km). Detailed photographs taken in Hα-light at Sacramento Peak show filamentary structures, in which the material streams upwards or downwards with speeds of the order of 10 to 20 km/sec. The solar magnetic fields clearly have an important influence on the structure of all prominences

good observing conditions, e. g. using a Lyot filter, has the appearance of a "burning prairie" of small prominances, the so-called spicules (Fig. 20.7) that move outwards or inwards with velocities of about 10 km/sec.

In Fig. 20.8 we show concisely the association of sunspots, prominences, corona etc. with reference to the 11.2 year solar cycle. Then we must ask ourselves above all else about its meaning. Obviously the picture of a static atmosphere in radiative equilibrium is no longer adequate; rather is everything in motion and in a state of flux.

We first enquire as to what thermodynamic engine, in which heat-energy flows from higher to lower temperatures in accordance with the second law of thermodynamics, generates the mechanical work needed to drive these motions. The so-called hydrogen convection zone (A. Unsöld 1931) performs this office. From the deeper photospheric layers, where $\log P_g \approx 5.2$ and $T \approx 6500°$ K, downwards to about log $P_g \approx 12$ and $T \approx 10^{6\circ}$ K the solar atmosphere is convectively unstable. This zone has thickness about one-tenth the solar radius. Above it, the most abundant element, hydrogen, is

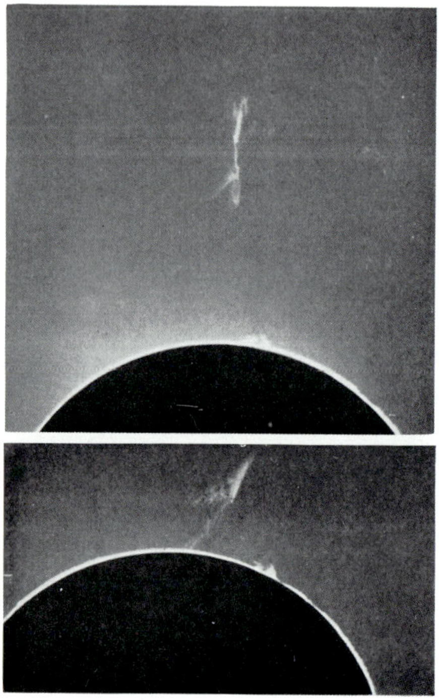

Fig. 20.6. *Eruptive or ascending prominence* 1928 November 19. The photographs are separated by an interval of $1^h\,11^m$. The greatest height observed is 900 000 km above the solar limb and the maximum velocity 229 km/sec

practically neutral, within it the hydrogen is partially ionized, below it the ionization is complete. If a volume element of the partially-ionized gas rises, the hydrogen begins to recombine and in each

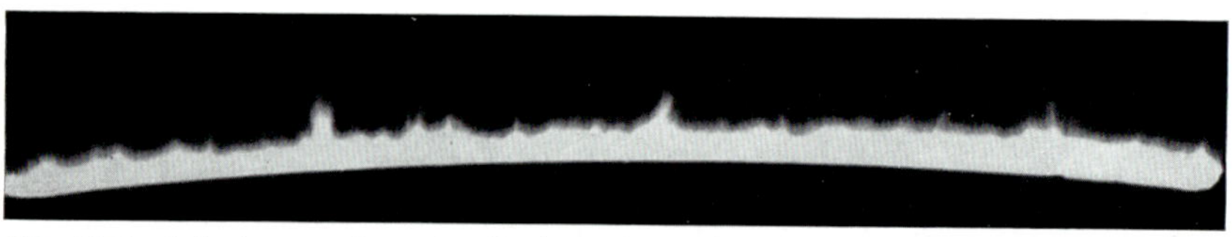

Fig. 20.7. *Spicules at the solar limb in* Hα (the disk has been covered). These structures, which were first described by A. Secchi, have a mean height above the solar limb of $\sim$ 8000 km and a thickness of approx. 500—1000 km. Moving with velocities of approx. 20 km/sec upwards or (more rarely) downwards, they have a lifetime of approx. 2 to 5 minutes. The direction taken by the spicules is that of the local magnetic fields

recombination process 13.6 eV (corresponding to 16 kT at 10 000° K) is added to the thermal energy of $\frac{3}{2}kT$ per particle. The adiabatic cooling is thereby so very much reduced that the effective ratio of the

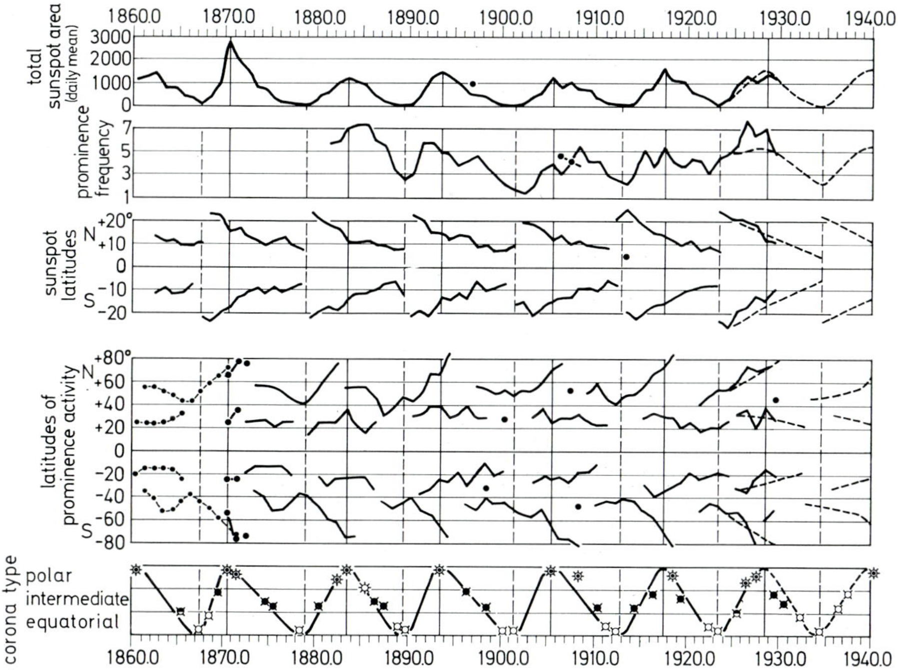

Fig. 20.8. *11.2-year cycle of solar activity* (after W. J. S. Lockyer). Area and distribution in latitude of sunspots and prominences, and variation in form of the corona

specific heats c_p/c_v approaches unity and the rising volume-element would become hotter than its new surroundings in radiative equilibrium. Thus the volume-element would rise higher still. Exactly the reverse happens to a sinking volume-element. This effect is further enhanced by the opposed influence of ionization upon the radiative temperature-gradient. Thus we obtain a zone of convection-currents. For $\log P_g \geqq 5.3$, convection takes over practically the whole of the energy-transport; in the deeper layers of the convective zone radiative energy-transport becomes insignificant.

While the thermodynamics of the hydrogen convective zone is fairly simple and clear, its *hydrodynamics* includes some of the most difficult problems of the subject. Only models using the rather crude mixing-length theory according to W. Schmidt and L. Prandtl have been worked out: a body of gas of dimensions l is supposed to describe a sort of free path of the same order l and then all at once to give up its temperature excess, its momentum, etc., by mixing with its surroundings — obviously a very crude schematization of highly complex convection-currents.

Solar granulation can now be ascribed to the specially marked instability of a surface layer, a few hundred kilometres thick, of the hydrogen convection-zone. The granules are of the order of magnitude of the thickness of this layer, but also not much greater than the equivalent height of the atmosphere there.

Besides the granulation, there is also a second coarser network of convection-cells of which the mesh has diameters between 15 000 and 40 000 km. We see this particularly clearly in the CaII-spectroheliograms (the calcium flocculi) which apply to a fairly high layer of the solar atmosphere. The flow travels upwards in the middle of a cell, with radial speed of about 0.4 km/sec outwards, and downwards round the edge. The lifetime of a cell (about equal to the time for it to turn over) is about 20 hours. Since its diameter is about equal to the thickness of the hydrogen convection-zone, it is tempting to associate the cells with flow through the *whole* convection-zone.

The currents so far considered of 0.5 to 2.5 km/sec influence the solar spectrum by giving the Fraunhofer lines a serrated structure and increasing their width on the average. So far as the moving volume-elements are optically thin, their doppler effects are simply superimposed on those of the thermal motion. We have then the so-called microturbulence which increases the purely thermal width $\Delta\lambda_D$ by factors between about 1.2 and 2. The doppler effects of the somewhat faster flow in the spicules of the upper chromosphere make themselves correspondingly evident in the cores of strong Fraunhofer lines.

We must not get lost in details but we must proceed at once to ask the disturbing question as to how in the transition region, the thickness of which is only some 20 000 km, between the chromosphere and the corona the temperature increases outwards from about 4000° to between about 1 to 3 million degrees. According to the second law of thermodynamics, this heating-up of the outermost atmospheric layers of the Sun against the "natural", entropy-generating, temperature drop from the inside towards the outside can occur only through the use of mechanical energy or other "ordered" forms of energy of greater negative entropy.

Following M. Schwarzschild and L. Biermann we can in fact show that in the upper layers of a partially convective atmosphere mechanical energy transport more and more supersedes radiative transport. As Proudman and Lighthill have shown more exactly, sound-waves first arise in the turbulent motions in the photosphere. When these propagate into the rarefied layers of the upper atmosphere they become* "steeper" and press on into the upper layers. There the mechanical energy** is dissipated and turned back into heat-energy, and since at low densities and high temperatures the radiative power of solar matter becomes progressively worse, the temperature rises until the atmosphere has found a new type of energy-transport. As H. Alfvén noticed, this arises from the circumstance that, at sufficiently high temperature and steep *inward* temperature-gradient, the heat-conduction by the free electrons of the plasma (like the high heat-conductivity of metals) suffices to convey the energy back to the inside where it is ultimately radiated away.

The transition layer where, within some 20 000 km, the transition from a few thousand to millions of degrees is accomplished has recently been further disclosed by space research. Hotter and hotter layers emit lines of higher and higher ionization and excitation potentials up to several hundred electron volts. These spectra lie in the short-wave ultra-violet or X-ray region in about 10 to 1000 Å, and so can be observed only from rockets and space-probes. In fact, for $\lambda < 1500$ Å the solar disk shows a spectrum of *emission lines* of high ionization-levels, the measured intensities of which ought to provide the data for the construction of a precise model of the transition layer.

The remarkable filamentary structure of prominences, the polar plumes of the corona at sunspot minimum and also the enormous coronal streamers over large centres of solar activity, and not least the flow-structure that is observed in Hα-spectroheliograms in the vicinity of spots (G. E. Hale's so-called hydrogen vortices) long ago prompted

* The energy-density of the sound waves is proportional to the density ϱ times the square of the velocity-amplitude v. If the waves advance with sound-speed c into a less dense medium, then v^2 increases correspondingly. If v reaches the order of magnitude of the sound-speed so that the Mach number $M = v/c$ tends to unity, then a shock-wave is produced.

** In the plasma of the solar atmosphere, in addition to the sound-waves and shock-waves mentioned, magneto-hydrodynamic waves may contribute to the energy-transport. In these a variable magnetic field is coupled to the vibrations of the material.

the thought that in solar physics hydrodynamics would not have the last word but that magnetic fields could also play an essential rôle. Thus in 1908 the ingenious G. E. Hale looked for the Zeeman effect in the spectrum of a sunspot. This effect is the splitting of, for example, the line FeI 6173 seen with the help of a polarization device that cuts out alternately the right- and left-hand circularly polarized outer Zeeman-components, the observation being made along the magnetic field-lines. He discovered magnetic fields that reached about 4000 gauss in the largest spots. Further, he showed that the two spots of a so-called bipolar spot-group show like a horseshoe magnet always one north and one south pole. Finally, a long series of observations by Hale and Nicholson showed that in such a group the leading spot (according to the Sun's rotation) has opposite polarity in the northern and southern hemispheres and that this alternates from one solar cycle to the next. Thus the true duration of a solar cycle amounts to 2×11.2 years.

Since all purely thermodynamic attempts to explain the low temperature of sunspots have failed, we must surely suppose that in the depth of the spot the convective energy-flow is strongly impeded by the magnetic field.

In 1952 H. W. Babcock and H. D. Babcock using much refined equipment succeeded in detecting Zeeman effects in the Sun corresponding to only one or two gauss, for which the splitting corresponds to only a tiny fraction of the line-width. It appears that the facular regions, or more precisely the *plages faculaires*, of the CaII-spectroheliograms in the surroundings of groups of sunspots ordinarily exhibit fields of 10 to 100 gauss, and they are apparently caused by these fields. Since the more intense regions of the corona, the so-called coronal condensations, occur above the "plages", we are driven to the concept that moderately strong magnetic fields bring about the mechanical energy-transport from the photosphere into the higher layers of the chromosphere and corona.

This may be bound up with the still baffling phenomenon of solar eruptions or flares (not to be confused with eruptive prominences) (Fig. 20.9). If we observe a spot-group (centre of activity) in the Hα-line or the calcium K-line, the facular "plages" or facular surfaces, between the spots and their surroundings, exhibit structures of the order of magnitude of a few thousand kilometres with small irregular brightness-fluctuations and motions. *Suddenly* many fairly bright structures coalesce, then, in the case of a 3^+ eruption a region of 2 to

3×10^{-3} of the solar hemisphere, corresponding to an entire spot-group, blazes up in the lines Hα, CaII K, ... (Eruptions are classified as of "importance" 1, 2, 3 or 3^+.) The Hα-emission line reaches a

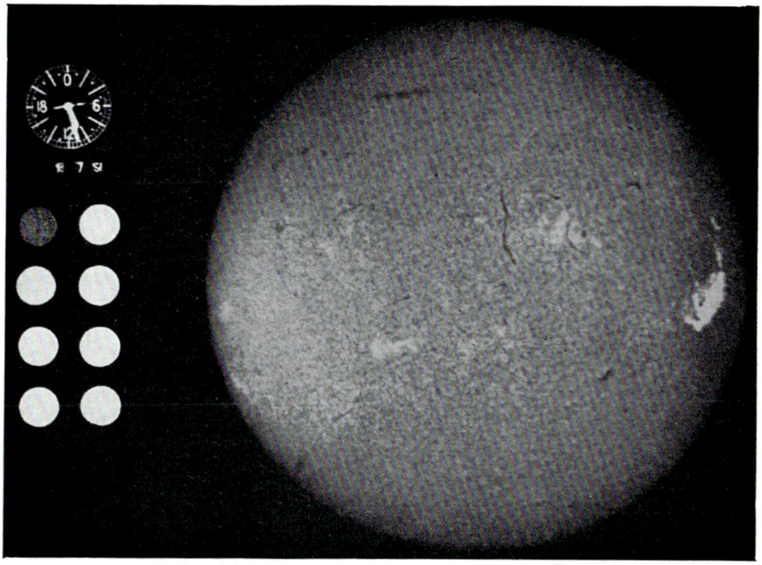

Fig. 20.9. *A large eruption* (3^+ flare, 1961 July 18, see right of picture) which was accompanied by a pronounced emission of cosmic rays. Just above the centre of the disk a long filament (i. e. prominence, compare Fig. 20.6) can be seen in absorption. Cape Observatory, Hα patrol photograph (intensity calibration marks and the time recording appear on the left)

central intensity up to about three times that of the normal continuum, and a width of many angstroms. The duration of the flare ranges from the order of a second for "micro-flares" of about 1″ diameter up to several hours for giant 3^+ flares. One often, somewhat vaguely, relates the occurrence of flares to some sort of instability in the magnetic field of the spot-group. The author prefers the view that the flow of mechanical energy, which among other things heats the faculae and corona, is so suddenly switched on that the energy cannot escape and an explosion occurs. An ordinary facular surface compared with a flare is somewhat like the burning of dynamite compared with its detonation.

Bound up with the flare, we have the emission of corpuscular radiation in a large range of energies. Plasma-clouds, which arrive at the Earth about one day later and so have a velocity about 2000 km/

sec, produce there magnetic storms ("sudden commencement") and polar lights. As Forbush and Ehmert found, great flares also make a solar contribution to cosmic rays with particle energies of 10^8 to 10^{10} eV. Their chemical composition largely matches that of normal solar material.

Connected with the flares, and also with less spectacular manifestations of solar activity, we have the non-thermal component of the *radio emission* from the Sun.

Its analysis with the aid of the radio-frequency spectrometer, which plots the intensity as a function of time over a large frequency range, led J. P. Wild and his collaborators in Australia to distinguish several types of *bursts* of radio-emission (Fig. 20.10). In order to understand this we must recall that in a plasma of electron-density N_e electromagnetic waves of frequency less than the so called critical or plasma frequency ν_0 cannot be propagated, where

$$\nu_0 = \left(\frac{e^2}{\pi m} N_e\right)^{1/2} \mathrel{\hat{=}} 9 \times 10^{-3} N_e^{1/2} \quad \text{MHz} \tag{20.5}$$

because then the refractive index would be negative. Thus radio-emission of a given frequency can originate in the corona only above a particular layer.

Now the Type II and Type III bursts show, respectively, (Fig. 20.10) a slow or fast displacement of their frequency band towards lower frequencies. Thence we can infer that the stimulating

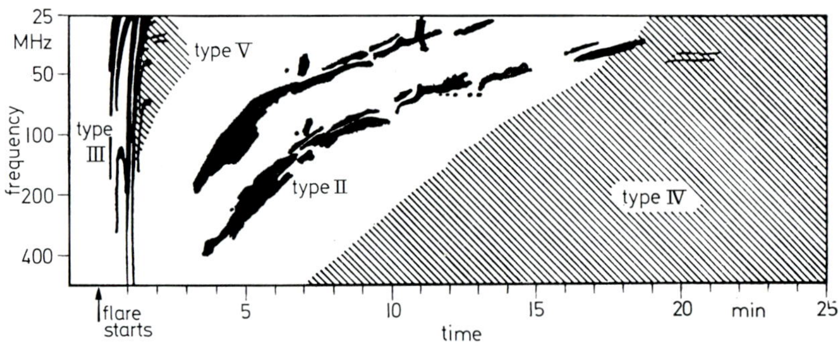

Fig. 20.10. *Dynamical (i. e. time-dependent) radio spectrum in the meter wavelength region, after* J. P. Wild. The diagram shows (schematic only) the development of "bursts" of various types following a large flare. The lower the frequency (i. e. uppermost part of diagram), the higher the coronal layer in which it arises. The upward speed of the exciting agency can be read off from the dynamical spectrum; thus, this is slow for the type II bursts and fast for the type III bursts. Type II and type III bursts frequently show a harmonic (2 : 1) overtone

agency sweeps through the corona with speeds of the order of 1000 km/sec in the case of Type II and up to about 40 percent of the speed of light in the case of Type III. Type IV occurrences emit a continuum in a wide frequency band over a longer time; here we very likely have to do with synchrotron emission by fast electrons.

We do not yet know how the whole super-thermal corpuscular emission from the Sun (or anywhere else in the cosmos) is accelerated. Proposed theories are fairly helpless when presented with an occurrence like the one in Fig. 20.6.

As the foregoing survey shows, one of the most important concerns of present-day astrophysics (as well as of the technique of nuclear fusion) is the theory of the flow of material of high conductivity in the presence of magnetic fields, i. e. *magneto-hydrodynamics* or *hydromagnetics*.

From the fundamental work of H. Alfvén, T. G. Cowling and others, going back over the past 20 years, we may infer (1) Almost all cosmic plasmas have very high electrical conductivity σ. (2) If we have a conductor at rest in a magnetic field H then the variation of the field generates induced currents which decay only slowly (as in a familiar experiment with eddy currents). In alternating-current practice the decay-time τ is given by the ratio of the self-induction L to the resistance R. The latter is proportional to $1/\sigma$. Since in electrostatic units σ has dimension [$\sec^{-1}$], a factor of dimension [$\sec^2$] must enter in forming τ. This can be only x^2/c^2, where x is a linear measure of the conductor and c is the speed of light. Actually we obtain from Maxwell's equations

$$\tau \approx \frac{\pi\,\sigma}{c^2}\,x^2. \tag{20.6}$$

The dependence of the sort $\tau \propto x^2$ shows that the propagation and decay of a magnetic field in a conductor *at rest* has the character of a diffusion process. (3) If we allow *motions* in the conducting medium we have to consider whether the magnetic field propagates more quickly by diffusion independently of the matter, or remains "frozen" in the material. Alfvén's condition for the latter

$$\sigma\,H\,x/c^2\,\sqrt{\varrho} > 1, \tag{20.7}$$

where ϱ is the density, is effectively always satisfied in cosmic plasmas. (4) Since the magnetic pressure (Maxwell stress) is of order of magnitude $H^2/8\,\pi$ this quantity is usually of the same order of magnitude as the dynamical pressure $\sim\frac{1}{2}\varrho v^2$ (in not too special magneto-hydrodynamic flows).

With the help of these preliminaries, we now understand: (a) The decay-time for the magnetic field in a stationary sunspot given by (20.6) is about $\tau \approx 1000$ years. Thus the magnetic field of the spot must have been present in the convection zone for a long time, and at its "coming into existence" it is only carried up to the surface with the material by magneto-hydrodynamic flows. (b) The flows take place mainly along the lines of force (giving zero Lorentz-force) of the magnetic fields "anchored" in the deeper layers of the chromosphere and corona. In this way we can account for the filamentary structures in the chromosphere, prominences and corona. (c) Magnetic field is also blown away with the plasma in the solar wind. In fact, measurements in space vehicles show markedly variable interplanetary magnetic fields of the order of 10^{-6} to 10^{-5} gauss, in order of magnitude agreement with the result (4).

[The further development of hydrodynamical, and particularly magneto-hydrodynamical, investigations presents the fundamental difficulty, from the mathematical standpoint, that the equations involved are *non-linear.* We can make progress with even the simplest problems only by means of special numerical computations from which it is often impossible to judge to what extent the result of the calculation is restricted to a particular model or to what extent it is possible to generalize it.]

21. Variable Stars: Motions and Magnetic Fields in Stars

The first observations of variable stars about the turn of the 16th to the 17th century formed at the time a powerful argument against the Aristotelian dogma of the immutability of the heavens. The observations by Tycho Brahe and Kepler of the supernovae of 1572 and 1604 have even in our own time contributed significantly to our knowledge of these puzzling objects, and they have made possible the identification of their remnants by radio-astronomy. We must also recall the discovery of Mira Ceti by Fabricus.

We have already spoken of the development of the technique of photometric measurements from Argelander's method of step-estimates, which is now used only by amateurs, through photographic photometry, which serves mainly in surveys of stellar clusters, Milky

Way fields, galaxies, etc., for variable stars, to photo-electric photometry, that yields the light-curve of an individual variable with an accuracy of a few thousandths of a magnitude. We need scarcely remind ourselves that to-day the measurement of colour-indices and the analysis of spectra play an essential part in addition to the measurement of magnitudes.

We denote variable stars by letters R, S, T, . . . , Z with the genitive of the name of the constellation, and we can use also RR, RS, . . . , ZZ; for particular stellar clusters, stellar fields, etc., we nowadays for the most part use simply the catalogue numbers of the stars.

It is clear from the outset that the investigation of *variable* stars promises deeper knowledge of the structure and evolution of the stars than the study of unchanging static stars. On the other hand, however, the observation and theory of variables present vastly greater difficulties. A warning against easy *ad hoc* hypotheses may not be out of place here.

Our concern is not to describe in outline the innumerable classes of variable stars. We leave aside the eclipsing variables, which we have already discussed, and select certain interesting and important types of intrinsically variable stars which we now survey from the standpoint of their physical significance.

Pulsating Stars

Pulsating stars include, amongst others, the following groups of variable stars; they are all giant stars (although there do exist in addition "dwarf Cepheids" more or less on the main sequence).

RR *Lyrae stars* or *cluster variables.* Stars with regular luminosity-change having periods of about 0.3 to 0.9 days, luminosity-amplitudes of about 1^m, spectral types about A to F. They belong to the halo and nucleus of the Milky Way and they are important in globular clusters.

Classical cepheids (δ Cephei-stars) are likewise entirely regular, with periods of about 2 to 40 days. They have about the same luminosity amplitudes as cluster variables, but their spectral types are F—G. They occur in the spiral arms of the Milky Way system.

W-Virginis variables, also regular, have similar periods, but are about two magnitudes fainter. They occur in the halo and nucleus of the Milky Way.

Then there are several groups of semi-regular variables with longer periods and later spectral types, in particular:

RV Tauri variables with periods of 60 to 100 days

Long period variables of several sorts, all of late spectral types, with quasi-periods of about 100 to 500 days; Mira Ceti is an example.

The radial-velocity curves of these sketchily described groups of variable stars give the first indication of their physical nature. These are closely related to the light-curves. At first people tried to refer all the regular velocity-changes of, say, a classical cepheid to double-star motion. Integration over the velocity, without further hypotheses, gave the dimensions of the "orbit", since if x is the coordinate in the direction of the sight-line

$$\int_{t_1}^{t_2} \frac{dx}{dt}\, dt = x_2 - x_1 \,. \tag{21.1}$$

It was found that the star in this orbit would have no room for a companion. Therefore H. Shapley in 1914 fell back upon the possibility of a radial pulsation of the star which A. Ritter had discussed in the 1880s as a purely theoretical problem. The pulsation theory of cepheids (and related variables) was then further developed by A. S. Eddington from 1917 onwards. This in turn provided the stimulus for Eddington's pioneering work on the *internal constitution of the stars* (Chapter 25).

A simple estimate embodies an important clue to the theory of pulsating stars, although the details are difficult: We think of the pulsation as a standing sound-wave in the star. Its velocity is $c_s = (\gamma\, p/\varrho)^{1/2}$, where $\gamma = c_p/c_v$ is the ratio of the specific heats, p is the pressure and ϱ the density. The mean pressure $\bar{p}$ inside a star is of the order of magnitude of the force of gravity upon a column of material of 1 cm² cross-section, that extends inwards from the surface of the star of mass M. Thus

$$\bar{p} \approx \underset{\text{mean density}}{\bar{\varrho}} \times \underset{\text{radius}}{R} \times \underset{\text{acceleration}}{G\,M/R^2} \tag{21.2}$$

The period of oscillation P is then, again as regards order of magnitude,

$$P \approx R/c_s \approx R\,(\gamma\, G\, M/R)^{-1/2}. \tag{21.3}$$

Since $M = \frac{4}{3}\pi R^3\, \bar{\varrho}$ we obtain the important relation between period P and mean density $\bar{\varrho}$ of the star

$$P \sim (\bar{\varrho})^{-1/2} \tag{21.4}$$

which has been well confirmed by extensive observational material.

W. Baade suggested a second test of the pulsation theory. The brightness of a star, other things being equal, is proportional to the area of its disk πR^2 times the radiation flux πF_λ at its surface. But

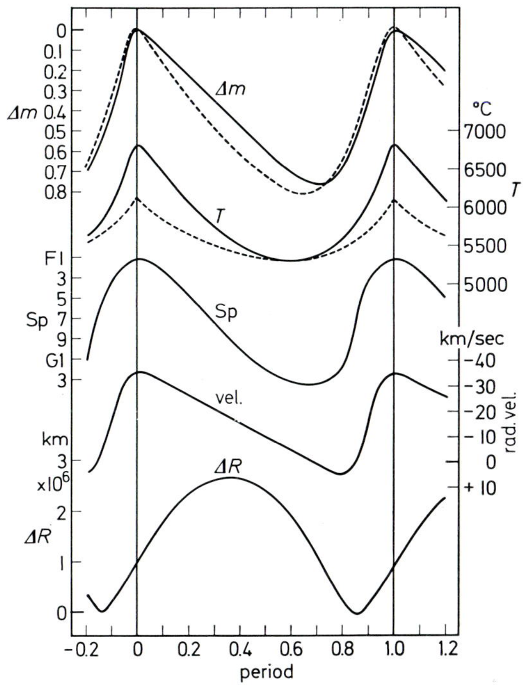

Fig. 21.1. *Delta Cephei* (after W. Becker). Periodic variations of, reading from the top a) Brightness, showing light curve — before and - - - after correction for the varying radius of the star. b) Temperature (— colour temperature, - - - radiation temperature). c) Spectral type. d) Radial velocity. e) $\Delta R = R - R_{min}$

we can obtain the time-variation of the stellar radius R directly by integration of the radial-velocity curve according to (21.1). Its value is usually about 5 percent of the stellar radius. On the other hand, we can evaluate the radiative flux πF_λ using the theory of stellar atmospheres from the colour-indices or other spectroscopic criteria. Some results for δ Cephei are collected in Fig. 21.1 following W. Becker. The required proportionality of the measured brightness with $R^2 F_\lambda$ is in fact well verified.

In 1912 Miss H. Leavitt at the Harvard Observatory discovered for the many hundreds of cepheids in the Magellanic Clouds a relation between the period P and, in the first instance, the apparent magnitude

m_v. Since all these stars have the same distance-modulus, this was the discovery of a *period-luminosity relation.* Using still rather meagre observational material on proper motions (Chapter 23), H. Shapley determined the zero-point of the scale of absolute magnitudes. Thence it was now possible (problems of interstellar absorption being then still unknown) to discover the distance of any cosmic system in which cepheids could be detected. With the help of Bailey's observations (1895) of many cluster variables (RR Lyrae stars) in globular clusters, H. Shapley was able in 1918 for the first time to evaluate their distances and thus to define the outlines of the galactic system in the modern sense. Using the same method, but applying it to classical cepheids (with longer periods) E. Hubble then in 1924 determined the distances of some of our neighbouring spiral nebulae, and he definitely showed that these are galaxies similar to our Milky Way system. We shall discuss this advance into cosmic space in Chapters 27 and 30. Here, however, we must recount an important correction to the basis of the cepheid-method, which W. Baade discovered in 1950. In fact, he was able to show that the zero-point of the period-luminosity relation is differently situated for different types of cepheids. In particular, for equal periods the classical cepheids of Population I (see Chapter 22 with reference to stellar population) are 1 to 2 magnitudes brighter than the W Virginis stars of Population II. The curve for cluster-variables or RR Lyrae stars lies about on the extension of the W Virginis curve.

The more exact study of light-curves permits the recognition of several sub-classes of the main groups here presented. Corresponding to these, there are also possibly small differences in the period-luminosity relation.

The cooler pulsating variables with longer periods of luminosity-change, like the RV *Tauri variables* and the *long-period variables,* have more and more irregular luminosity-variations. The theory of the internal constitution of the stars (Chapter 25) shows that in cooler stars the hydrogen convection-zone becomes ever more extensive. Therefore it is natural to see a coupling of the pulsation with the turbulent convection currents as the cause of the observed semi-regular luminosity-change.

The amplitude of the variations, measured for example in visual magnitude m_v, increase systematically towards cooler stars. This depends essentially upon the known form of Planck's law of radiation. If we write the visual magnitude m_v in Wien's approximation, as in

(13.12), we have

$$m_V = \frac{1.562}{\lambda_V T} + \text{const.}_V\,. \tag{21.5}$$

A given temperature fluctuation ΔT thus corresponds to a luminosity amplitude

$$\Delta m_V = -\frac{1.562}{\lambda_V T^2}\,\Delta T, \tag{21.6}$$

which increases like $1/T^2$ for cooler stars.

The maintenance of the pulsation presents an interesting theoretical problem: What "valve" ensures that the oscillation of the star — like the piston of a heat-engine — always receives its push in the correct phase? The generation of thermonuclear energy near the centre of the star is practically unaffected by the pulsation. We have to do much more with the temperature- and pressure-dependence of the *opacity*, which regulates the flow of radiative energy and so determines the temperature to which any particular layer adjusts itself at any instant. In combination with the change in the adiabatic temperature gradient, this effect shows itself to be specially important in the region of the second ionization of helium. The relevant calculations — very difficult ones — also show and make theoretically comprehensible the combination of parameters of the star, i. e. the region of the colour-magnitude diagram, for which pulsation can occur.

Here we conclude our very incomplete survey of the pulsation-variables and turn to a second group of stars (that also possesses numerous sub-classes) of which the variability may ultimately have its location in the hydrogen convection zone, which naturally leads to many analogies and connections with the phenomena of solar activity.

Variable Stars with Convective Mechanism — Stellar "Activity"

The theory of the internal constitution of the stars, as we have already remarked, shows that the importance of the convection-zone becomes ever greater in the direction of cooler stars, while the calculations of E. Böhm-Vitense show that it gradually dies out above effective temperatures of about 7500° for main-sequence stars, 6500° for giants, and 5000° to 4000° for supergiants. As in the case of the Sun, we must also expect some kind of magnetic fields to be associated with the convection currents. To this group of stars there belong:

Spectrum-variables. These stars show a periodic change in the intensity of certain spectral lines, different lines behaving in different ways. The prototype is α^2 Canum Venaticorum, with period 5.5 days, in which the lines of EuII and CrII vary with opposite phases, while, for example, SiII and MgII remain almost constant. Luminosity changes of about $0^m.1$ often accompany the spectral changes. By measuring the Zeeman effect, H. W. Babcock was able to show that these stars possess magnetic fields of several thousand gauss, of which the strength and often also the sign show periodic changes. Following A. Deutsch, we can account for at least a considerable part of the observations by supposing these stars to exhibit gigantic magnetic *spots* in which, according to the polarity, one set or the other of spectral lines is enhanced. The variations are ascribed to the rotation of the star. In fact, the measured rotational broadening of the lines agrees well with the periods of the stars, i. e. rapidly varying stars mostly have broad lines, and conversely. However, it is still a complete mystery as to why the line intensities should vary with the polarity and strength of the magnetic fields.

Metallic-line stars (Am) are related to spectrum-variables or Ap-stars (peculiar A stars; but there are other neighbouring types as well) but they are mostly not variable. They lie on the main sequence and follow after the Ap-stars on the side of cooler temperatures. According to their hydrogen lines we classify the metallic-line stars as about A5 to F2. Most of the metallic lines are then too strong for these "hydrogen" types; on the other hand, the H and K lines of CaII are too weak. Magnetic fields are detected in some cases. People have tried, but with only partial success, to account for the spectra of metallic-line stars by some anomalous stratification of their atmospheres; in any case the cause of the latter would still require explanation.

Flare-stars or flash-stars are cool main-sequence stars. At unpredictable longer or shorter intervals they blaze up within a few seconds, often getting several magnitudes brighter, and then return to their initial state after minutes or up to about an hour. The emission lines observed in the outburst, as well as the recently discovered simultaneous radio-bursts, leave no doubt that we are witnessing giant flares, which are wholly analogous to solar flares, but they contrast much better with the weak continuum of these red stars.

T *Tauri-stars* or UV *Ceti-stars,* so called after their prototypes, which have been intensively observed both by G. Haro and by G. Herbig, are obviously related to flare- or flash-stars. We find them in the sky always in association with dark clouds of interstellar material. They show irregular luminosity fluctuations, which cannot, however, be explained (as was earlier thought) simply as darkening by dust-clouds passing in front of them. In addition to the usual Fraunhofer spectrum, the spectra show also emission lines, especially the H and K lines of Ca II and the hydrogen Hα-line as well as forbidden lines, which arise in a kind of nebular envelope. As for flare-stars and flash-stars, the emergence of a not yet definitely identified ultraviolet continuum is tied up with the outbursts of luminosity.

As we shall see in Chapter 26, flare-stars and flash-stars, and variables of the type of T Tauri and UV Ceti, etc., all belong to young galactic clusters or associations. In the colour-magnitude diagrams of these systems, they lie above the main sequence in the range of spectral types G, K and M. We have to do with fairly cool stars that have "just" been formed from interstellar material and are still evolving by contraction towards the main sequence.

Many stars of spectral types above G to M also show emission-components * in their Hα lines as well as in the H and K lines of Ca II. We may ascribe them to stellar "plages faculaires", and we may regard them as signs of some form of "activity" and of extended, but weaker, stellar magnetic fields. As O. C. Wilson has discovered, the widths of these emission lines are astonishingly well-correlated with the *absolute magnitude* of the stars. Therefore we can use them for the determination of spectroscopic parallaxes. This can only mean that the turbulent velocities in the chromospheres of these stars depend intimately upon the acceleration due to gravity, or the pressure distribution. As regards understanding these phenomena from the hydrodynamic standpoint we still lack above all an exact theory of "mechanical" energy-transport (which is also so important for the heating of the corona). The *intensities* of the emission lines differ greatly from star to star and to some extent they also show changes with time.

In certain cooler dwarf stars G. E. Kron observed small periodic luminosity changes, that can be attributed to "starspots" analogous to sunspots. For example, Ross 248 shows fluctuations with amplitude $0^{m}.06$ and period about 120 days, which we may well interpret as the period of rotation of the star.

---

* According to O. Struve, Be-stars, i. e. B-stars with hydrogen emission-lines, are totally different. Mainly in connexion with fast rotation, they have extended gaseous envelopes or rings, which are excited to fluorescence by the ultraviolet radiation of the star itself.

It would obviously be very interesting, but also very laborious, to learn from the phenomena of activity in stars of late spectral types more about their periods of rotation and their cycles of activity, corresponding to the 11-year solar cycle.

All these evidences of stellar activity are ultimately of a magneto-hydrodynamic character. It is noteworthy that according to the extensive investigations of G. Haro and of O. C. Wilson they are observed only in stars that, according to the criteria of stellar astronomy, must be considered young, and indeed younger than some hundred million years. But this length of time corresponds in turn, as regards order of magnitude, with the time estimated from (20.6) for the ohmic dissipation of the magnetic field in the hydrogen convection-zone of a star of mean or later spectral type. Thus we are driven to the idea that the star itself does not generate the "frozen-in" magnetic field-lines, but acquires them in its formation from interstellar matter *.

R Coronae Borealis stars are another class of relatively cooler stars. From time to time the luminosity suddenly drops by several magnitudes and then slowly recovers. Since, judging by the spectrum, these relatively cool stars appear to be rich in carbon and poor in hydrogen, we think of them as seen through a covering of colloidal carbon, i. e. a kind of soot.

Novae and Supernovae

The next group including *novae*, and *nova-like variables* which undergo similar but feebler outbursts in a more or less regular sequence, P Cygni stars, U Geminorum = SS Cygni stars, Z Camelopardis stars, etc., present quite different features. As we must state in advance, their theoretical interpretation is still altogether uncertain.

A *nova-outburst* proceeds about as follows. The initial state, the *pre-nova*, is a hot star with absolute magnitude $M_v \approx +5$ lying in the HR diagram between the main sequence and the white dwarfs. Within at most 2—3 days the luminosity increases to a maximum of about $M_v \approx -6$ to -8.4, and so by a factor of 4 to 6 powers of ten. The spectrum then resembles that of a supergiant like α Cygni (cA 2). The luminosity-increase does not depend upon a rise in temperature but, as confirmed by the radial velocities, upon an enormous expansion of the star. After passing through the luminosity-maximum (and during the

* The field-lines would also be compressed along with the material and so the field-intensity would be considerably increased.

subsequent decrease the luminosity often shows cepheid-like fluctuations) broad emission-lines make their appearance. Their doppler effects show that the nova is now ejecting shells with velocities of the order of magnitude of 2000 km/sec. In several cases, e. g. Nova Aquilae 1918, such an envelope and its expansion could be followed in direct photographs for more than a decade. The measured radial velocity of 1700 km/sec corresponded to the expansion of the envelope by $1''$ per year; the comparison of these numbers gives the distance and absolute luminosity of the nova as well as the dimensions of the envelope.

In the course of several years the nova returns to its initial luminosity and presumably to its initial state. We see this most clearly in the case of recurrent novae, like T Pyxidis which has undergone several nova-like outbursts ($\Delta m \sim 7^{m}$) at intervals of about 10 years and which shows almost constant luminosity in between. According to recent observations, many novae, if not all, appear to belong to binary systems. In our Milky Way Galaxy and in similar galaxies some 30 to 50 outbursts take place per year.

U Geminorum or SS Cygni stars, again so called after their prototypes, show much less violent outbreaks at irregular intervals of the order of a month to a year.

On the other hand, P Cygni has remained quiet since a lively performance at the beginning of the 17th century. The emission lines, especially those of the Balmer series, with their absorption components shifted to the violet, show that nevertheless an envelope is being continually ejected. More than a dozen so-called P Cygni stars behave similarly.

Even though its estimation is rather uncertain because of the need to take account of the bolometric correction, the *energy* given up during a nova-outburst is of the order of magnitude 10^{45} erg. This corresponds to the thermal energy-content of a thin layer at for example 5×10^{6} °K and only $^{1}/_{1000}$ solar mass. Everything therefore points to the nova-outburst being a "skin disease" of the star. The hypothesis proposed by the author in 1930, that an ionization zone which has become convectively unstable (because of the high temperature this is certainly not the hydrogen convection-zone) suddenly collapses, in the manner in which delayed boiling can lead to the explosion of a kettle, is perhaps not altogether astray.

Supernovae represent cosmic explosions of quite a different magnitude, their special status having been recognized by W. Baade and

F. Zwicky in 1934. At its maximum, a supernova can outshine the entire galaxy to which it belongs. We distinguish two types * of supernovae:—

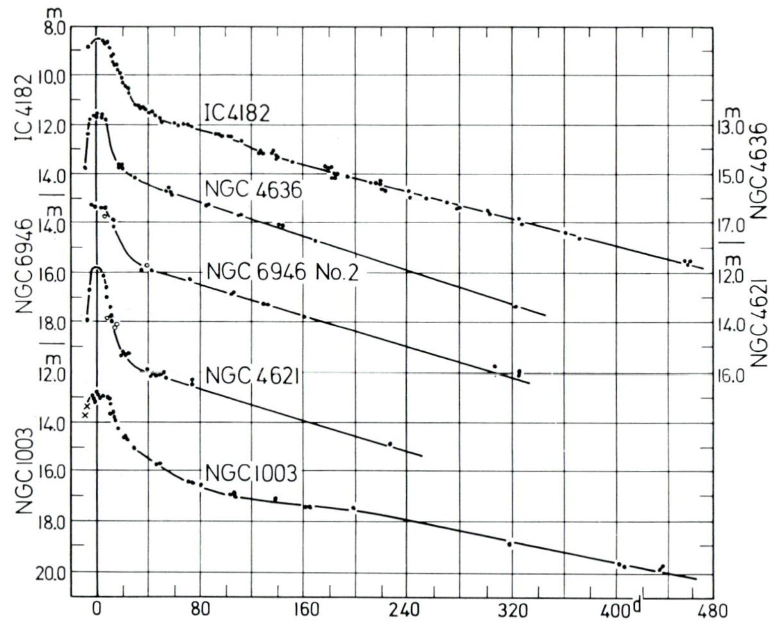

Fig. 21.2. *Photographic light curves of type I supernovae in different galaxies*

a) Type I supernovae reach at maximum a mean photographic absolute magnitude (corrected for interstellar absorption) of $M_{pg} = -19.0$. Their light-curves (Fig. 21.2) are very standard. In the first 20—30 days following maximum the luminosity drops by 2 or 3 magnitudes; from then on the brightness falls off exponentially with time. The spectra of supernovae I show broad bands, which shift with time. At present we do not yet know whether we should regard them as absorption or emission lines — to say nothing about identifying them.

b) Type II supernovae reach "only" $M_{pg} = -17.7$. All the same, this corresponds to about 10 000 times the brightness of an ordinary nova! The fall in brightness after the maximum at first proceeds more steeply, and later, on the other hand, more slowly than in Type I. Further, the light-curves show larger individual differences. The spectrum of a Type II supernova, and its development with time, resembles to an astonishing and far-reaching degree that of ordinary novae.

* Recently F. Zwicky proposed a division into 5 types.

In our Milky Way system the following could be identified as Type I supernovae: The nova of 1054 AD described in old Chinese and Japanese records, from which the Crab nebula has arisen, and Bolton identified this with the radio source Taurus A. By comparing the expansion speed of $0\overset{''}{.}21$ per year with the spectroscopically measured value of 1300 km/sec we obtain a distance of about 1300 pc. Incidentally, as regards the history of ideas, it is remarkable that in contemporary Europe there was no syllable of a mention of this supernova, which at maximum was brighter than Venus. At the beginning of the chapter we recalled Tycho Brahe's supernova of 1572 in Cassiopeia and Kepler's supernova of 1604.

Up to 1964, mostly as a result of F. Zwicky's Palomar survey, some 140 supernovae have been discovered in distant galaxies.

We may attempt to estimate the total energy-output E of a Type I supernova by integrating under the light curve and assuming that the ratio of total radiation to photographic (or — what makes scarcely any difference — to visual) radiation is the same as for the Sun. Under this assumption (BC ≈ 0) we obtain $E \approx 3.6 \times 10^{49}$ erg. The true value could be greater, say about 10^{51} erg as an estimate. For comparison we easily calculate that, for example, the amount of energy liberated by the conversion of one solar mass of hydrogen into helium is

$$0.0072 \times M c^2 = 1.3 \times 10^{52} \text{ erg.}$$

Thus we must assume that in supernova explosion the star releases an appreciable fraction of its available nuclear energy.

Part III

Stellar Systems

Milky Way and Galaxies: Cosmogony and Cosmology

22. Advance into the Universe

*Historical Introduction to Astronomy * in the Twentieth Century*

In the second and third decades of our century, there began a development of astronomy that in its significance does not rank behind the almost contemporaneous discoveries of relativity theory and quantum theory. *The universe as a whole, the cosmos in its spatial structure and in its temporal evolution, became an object for exact scientific investigation.* We pick up the thread of Chapter 10, and seek to follow the unfolding of the new ideas first in their historical association. This brief survey must at the same time elucidate the understanding of the detailed presentation which we must arrange from the standpoint of the topics concerned.

About the turn of the century — following upon the "star gauging" of W. Herschel and J. Herschel — H. v. Seeliger (1849—1924), J. Kapteyn (1851—1922) and others tried to investigate the structure of the Milky Way system by the methods of stellar statistics. Even if the goal was not attained, the incredible labour of these undertakings has nevertheless proved to be very valuable in other connexions.

H. Shapley's (1918) method of photometric distance determinations using cepheids (cluster variables) brought the decisive advance. The period-luminosity relation, that is the relation between the period P of the luminosity-variation and the absolute magnitude M_V, made it possible to measure the distance of every cosmic system in which one could detect any sort of cepheids.

* In keeping with modern usage, we regard astrophysics as part of astronomy, that is of the comprehensive study of the stars.

More exact investigation of the assumptions of this procedure, (1) the absence of interstellar absorption and (2) the applicability of the same period-luminosity relation for all types of cepheids, subsequently made considerable corrections necessary. In 1930 R. J. Trumpler discovered general interstellar absorption and reddening, and about 1952 W. Baade recognized that the period-luminosity relations of classical cepheids and the W Virginis stars, i. e. the pulsating variables in stellar populations I and II (see below) differ from each other by some 1 to 2 magnitudes. In the following, in numerical statements we shall from the outset take full account of these new corrections, and so to that extent we depart from a purely historical standpoint.

The distance of the globular clusters determined by H. Shapley made it known that these clusters form a slightly flattened system whose centre lies at a distance of about 10 kpc, or about 30 000 light years, in the direction of Sagittarius.

The present picture of our Milky Way system developed rapidly from these beginnings. The main body of stars form a flat *disk* of about 30 kpc diameter, the spiral arms being contained in this disk. As regards the *nucleus* of the system, we catch sight of its outer parts in the form of bright star-clouds in Scorpio and Sagittarius. The *galactic centre* itself is hidden from us by clouds of dark interstellar matter. Radio astronomy has made its direct observation possible for the first time. We ourselves are situated far out in the disk, at some 10 kpc from its centre. The disk is surrounded by the much less flattened *halo*, to which the *globular clusters* and certain classes of field-stars belong.

So long ago as 1926/27 B. Lindblad and J. Oort were able to a great extent to elucidate the *kinematics and dynamics of the Milky Way system*. The stars of the disk revolve round the galactic centre under the gravitational attraction of the masses that are fairly strongly concentrated there. In particular, the Sun describes its circular path of radius about 10 kpc with a speed of about 250 km/sec in about 250 million years. However what we first notice is only the *differential rotation*. Stars further out from the centre (as in the case of planets going round the Sun) move somewhat more slowly, stars closer in move faster, than we do. Thence we easily estimate the mass. After numerous adjustments the mass of the whole system is found to be about 2×10^{11} solar masses.

The globular clusters and the stars of the halo move round the centre in elongated orbits similar to ellipses. Their speeds relative to the Sun are consequently mostly of the order of 100 to 300 km/sec; these are the so-called high-velocity stars.

Since 1951 radio-astronomical observations using the 21-cm line of neutral interstellar hydrogen have given a much more detailed picture of the Milky Way. As early as 1944, H. C. van de Hulst had remarked that observations of this radio-spectral line would yield both the density and the velocity of interstellar hydrogen. In fact the groups of investigators in Leiden and Sydney in this way clarified very extensively the spiral structure of the Milky Way and its kinematics. However, we are running ahead of the historical development!

After considerable refinement of photographic technique used with the 100-inch Hooker telescope of the Mount Wilson Observatory, in 1924 E. Hubble succeeded, to a great extent, in resolving the outer parts of the Andromeda nebula (and of others of our neighbouring galaxies) into individual stars. He was able to discern (classical) cepheids, novae, bright blue O and B stars, etc. These made it possible by photometric means to determine the distance. This came to about 700 kpc or about 2 million light-years (and here we have taken account of more recent corrections). After wearisome controversies, it was now made clear that the Andromeda nebula and our Milky Way system are to a great extent similar cosmic structures. W. Baade's subsequent investigations confirmed this up to fine details. Today we can therefore largely combine the studies of the Andromeda nebula (M 31 = NGC 224) and the Milky Way into a single picture; some observations are better made "from outside" and others "from inside". Since the work of Hubble it has become customary to designate the "relatives" of our Milky Way as *galaxies,* and so far as possible to reserve the term nebula for bodies of gas or dust *within* the galaxies.

In 1929 E. Hubble made a second discovery of the greatest importance: the spectra of galaxies showed a *redshift* proportional to their distance. We interpret this as a uniform *expansion of the universe;* we easily realize that inhabitants of other galaxies would observe exactly the same as what we do. If we extrapolate, somewhat formally, the flight of the galaxies backwards, then at a time $T_0 = 10^{10}$ to 1.5×10^{10} years ago the whole cosmos would have been tightly packed together. We call the time T_0 the *age of the universe.* What might lie further back is outside the scope of our enquiry, and in any case at time T_0 years ago the universe must have been entirely different from today. We may mention only briefly the forerunners of Hubble's discovery, V. M. Slipher, C. Wirtz and others, as well as M. Humason's collaboration with Hubble on the 100-inch telescope. Here for the first time the *universe as a whole* had become the object of exact observational scientific investigation. Within the framework of the theory of general relativity, which first sought to treat gravitation and inertial forces together, from 1916 A. Einstein, and after him de Sitter, A. Friedman, G. Lemaître and others, had developed the principles of theoretical cosmology.

The recognition of an age of the universe of 10 to 15×10^9 years, not very much greater than the age of the Earth inferred from radioactive data as 4.5×10^9 years, gave a powerful stimulus to the study of the evolution of stars and stellar systems, i. e. of cosmogony.

In connexion with the age of the Earth and of the Sun, J. Perrin and A. S. Eddington had as long ago as 1919-20 conjectured that the energy radiated by the Sun is generated by the conversion of hydrogen into helium. On the basis of the nuclear physics meanwhile developed, by 1938 H. Bethe and C. F. v. Weizsäcker were then able to show which thermonuclear reactions are capable of slowly "burning up" hydrogen to produce helium in, for instance, the interior of a main-sequence star at temperatures* of about 10^7 °K. Bethe and then A. Unsöld in 1944 thence showed that nuclear energy-sources would suffice for a time of the order of the age of the universe only for the cooler main-sequence stars of types G, K, M. For hot stars shorter lifetimes resulted, which in the case of O and B stars of great luminosity went as low as about 10^6 years. In such a short time, the stars could not travel far from their place of origin. So they must have been formed close to where we now see them. Apart from any speculative hypotheses, W. Baade in particular repeatedly remarked that the close association in space between blue OB stars ** and dark clouds, e. g. in the Andromeda galaxy, pointed to the formation of these stars from interstellar matter.

Again to a great extent traceable to the stimulus of W. Baade, investigations by A. R. Sandage, H. C. Arp, H. L. Johnson and others

* H. N. Russell had already called attention to the circumstance that stars all along the whole main sequence, in spite of their very different effective (or surface) temperatures, have nearly the equal central temperatures of about $10^{7\circ}$ K.

** With spectra of small dispersion, in the case of very distant stars it is often no longer possible to distinguish between the rather similar spectra of types O and B. Then we speak simply of OB stars.

on the colour-magnitude diagrams of globular clusters and galactic clusters led us further. Along with the theory of the internal constitution of the stars, they produced something like the following picture. A star, which has been formed from interstellar material, first runs through a relatively short contraction-phase. Hydrogen burning starts on the main sequence; the star stays there until about 10 percent of its hydrogen has been used up. It then moves off to the right in the colour-magnitude diagram (M. Schönberg and S. Chandrasekhar, 1942) and becomes a red giant. The place in the diagram where the curve for a cluster turns off to the right from the main sequence — the so-called "knee" (Fig. 26.1) — shows which stars have consumed about 10 percent of their hydrogen since the cluster was formed. Clusters possessing bright OB stars, like h and χ Persei, are therefore very young, while in old stellar clusters the main sequence is still present only below G0. Theoretical researches on stellar evolution, beginning with F. Hoyle and M. Schwarzschild in 1955, denoted also a fundamental extension of A. S. Eddington's theory of the internal constitution of the stars (Chapter 25) because the earlier assumption of a continual mixing of the material inside a star had to be given up in face of the observations. The new concept is that in the centre of the star a burnt-out helium-zone is formed. The resulting outward advance of the nuclei-burning zone causes the inflation of the star into a *red giant*. We cannot here go into the subsequent phases of stellar evolution except to remark that most stars finish up as exceedingly condensed degenerate matter, that is as *white dwarfs*. In this state the store of thermal energy is sufficient to maintain the output of radiation over times of the order of the age of the universe.

In what way are the colour-magnitude diagrams of globular clusters different from those of the old galactic stellar clusters such as M 67? In 1944, W. Baade had already shown that various regions of our Milky Way can be different from one another, not only in dynamical respects, but also in their colour-magnitude diagrams. The concept of *stellar populations* arose in this way. It soon appeared that these differ from each other essentially in their *ages* and in the *abundances of the heavier* * *elements relative to hydrogen* (we usually say for short: metal-abundance).

Apart from sub-divisions and intermediate types, we have:—

---

* Carbon is included among the heavier elements for this purpose. Unfortunately, we still know rather little about helium.

1. *Halo population II.* Globular clusters, subdwarfs, etc., forming a slightly flattened system, but with strong concentration towards the galactic centre. Elongated orbits about the centre. Small metal abundance, about $^1/_{200}$ to $^1/_5$ of "normal". Almost no interstellar matter. Oldest part of the Galaxy.

2. *Disk population.* Most stars in our neighbourhood belong to this; they form a strongly flattened system with strong concentration towards the galactic centre. Almost circular orbits. "Normal" metal-abundance, e.g. as in the Sun.

3. *Spiral-arm population I.* Within the disk, interstellar material is denser in the spiral arms; here are the galactic star-clusters and the associations of young stars. Orbits and metal-abundance are not markedly different from those of the disk-population. Here the production of stars from interstellar matter is still in progress.

The colour-magnitude diagram of globular clusters is thus that of old metal-poor population II stars. As their colour-magnitude diagrams show, the old galactic clusters have indeed only a little more recent origin, but the material of these population I stars already has the same metal-content as for example the Sun.

In order to account for the age-sequence of stellar populations the picture has been formed that our Galaxy at first consisted of an almost spherically symmetric mass on nearly pure hydrogen. The globular clusters and the first generation of metal-poor stars were formed in this halo; after the lapse of their lifetime, the material of the stars became dispersed again in interstellar space. In the course of a few generations, more and more heavy elements were formed (see below). At the same time, the greater part of the material condensed towards the galactic disk. The stars of the disk-population were formed there, in circular paths, from dispersed material whose chemical composition had meantime attained nearly its present state. Magneto-hydrodynamic processes in the disk then presumably led to the formation first of arms of gas and dust. From this dispersed material were formed, and are still being formed, the spiral-arm population of stars, of which the chemical composition is indistinguishable from that of the stars of the disk.

Weighty considerations have nevertheless been advanced in recent times, especially by V. A. Ambartsumian, against the explanation of the halo-population II as described. According to Ambartsumian's interpretation, the halo did not come into being by the condensation of intergalactic material from without, but from within by a gigantic explosion of the condensed proto-galaxy (see Chapter 28).

The classification of galaxies laid down by E. Hubble in 1926 (we shall get better acquainted with it later on) was first made according to their forms but showed itself later to be basically a classification according to the preponderance of population II-features or of population I-features. The massive elliptical galaxies contain only a little interstellar matter, so they comprise mainly a stellar population II similar to globular clusters. At the other end, the structurally very different Sc and Irr I galaxies contain much gas and dust giving well-defined spiral arms or other structures as well as bright blue O and B stars, all features of stellar population I.

The origin of heavy elements presents a difficult problem in cosmogony. Until the 1950s, the similarity of their abundance distribution (Table 19.1) in most stars was considered to be an important argument in favour of their almost simultaneous origin. This would have occurred when the formation of the present expanding universe started some 10 to 15 thousand million years ago with the explosion of a primeval mass of extremely high density and temperature. We speak therefore of the "big bang" theory. However, the discovery of metal-poor subdwarfs and other stars, of which the anomalous spectra *could* indicate the intervention of nuclear processes, was a powerful argument for the origin of heavy elements being bound up with the origin, evolution and partial dissolution of the stars in the interstellar medium. In 1957 E. M. Burbidge, G. R. Burbidge, W. A. Fowler and F. Hoyle (often quoted as B^2FH) first investgated which kinds of nuclear processes under which physical conditions could be made responsible for the production of the various chemical elements. It has (still?) not been possible to accommodate all these varied conditions in a theory of stellar evolution. On the other hand, it is astonishing how alike the relative-abundance distribution of all the heavy elements is in all ordinary stars in population I and in population II. The exact analysis of the spectra of "pathological objects" such as helium-stars, carbon-stars, . . . , will probably help us further in this regard.

Or may the heavy elements have originated mainly in the original *galactic* explosion demanded by V. A. Ambartsumian?

In another direction, the new subject of high-energy astronomy, or the astronomy of super-thermal radiations, opens up entirely new kinds of possibilities. *Cosmic rays* have introduced us to cosmic material in the energy-range of about 10^8 to 10^{20} eV. Straightforward observations of the directions in which they reach us give no information about the question of their origin, since the charged particles are strongly deflected on the way by cosmic magnetic fields. The non-thermal component of radiation in *radio frequencies* (particularly in the meter-wavelength range) is *synchrotron radiation.* K. G. Jansky discovered this radiation in the Galaxy in 1931; numerous galactic and extra-galactic radio-sources have later been identified. As H. Alfvén and H. Herlofson, and then K. O. Kiepenheuer, conjectured and as the Soviet astronomers I. S. Shklovsky, V. L. Ginzburg and others were able to render more certain, it occurs if electrons of high energy move in spiral paths round the lines of force of cosmic magnetic fields. In this way we can locate the site of high-energy processes. Also X-ray astronomy and γ-ray astronomy from rockets and space-probes give promise of important contributions here. The explosions of whole galaxies, or of large parts of them, which were discovered in 1964 by radio-astronomy and then studied optically, may well prove important for the understanding of other features of galaxies as well. Is this where the sources of the most energetic cosmic-ray particles are to be found? May the heavy elements also have their origin in such conditions? Before we can even talk about answering these questions, there is appearing over the horizon of our exploration the possibility of *neutrino astronomy.* Neutrinos, which occur as a by-product of stellar energy-generation and in other "high-energy regions", must (like cosmic rays) account for a considerable part of the energy-density of interstellar space. On account of their tiny cross-sections, however, they have no chance at all of being annihilated again by absorption processes. Thus they might make quite novel contributions to cosmological knowledge, namely that concerning the *energy-budget* of the universe.

23. Constitution and Dynamics of the Galactic System

W. Herschel (1783—1822) was the first to seek to penetrate the structure of the galactic system, in his "star gauging", by counting how many stars down to a given limiting magnitude he could see in various directions.

What should we expect, were space fully transparent and uniformly populated with stars? For stars of given absolute magnitude the apparent luminosity decreases with distance like $1/r^2$ [see (14.2),

(14.3)]. Stars brighter than m thus occupy a sphere having $\log r = 0.2\,m + \text{constant}$. Their number $N(m)$ is proportional to r^3, so we must have

$$\log N(m) = 0.6\,m + \text{const.} \tag{23.1}$$

In Fig. 23.1 we compare the numbers of stars per square degree according to F. H. Seares (1928) for the galactic plane and for the direction of the galactic pole ($b = 0°$ and $b = 90°$) with the numbers

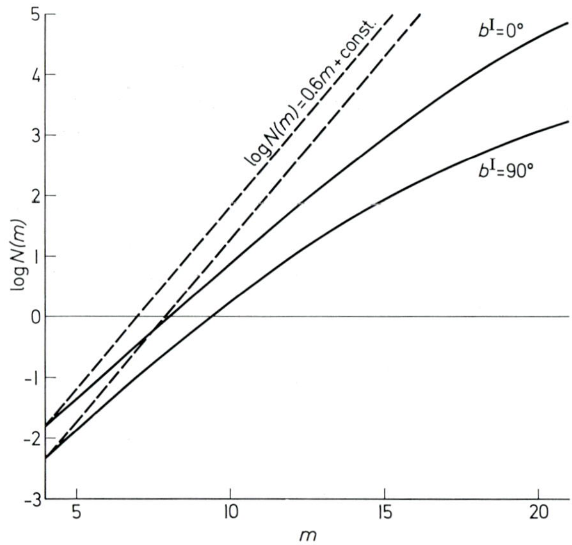

Fig. 23.1. *Number of stars N(m) brighter than a certain magnitude m* per square degree. Here m is on the international photometric scale. The solid curves are from star counts at the galactic equator ($b^{\rm I} = 0°$) and the galactic pole ($b^{\rm I} = 90°$) by F. H. Seares (1928). The broken-line curves denote the calculated functions $\log N(m) = 0.6\,m + \text{const}$ for constant star density, in the absence of galactic absorption. The constant was obtained by matching the calculated functions to the observations at $m = 4$

to be expected according to (23.1). The much slower increase of $N(m)$ for faint magnitudes in actuality can have only two causes, (1) decrease of star-density at large distances, (2) interstellar absorption, or both. H. v. Seeliger (1849—1924) and J. Kapteyn (1851—1922) took account of the dispersion of absolute magnitudes of the stars. They showed how we may represent the star-numbers $N(m)$ basically by superposition (mathematically speaking, convolution) (1) of the density function $D(r)$ = number of stars per pc^3 at distance r in a given direction with (2) the luminosity function $\Phi(M)$ = number of stars per pc^3 in the interval of absolute magnitude from $M-\frac{1}{2}$ to $M+\frac{1}{2}$, taking account of (3) an interstellar absorption of $\gamma(r)$ magnitudes per parsec. Even if we assume the luminosity-function $\Phi(M)$ to be everywhere the same, and if we determine it with the help of stars of known parallaxes in a region of 5 or 10 pc, we can nevertheless not separate the functions $D(r)$ and $\gamma(r)$. We can therefore pass over the results of older stellar statistics. The concepts introduced in this way, however, do remain important, as does the great sampling survey of the whole sky (magnitudes, colour indices, spectral types) in the Kapteyn-fields.

The study of the motions of the stars then leads further. We have already mentioned the spectroscopic measurement of *radial velocities* V (km/sec) using the *doppler effect* (H. C. Vogel, 1888; W. W. Campbell and others) *.

The *proper motions* μ of the stars on the celestial sphere, discovered much earlier by E. Halley (1718), have then to be considered. They are usually expressed in seconds of arc per year. We measure relative proper motions (referred to faint stars of small proper motion) by comparing ("blinking") two plates taken at an interval of 10 to 50 years apart, if possible with the same instrument. Reduction to absolute proper motions is achieved by using meridian instruments to measure absolute positions of certain stars at different epochs. Recently C. D. Shane has used remote galaxies to furnish an extragalactic system of reference.

The transverse component T (km/sec) of stellar velocity is related to the proper motion μ as follows:

If p is the parallax of the star in seconds of arc, then μ/p is equal to T in astronomical units per year. This unit of speed is equal to $1/2\,\pi$ times the orbital speed of the Earth and so is 4.74 km/sec. Therefore

$$T = 4.74\ \mu/p \qquad \text{(km/sec)} \tag{23.2}$$

and the space velocity of the star is

$$v = (V^2 + T^2)^{1/2}. \tag{23.3}$$

The angle ϑ between the space velocity and the sight-line is determined

* The sign is defined so that a positive radial velocity means a red-shift, i. e. recession from the Sun; a negative radial velocity means a blue-shift, i. e. approach towards the Sun. We reduce the values to the Sun as point of reference by removing the varying part depending upon the orbital and rotational motion of the Earth.

by the relations

$$V = v \cos \vartheta \qquad T = v \sin \vartheta \,. \tag{23.4}$$

Since, for example, a star of the sixth magnitude has on the average a parallax of 0.″012 but a proper motion of 0.″06 per year, by using proper motions over, say, 20 years we can penetrate about 100 times further into space than by means of parallax-measurements.

If we first make the simplifying assumption that the stars are at rest and that only the Sun is in motion relative to them with velocity $v_\odot$ in the direction of the apex, then we expect the distribution shown in Fig. 23.2 of radial velocities V and transverse velocities T (or proper motions) with the angular distance λ of the star from the apex.

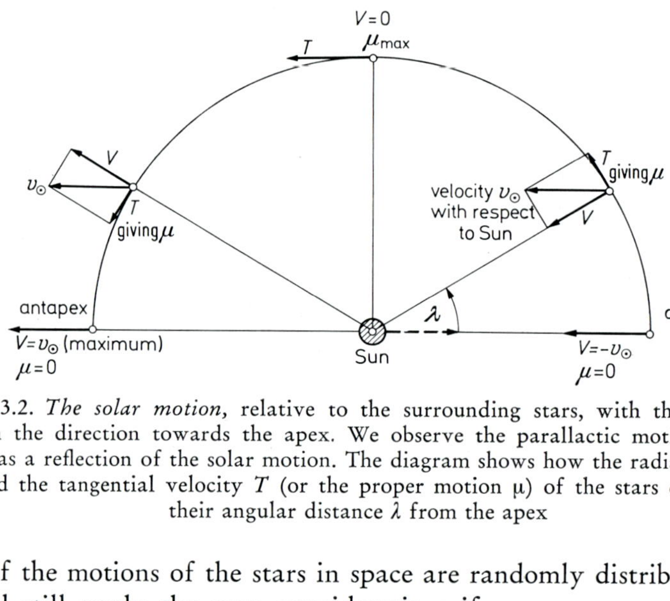

Fig. 23.2. *The solar motion,* relative to the surrounding stars, with the velocity $v_\odot$ in the direction towards the apex. We observe the parallactic motion of the stars as a reflection of the solar motion. The diagram shows how the radial velocity V and the tangential velocity T (or the proper motion μ) of the stars depend on their angular distance λ from the apex

If the motions of the stars in space are randomly distributed, we could still apply the same considerations if we average over many stars. Thus we obtain from the proper motions and radial velocities of the stars in our neighbourhood

$$\begin{aligned} &\text{solar motion} \quad v_\odot = 20 \text{ km/sec} \\ &\text{towards apex RA } 18^{\text{h}}00^{\text{m}}, \ \delta = +30^\circ. \end{aligned} \tag{23.5}$$

Using a few proper motions, W. Herschel carried out the first determination of the solar apex as long ago as 1783. Later it was found that, to be precise, the solar motion depends upon which stars we use for its determination; this was the first indication of a systematic constituent of stellar motions. Equation (23.5) gives the so-called standard solar motion which we generally use in order to refer stellar motions to a local standard of rest.

We can now use the knowledge of the solar motion (23.5) in order to isolate the statistical part of the proper motions (the so-called *peculiar motions),* in the measured proper motions of a significant selected group of stars, from the reflexion of the solar motion or the parallactic motion. The latter obviously depends upon the mean parallax $\bar{p}$ of the stellar group. Corresponding to (23.4), the part of the transverse velocity caused by the solar motion is $T = v_{\odot} \sin \lambda$, where λ again denotes the angular distance of the stellar group * from the apex. Accordingly we can now use (23.2) if, in averaging over the proper motions, we restrict ourselves to their components in the direction of the apex, the $\bar{v}$ components. The mean, or secular, parallax of the group of stars is thus

$$\bar{p} = -\frac{4.74\,\bar{v}}{v_{\odot} \sin \lambda}\,. \tag{23.6}$$

The hypothesis of statistically distributed peculiar motions is to be used with much caution.

In 1908 L. Boss discovered that, for instance, in the case of a comprehensive group of stars in Taurus, which arrange themselves around the Hyades galactic cluster, the proper-motion vectors as drawn on the sphere or in a star-chart point towards a convergent point at $\alpha = 93^{\circ}$, $\vartheta = +7^{\circ}$. The members of this star stream, or moving cluster, thus describe parallel motions in space, like a shoal of fish, whose direction points towards the convergent point. Let the velocity of the cluster relative to the Sun be v_{H}. Now if we know

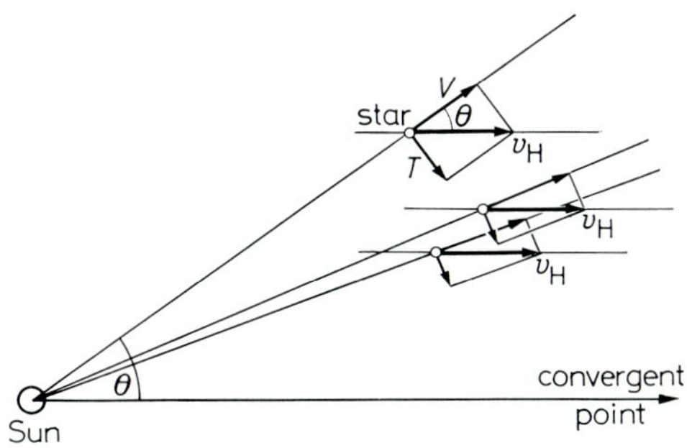

Fig. 23.3. *Moving cluster or star stream*

* Here we assume that the stellar group occupies a relatively small region of the sky, so that a single mean value of λ suffices.

(Fig. 23.3) the proper motion μ and the radial velocity V relative to the Sun for a star of the cluster, and if ϑ is the angle on the sky from the star to the convergent point, then we can take over the notions in (23.2), (23.4). We find

$$V = v_{\mathrm{H}} \cos \vartheta \quad T = v_{\mathrm{H}} \sin \vartheta = 4.74\, \mu/p$$

whence we immediately obtain the parallax of the star

$$p = \frac{4.74\, \mu}{V \tan \vartheta}\,. \tag{23.7}$$

The Taurus cluster, for example, is about 42 parsecs away and its velocity amounts to 31 km/sec; most of its stars lie within a region of some 10 parsecs diameter. This method of stream parallaxes surpasses the method of trigonometric parallaxes in scope and often also in accuracy.

The common small proper motions of stars in galactic clusters (Pleiades, Praesepe, etc.,) permit to some extent the determination of their parallaxes. However, they are important chiefly in order to test individual stars for membership of the cluster.

According to our current ideas, the diameter of the Galaxy is of the order 30 000 pc; the other galaxies are comparable structures at distances ranging from hundreds of thousands to thousands of millions of parsecs. The whole of the knowledge we have of the size and structure of the "new cosmos" depends essentially upon the methodology of photometric measurement of distance. According to the well known inverse-square law of photometry, we see a star of absolute magnitude M and parallax p, or of distance $1/p$ parsec, to be of apparent magnitude m, where from (14.2) its distance modulus is given by

$$m - M = -5\,(1 + \log p). \tag{23.8}$$

Here we have not taken account of interstellar absorption. We shall return to this in the succeeding Chapter 24, but for the time being we shall quote values of distances, etc., in which the required correction has been made.

Ultimately we must always refer back to absolute magnitudes of particular objects, which have been derived using trigonometric, secular, moving-cluster . . . parallaxes. It was on account of their basic significance that we first explained the methods of geometrical distance determination so fully.

In order to gain insight into the structure of the Galaxy, it is well to start from aggregates of stars whose structure is immediately recognizable to the naked eye or in suitable photographs: *globular clusters* in which the stars seem to be crowded together like a swarm of bees, the less condensed *galactic* or *open clusters* of stars, stellar *associations* and *galactic star clouds*. The classical position-catalogues, for all "nebulae" which nowadays we distinguish on the one hand as galaxies and on the other hand as galactic nebulae and as planetary nebulae, are Messier's catalogue (M) of 1784, and Dreyer's New General Catalogue (NGC) of 1890 with its continuation in the *Index Catalogue* (IC) 1895 and 1910.

a) *Globular clusters.* The two brightest globular clusters, ω Centauri and 47 Tucanae, are situated in the southern sky. In the northern sky M 13 = NGC 6205 in Hercules is still a naked-eye object;

Fig. 23.4. *Globular cluster* M 13 (NGC 6205) in Hercules. Distance 8 kpc

its brightest stars are about $13^{m}.5$. Photographs with large reflecting telescopes (Fig. 23.4) show in the brighter globular clusters more than 50 000 stars; in the centre the individual stars cannot be distinguished from each other. The globular clusters show a strong concentration in the sky in the Scorpio-Sagittarius direction.

b) *Galactic or open stellar clusters* (Fig. 23.5) occur in the sky right along the bright band of the Milky Way. Many are relatively

Fig. 23.5. *The galactic double cluster h and χ Persei*

rich in stars and contain hundreds of members, but still many fewer than the globular clusters; others are poor in stars with only a few dozen members. Also the concentration of the stars towards the centre, the compactness of the cluster, varies a great deal. The best known are the Pleiades and the Hyades in Taurus, the double cluster h and χ Persei, and so on.

c) The OB *associations* are relatively loose groups of bright O and B stars, which often surround a galactic cluster as, for example, the ζ Persei association surrounds the $h+\chi$ Persei cluster. On the other hand, the T *associations* are corresponding groups of T Tauri or RW Aurigae variables and other stars near the lower part of the main sequence. In 1947 V. A. Ambartsumian called attention to the cosmical significance of OB and T associations as very young systems.

d) The *star-clouds* of the Galaxy, for instance in Cygnus, Scutum, Sagittarius . . . are much more extended systems which we may tentatively regard as analogues of the bright "knots" in the arms of distant spiral galaxies.

We here conclude the first survey and again turn our attention to the cardinal problem of distance determination.

In 1918 H. Shapley produced a turning point in modern astronomy when he determined the distances of numerous *globular clusters* by means of the cluster variables (RR Lyrae stars) contained in them. As we have said, the essential difficulty here lay in deriving the absolute magnitudes, or in calibrating the period-luminosity relation. In the case of globular clusters that contained no variables, Shapley used as a secondary criterion the *brightest stars* of the cluster. In applying this, one purposely leaves aside the five brightest stars, as

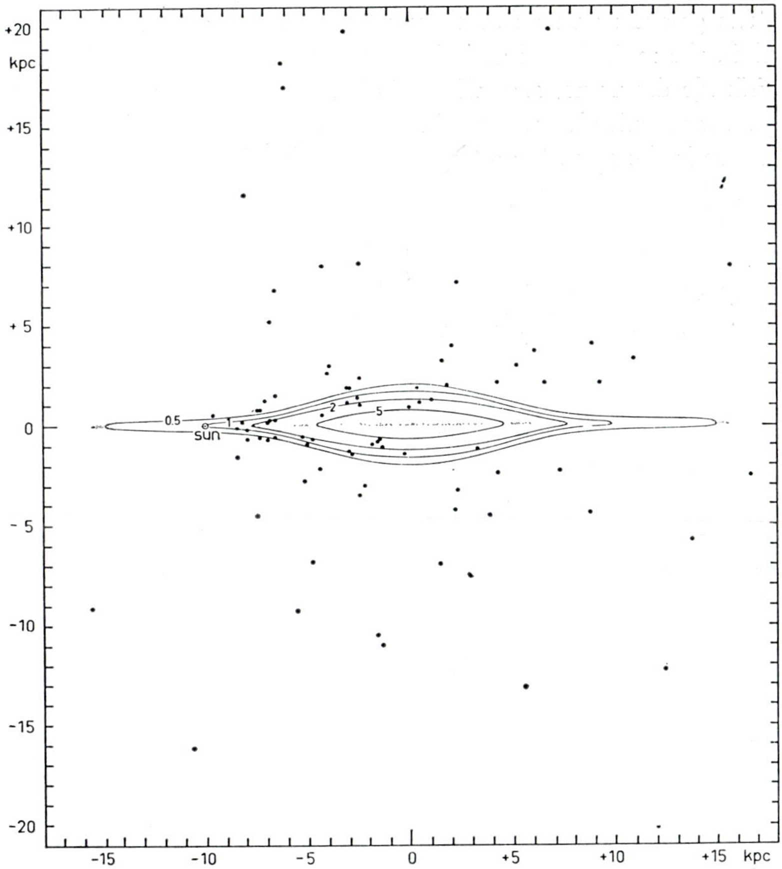

Fig. 23.6. *The Milky Way system.* The diagram shows the spatial distribution of globular clusters projected on to a plane perpendicular to the galactic plane and passing through the Sun and the galactic centre. Lines of equal mass density are shown, the numbers giving the density with respect to that in the solar neighbourhood. The thin layer of interstellar matter lying in the galactic plane and associated with the extreme (spiral arm) Population I is shown dotted. (After J. H. Oort)

possible foreground stars; one takes as well-defined the absolute magnitudes of the next brightest down to say the 30th. With due precautions, one may next employ as a criterion the *total brightness* of the cluster, or its angular diameter. From his observational material, which at that time concerned 69 globular clusters, Shapley was able to draw the conclusion that these form a slightly flattened system (Fig. 23.6) whose centre 10 kpc from us lies in Sagittarius (the distance being at that time evaluated as 13 kpc). Thus the stage was set for all subsequent investigations concerning the Galaxy.

The galactic clusters that have been studied are much nearer to us than the majority of globular clusters. We can plot their colour-magnitude diagrams, e. g. with $B-V$ colour indices as abscissa and apparent magnitude m_V as ordinate. If we may assume that the main sequence is the same for all systems and for the so-called field stars

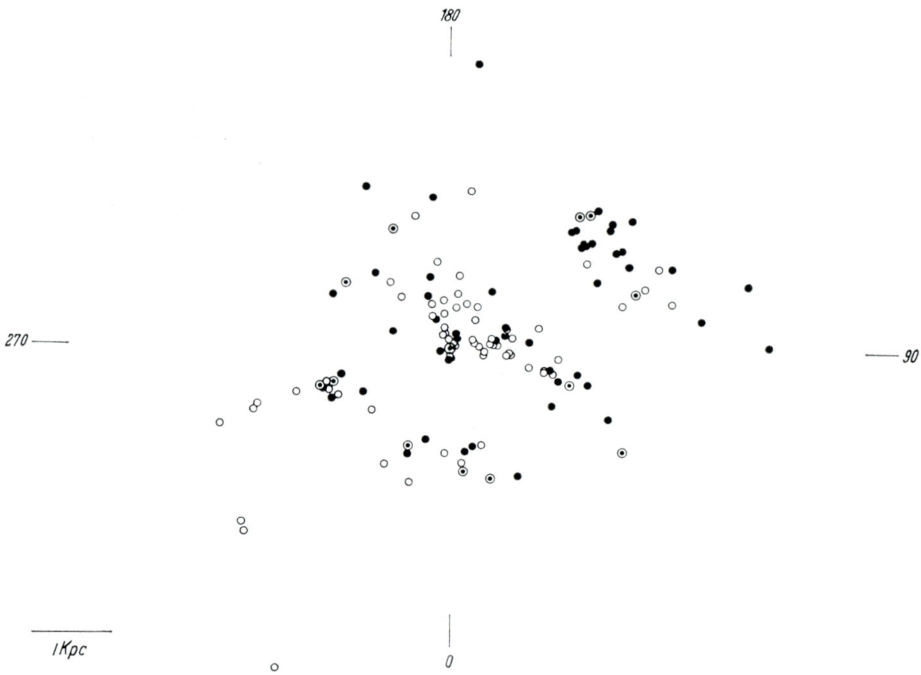

Fig. 23.7. *Distribution of young galactic clusters* ● (i. e. those which still contain O-B2 stars), of H II regions O and of clusters with H II regions ⊙ in the plane of the Milky Way (after W. Becker). The galactic longitude $l^{\mathrm{II}} = 0$ (below) defines the direction towards the galactic centre, the Sun being situated at the centre of the system of coordinates. All the objects included in the diagram arrange themselves along the *spiral arms;* they belong to extreme Population I

of our neighbourhood, which do not belong to any recognizable system, then the vertical separation between the $(B-V, m_v)$ diagram of a cluster and the $(B-V, M_v)$ diagram for our neighbourhood yields directly the distance modulus $m_v - M_v$ and hence the distance of the cluster. (We shall return to consideration of finer differences between the colour-magnitude diagrams and to the effect of interstellar absorption; the following results have been corrected for these effects.) We can only mention the fundamental work of R. Trumpler. In Fig. 23.7 we show the distribution plotted by W. Becker 1964 of galactic clusters, that contain O to B2 stars as their earliest spectral types, together with HII regions, i. e. bodies of hydrogen gas which have been ionized and excited to give Hα emission by the radiation of the O or B stars in clusters or in OB associations or by single O or B stars (Chapter 24). We notice at a glance the arrangement into elongated regions in which we may perceive the neighbouring portions of spiral arms.

At this point we may introduce a system of *galactic coordinates* that are useful in the study of the Milky Way, namely the galactic longitude l in the plane of the Galaxy and galactic latitude b perpendicular to the plane, positive to the north and negative to the south. The older system l^{I}, b^{I} used the galactic north pole RA $12^{\mathrm{h}}\,40^{\mathrm{m}}$, $\delta + 28^{\circ}$ (1900.0) and measured l^{I} from the intersection (ascending node) of the galactic plane with the celestial equator 1900. An improved system of galactic coordinates was introduced in 1958 taking particular account of radio-astronomical observations:

$$l^{\mathrm{II}},\ b^{\mathrm{II}} \text{ with the galactic north pole RA } 12^{\mathrm{h}}\,49^{\mathrm{m}},\ \delta + 27\overset{\circ}{.}4 \text{ (1950.0)} \tag{23.9}$$

where l^{II} is now measured from the

$$\text{galactic centre RA } 17^{\mathrm{h}}\,43^{\mathrm{m}},\ \delta - 28\overset{\circ}{.}9 \text{ (1950.0)}.$$

Old and new galactic longitudes (both of which are still often used in similar work) in the vicinity of the galactic plane are therefore connected by the relation

$$l^{\mathrm{II}} \approx l^{\mathrm{I}} + 32\overset{\circ}{.}3 \quad \text{or} \quad l^{\mathrm{I}} - 327\overset{\circ}{.}7. \tag{23.10}$$

The "old" position of the galactic centre $l^{\mathrm{II}} = 0$, $b^{\mathrm{II}} = 0$ is

$$l^{\mathrm{I}} = 327\overset{\circ}{.}7 \quad b^{\mathrm{I}} = -1\overset{\circ}{.}4. \tag{23.11}$$

The Lund Observatory has published (in 1961) tables for the mutual conversion of equatorial coordinates and old, as well as new, galactic coordinates.

We define the galactic components of the space velocity of stars (with positive senses)

U radially outwards from the galactic centre
V in the direction of galactic rotation ($l^{\mathrm{II}}=90°$)
W perpendicular to the galactic plane towards the galactic north pole. (23.12)

It has to be stated whether the solar motion has been subtracted.

After these rather formal preliminaries, we turn to the kinematics and dynamics of the galactic system, as B. Lindblad and J. H. Oort developed them in their theory of the differential rotation of the Galaxy 1926-27.

We first assume that all motions take place along circular orbits around the galactic centre in the galactic plane (Fig. 23.8). Let the

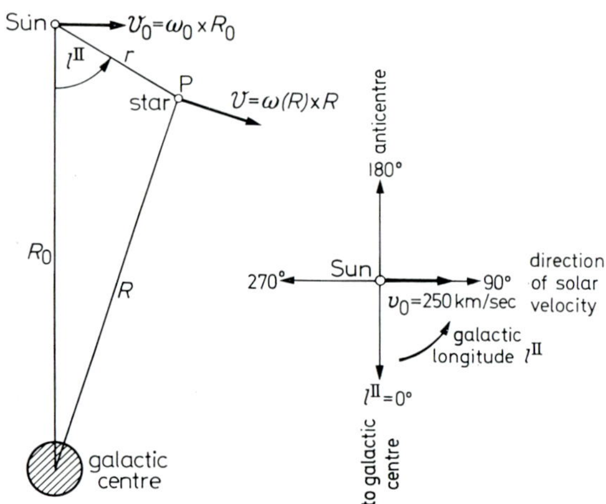

Fig. 23.8. *Galactic rotation and galactic longitude* l^{II}

angular velocity ω of a star P as a function of the distance R from the centre be $\omega(R)$. For the Sun, let $R=R_0$ and $\omega(R_0)=\omega_0$, so that $\omega_0 R_0=\mathcal{V}_0$ is the velocity of the Sun in its galactic orbit. In what follows, we shall concern ourselves with the solar neighbourhood, and we shall remove the solar motion (23.5) from all observations.

The velocity of the star P relative to the Sun is

$$R_0\{\omega(R)-\omega_0\} \tag{23.13}$$

as we see immediately from Fig. 23.8, if we first think of P as held fixed and consider the motion of the Sun relative to P.

The corresponding radial velocity V_r of the star P (positive sign implies recession) is then

$$V_r = R_0 \{\omega(R) - \omega_0\} \sin l^{II}, \qquad (23.13\text{ a})$$

where l^{II} is the galactic longitude. If the distance of the star from the Sun $\odot P = r \ll R_0$, we can write to a first approximation (the subscript 0 referring always to $R = R_0$)

$$V_r = -r R_0 \left(\frac{d\omega}{dR}\right)_0 \sin l^{II} \cos l^{II} = -\frac{1}{2} r R_0 \left(\frac{d\omega}{dR}\right)_0 \sin 2 l^{II}. \qquad (23.14)$$

The corresponding transverse component T of the velocity of P relative to the Sun is

$$T = R_0 \{\omega(R) - \omega_0\} \cos l^{II}. \qquad (23.15)$$

Thence we obtain the proper motion of P (positive in the sense of l^{II}) if we subtract from T/r the galactic rotation of the Sun ω_0. If we introduce the series-expansion already used, we find in circular measure

$$-R_0 (d\omega/dR)_0 \cos^2 l^{II} - \omega_0 \qquad (23.16)$$

or using $2\cos^2 l^{II} = 1 + \cos 2 l^{II}$ and converting to seconds of arc,

$$\text{proper motion} = \frac{1}{4.74}\left\{-\frac{1}{2} R_0 \left(\frac{d\omega}{dR}\right)_0 \cos 2 l^{II} - \frac{1}{2} R_0 \left(\frac{d\omega}{dR}\right)_0 - \omega_0\right\}. \qquad (23.17)$$

We call the coefficients in (23.14), (23.17) the *Oort constants of differential galactic rotation.* We also express them in terms of the orbital velocity $\mathcal{V}(R)$ where

$$\omega = \frac{\mathcal{V}(R)}{R} \quad \text{so that} \quad \frac{d\omega}{dR} = \frac{1}{R}\left(\frac{d\mathcal{V}}{dR} - \frac{\mathcal{V}}{R}\right) \qquad (23.18)$$

and obtain

$$A = -\frac{1}{2} R_0 \left(\frac{d\omega}{dR}\right)_0 = \frac{1}{2}\left\{\frac{\mathcal{V}_0}{R_0} - \left(\frac{d\mathcal{V}}{dR}\right)_0\right\} \qquad (23.19)$$

$$B = -\frac{1}{2} R_0 \left(\frac{d\omega}{dR}\right)_0 - \omega_0 = -\frac{1}{2}\left\{\frac{\mathcal{V}_0}{R_0} + \left(\frac{d\mathcal{V}}{dR}\right)_0\right\}$$

or

$$A + B = -(d\mathcal{V}/dR)_0 \quad \text{and} \quad A - B = \mathcal{V}_0/R_0. \qquad (23.20)$$

Finally, the radial velocity V_r and the proper motion PM for our neighbourhood, as functions of galactic longitude l^{II}, take the simple forms (see Fig. 23.9)

$$V_r = A\, r \sin 2\, l^{\mathrm{II}} \tag{23.21}$$

$$\mathrm{PM} = \frac{1}{4.74}\, \{A \cos 2\, l^{\mathrm{II}} + B\}. \tag{23.22}$$

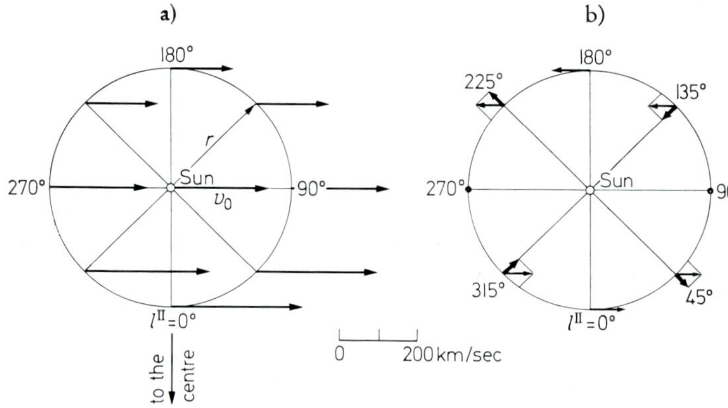

Fig. 23.9. *Differential galactic rotation.* a) Absolute velocities of stars at a distance r from the Sun. In the diagram $r = 3$ kpc; the lengths of the velocity vectors give the motions of the stars in 10 million years. b) Velocities of the same stars with respect to the Sun and their radial components (thick arrows), the *double wave of radial velocities* given by equation 23.21

After disposing of peculiar motions, observation confirms well the "double wave" ($\sin 2\, l^{\mathrm{II}}$) of the radial velocity and of the proper motion. While the amplitude of V_r is proportional to r, the proper motion is independent of r. After discussion of very extensive observational material, which we cannot review here, in 1964 the International Astronomical Union proposed the following values

$$A = 15 \text{ km/sec/kpc} \quad B = -10 \text{ km/sec/kpc}. \tag{23.23}$$

If we combine these with the distance of the Sun from the galactic centre

$$R_0 = 10 \text{ kpc} \tag{23.24}$$

we obtain using (23.20) the orbital velocity $\mathcal{V}_0$ and period P_0 of the Sun, or rather of the solar neighbourhood, as

$$\mathcal{V}_0 = 250 \text{ km/sec} \quad P_0 = 250 \text{ million years}. \tag{23.25}$$

Thus we have travelled once round the galactic system since the end of the carboniferous age.

Were the entire mass M, under whose influence the Sun describes its circular orbit, concentrated at the galactic centre, then as in the case of planetary motion, we must have from (6.35)

$$\mathcal{V}^2 = G\,M/R. \tag{23.26}$$

In this way we estimate the mass of the Galaxy as

$$M = 2.9 \times 10^{44}\,\mathrm{g} = 1.5 \times 10^{11}\ \text{solar masses}. \tag{23.27}$$

More exact calculations also lead to a total mass of the same order of magnitude. By logarithmic differentiation of (23.26) we get $\mathrm{d}\mathcal{V}/\mathcal{V} = -\frac{1}{2}\,\mathrm{d}R/R$ and thus from (23.20) $[(A-B)/A+B)]_{\text{calc}} = 2$ while the observed rotation constants (23.25) lead to $[(A-B)/(A+B)]_{\text{obs}} = 5$. Thus the assumption of a $1/R$ potential can be only poorly satisfied *.

What is then the situation in regard to our assumption up to now of circular galactic orbits? As soon as we admit appreciable eccentricities e of the stellar orbits, radial velocities of the order of 100 km/sec and more make their appearance in the immediate vicinity of the Sun. As J. H. Oort remarked in 1928, we can in this way understand the phenomenon of high velocity stars.

If we limit ourselves first to orbits of stars in the galactic plane then their galactic velocity components U, V or the corresponding components relative to the solar neighbourhood

$$U' = U \quad \text{and} \quad V' = V - 250\ \text{km/sec}, \tag{23.28}$$

determine their galactic orbits. With sufficient accuracy, for the stars that are accessible to exact observation, we can set the positional coordinates equal to those of the Sun. In a diagram with coordinates U', V' we can then for an assumed galactic force- or potential-field, plot curves of constant eccentricity e, curves of constant apogalactic distance R_1, and so on. In 1932 F. Bottlinger was the first to perform such calculations for a $1/R^2$ field of force. Figure 23.10 shows such a Bottlinger diagram for a field of force more appropriate to the actual Galaxy. The velocity vectors of the high-velocity stars demonstrate that these stars move in orbits of large eccentricity, partly with direct motion and partly with retrograde motion round the galactic centre. On the

* The values of the constants $A = 19.5$, $B = -6.9$ km/sec/kpc with $R_0 = 8.2$ kpc, which were preferred until a few years ago, giving $(A-B)/(A+B) = 2.1$ would have agreed much better with our certainly much too crude theoretical result!

other hand, the "normal" stars of our neighbourhood have small U' and V', i.e. like the Sun they all move with direct motion in nearly circular orbits.

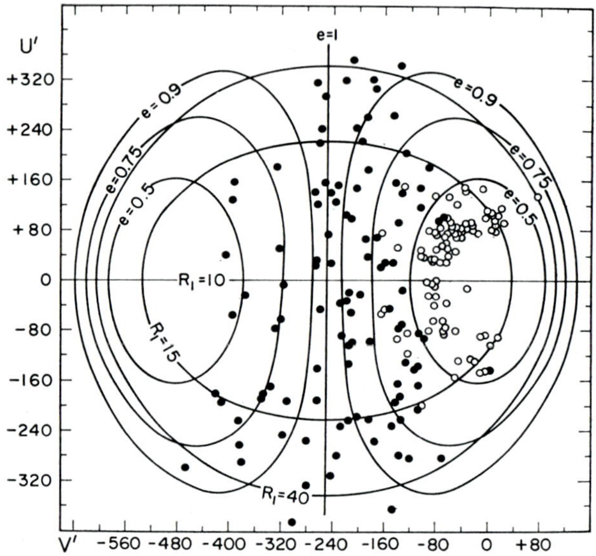

Fig. 23.10. *Bottlinger diagram.* The coordinates are the galactic velocity components U' (towards the anticentre) and V' (in the direction of rotation) with respect to the solar neighbourhood. The absolute velocity components U and V may be read by reference to the coordinate axes. The orbital eccentricity e and orbital apogalactic distance R_1 in kpc may be read off from the two intersecting sets of curves. For later reference (Chapter 27) we note that the filled circles represent stars with an ultraviolet excess $\delta(U-B) > +0\overset{\mathrm{m}}{.}16$ i.e. metal-poor stars of the halo population II. These are high velocity stars. Open circles denote stars with $\delta(U-B) < 0\overset{\mathrm{m}}{.}16$. These stars represent the transition from the halo population II to the disk population i.e. the transition towards stars with more circular orbits. (After O. J. Eggen)

As regards the motions of the stars of our neighbourhood perpendicular to the galactic plane (the W-components in (23.12), which generally used to be called Z-components), according to J. H. Oort (1932 and 1960) we can largely explain them if we regard the distribution of mass-density ϱ in our region of the galactic disk as a plane-stratified distribution. Thus we here take account of a dependence of mass-density, etc., only upon the distance z from the galactic plane. Then we may first treat the W-components of the stellar velocity vectors on their own account, independently of the motion parallel to the galactic plane (U and V). The stars perform perpendicularly to the galactic plane oscillations about it with periods of about 10^8 years. The density-distribution of stars at right angles to the galactic plane is related to the gravitational field of the galactic disk on the one hand, and to the velocity-distribution of the W-components on the other hand in a way analogous to that in which the density-distribution of molecules in an atmosphere is related to the gravitational field and the Maxwell velocity-distribution i.e. the temperature. However, here in addition the gravitational field depends upon the mass-density ϱ through Newton's law of attraction (or through Poisson's equation). Therefore, Oort was able to derive the total density of matter in the galactic plane near the Sun; a recent discussion (1960) yields with an uncertainty of about $\pm$ 10 percent

$$\varrho = 10.0 \times 10^{-24}\,\mathrm{g\,cm^{-3}} = 0.15\,\mathfrak{M}_\odot/\mathrm{pc}^3 \qquad (23.29)$$

According to W. Gliese (1956) the total density of observed *stars* within a region of 20 pc radius is $5.9 \times 10^{-24}\,\mathrm{g\,cm^{-3}}$ to (taking somewhat stronger interstellar absorption) about $6.7 \times 10^{-24}\,\mathrm{g\,cm^{-3}}$. The density of interstellar *gas* may correspond to about one hydrogen atom $\mathrm{cm^{-3}}$, that is $1.7 \times 10^{-24}\,\mathrm{g\,cm^{-3}}$. Thus dark material in the form of large fragments, whose presence we could be aware of only from Oort's method, need not play any very important part. The analysis of the W-velocity distribution again allows the two sorts of stars already mentioned to be distinguished: the disk-stars with $|\overline{W}| \approx 12$ km/sec and the high-velocity stars with $|\overline{W}| \approx 24$ km/sec (but with the occurrence of considerably higher individual values). Thus while in fact disk-stars describe almost circular orbits almost in the galactic plane, high-velocity stars mostly move in orbits of high eccentricity and of high inclination to the galactic plane.

In 1944 W. Baade recognized that these dynamically distinguishable groups of stars also have different HR or colour-magnitude diagrams (Chapter 25). Thus he was the first to distinguish between stellar populations I and II. It soon became necessary to refine the subdivision; at the colloquium on stellar populations held in Rome in 1957 the population groups shown in the table p. 342 were isolated. Spectrum analysis (Chapter 19) shows that the abundance of the heavy elements decreases from right to left. We shall infer (Chapter 25) from the colour-magnitude diagrams that the age of the structures increases from right to left. Thus we can hope to gain a genetic interpretation of the structure of the Galaxy.

24. Interstellar Matter

Diffuse material between the stars of the Galaxy first entered the astronomers' awareness in the form of *dark clouds* whose absorption weakens and reddens the light from stars seen through them. However, in 1930 R. J. Trumpler was first able to show that, in addition to detectable dark clouds, interstellar absorption and reddening are by no means negligible anywhere in the Galaxy in regard to the photometric measurement of distances beyond a few hundred parsecs. In 1922 E. Hubble had already realized that (diffuse) galactic *reflection-nebulae,* like those around the Pleiades, are produced by the scattering of the light of relatively cool stars by cosmic dust-clouds, while in (diffuse) galactic *emission-nebulae* interstellar gas is excited by the radiation of hot stars so as to emit a line spectrum. The study of the interstellar gas followed quickly in 1926-1927. Indeed, as long ago as 1904 J. Hartmann had discovered the stationary lines of CaII which do not show any orbital displacement in the spectra of double stars. However, in 1926 A. S. Eddington from theory, and O. Struve, J. S. Plaskett and others from the observations, first developed the concept that the interstellar lines of CaII, NaI, etc., arise in a layer of gas which is partly ionized by stellar radiation, which fills the whole disk of the Galaxy, and which shares in the differential rotation. On another aspect, in 1927 I. S. Bowen succeeded in finding the long-sought identification of the "nebulium lines" in the spectra of the gas-nebulae, as the forbidden transitions in the spectra of [OII], [OIII], [NII], etc., and H. Zanstra worked out the theory of nebular radiation. It was some ten years later that astronomers recognized *hydrogen* as being by far the most abundant element in the interstellar gas, as it is in stellar atmospheres. O. Struve and his collaborators first remarked that the assumption of large hydrogen abundance considerably diminished the quantitative difficulties in regard to the ionization of the interstellar gas. With the help of nebular-spectrographs of great light-gathering power, they then discovered that many O and B stars, or groups of these, are surrounded by a fairly well-defined region which radiates in the red Balmer line Hα. So here interstellar hydrogen must be ionized. B. Strömgren then in 1938 developed the theory of these *HII regions.*

Neutral hydrogen — we speak of HI regions — at first seemed not to be directly observable until in 1944 H. C. van de Hulst calculated that the transition between the two levels in the hyperfine structure of the ground state of hydrogen must yield an emission line of interstellar hydrogen with a radio frequency and with measurable intensity. The transition is from $F = 1$ (nuclear spin parallel to the electron spin) to $F = 0$ (nuclear spin anti-parallel to the electron spin) giving

$$\lambda_0 = 21.1 \text{ cm} \quad \text{or} \quad \nu_0 = 1420.4 \text{ MHz}.$$

Its observation for the first time in 1951 — almost simultaneously at the Harvard Observatory, in Leiden and in Sydney — gave such an impetus to the study of interstellar matter and to galactic and extragalactic research, that, as against the historical development, we may begin here with an introduction to 21-cm radio astronomy.

We can calculate the absorption coefficient $\varkappa_\nu$ of the 21-cm line from quantum mechanics. Its frequency dependence is entirely determined by the doppler effect of the motion of the interstellar hydrogen. If there are $N(V)\,\mathrm{d}V$ hydrogen atoms per cm³ with radial velocity between V and $V + \mathrm{d}V$ cm/sec, then we find

$$\varkappa(V) = \frac{N(V)}{1.835 \times 10^{13}\, T} \tag{24.1}$$

where T is the temperature that determines the velocity distribution of the hydrogen atoms and so also (as a result of collisions) the distribution of the atoms between the two hyperfine-structure levels. Then the intensity of the radiation emitted by an optically thin layer, in accordance with Kirchhoff's law, is

$$I(V) = \int \varkappa(V)\, B_\nu(T)\, \mathrm{d}l \tag{24.2}$$

where, since $h\nu/kT \ll 1$, the Kirchhoff-Planck function $B_\nu(T)$ is given sufficiently accurately by the Rayleigh-Jeans approximation (11.25)

$$B_\nu(T) = 2\,\nu^2\, k\, T/c^2. \tag{24.3}$$

The integration is to be taken over the whole sight-line.

The case of an optically thick layer arises in only a few places in the Galaxy. We recognize these by the fact that over a large range of frequency, or of velocity, the intensity reaches a constant maximum value

$$I(V) = B_\nu(T) = 2\,\nu^2\, k T/c^2 \tag{24.4}$$

from which we can at once derive the temperature of the interstellar hydrogen

$$T \approx 125^\circ \text{ K}. \tag{24.5}$$

Fortunately, a more exact value is of no great importance, since T cancels out of our formula (24.2) for an optically thin layer.

From (24.1), (24.2), the measurement of the line-profile $I(V)$ by means of a high-frequency spectrometer with a radio telescope gives the number of hydrogen atoms in a column of unit cross-section along the sight-line, the radial velocity V or frequency ν of which lie in the range

$$V \text{ to } V+\mathrm{d}V \quad \text{or} \quad \nu_0-\frac{V}{c}\nu_0 \quad \text{to} \quad \nu_0-\frac{V+\mathrm{d}V}{c}\nu_0 . \tag{24.6}$$

The velocity distribution of interstellar hydrogen is made up of two contributions: First we have statistically distributed velocities, a kind of turbulence, whose distribution-function is similar to Maxwell's. From the interstellar lines of CaII, we already know the mean

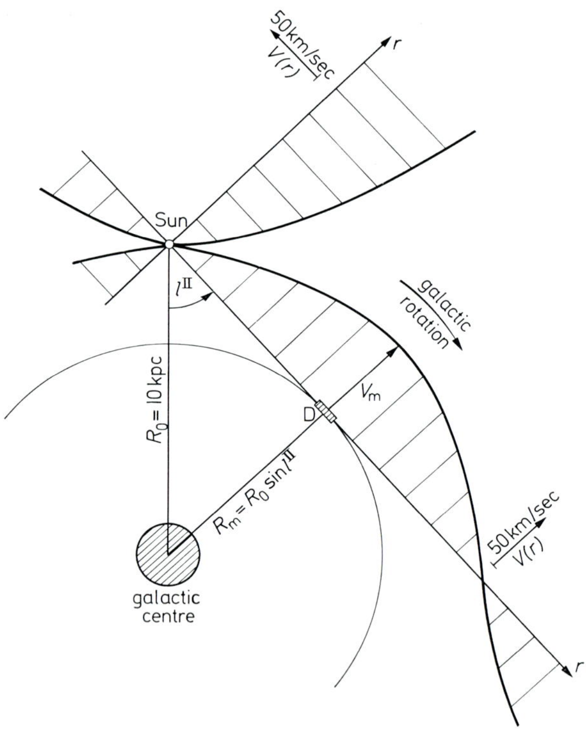

Fig. 24.1. *Differential galactic rotation* of interstellar hydrogen. The radial velocity $V(r)$ of the interstellar hydrogen with respect to the solar neighbourhood, observed along a line of sight of galactic longitude l^{II}, is plotted as a function of the distance r. This velocity reaches a maximum V_m at D, where the line of sight just touches a circular orbit. The above illustration should be compared with Fig. 23.9, which corresponds to the approximation $r \ll R_0$

value to be about 6 km/sec. The contribution of the differential rotation of the Galaxy is more important. After taking account of the solar motion, in our neighbourhood, according to (23.21), the radial velocity V increases linearly with distance; its dependence on direction shows the characteristic double wave proportional to $\sin 2\, l^{\mathrm{II}}$. For greater distances from the Sun we must resort to the exact equation (23.13). From Fig. 24.1 we see that the radial velocity V attains its maximum value V_{m} for a sight-line in the direction of galactic longitude l^{II} ($|l^{\mathrm{II}}|<90°$), such that the ray is tangent at D to a circle about the galactic centre having radius $R_{\mathrm{m}}=R_0 \sin l^{\mathrm{II}}$. Thus the line profile falls away steeply at V_{m} towards greater radial velocities. By combining the values of V_{m} for different galactic longitudes l^{II}, we can then derive the *rotational velocity* $\mathcal{V}(R)$ as a function of distance from the galactic centre. From this, conversely, we can calculate the distribution of radial velocity along each sight-line and, using the resulting curves together with the measured line profiles, we can determine the density distribution of the hydrogen. For $|l^{\mathrm{II}}|<90°$ the ambiguity as to whether a particular value of $\mathcal{V}$ belongs to the corresponding point in front of D or behind D can often be resolved by the consideration that the more remote object will in general show the smaller extension at right angles to the galactic plane, i. e. in b^{II}. For the rest, in assessing the 21-cm measurements we must all the time have regard to the internal consistency of the resulting picture.

Fig. 24.2 shows the distribution of neutral hydrogen in the galactic plane derived in this way from the observations of the Netherlands and Australian radio astronomers. We see clearly its concentration into spiral arms. More recent observations have shown that within a distance of 3 kpc from the galactic centre there is a spiral arm that, surprisingly, expands away from the centre at 50 km/sec. The dynamics of the innermost region of the Galaxy, especially that of the expansion of the 3 kpc arm, is still largely unexplained.

The distribution of neutral hydrogen perpendicular to the galactic plane has also been studied. In the mean, it forms a flat disk. The distance between the surfaces, where the density has fallen to half its average value in the galactic plane, in the range $3<R<10$ kpc is about 220 pc; in the innermost 3 kpc the disk is even somewhat thinner.

In the surroundings of bright O and B stars, the interstellar gas, and particularly the hydrogen, is ionized and is rendered luminous; we then see a diffuse nebula or an HII region. As H. Zanstra realized in 1927, this comes about as follows: If a neutral atom absorbs radiation from the star in the Lyman continuum $\lambda < 912$ Å it becomes ionized (see Fig. 17.1). The resulting photo-electron is later recaptured

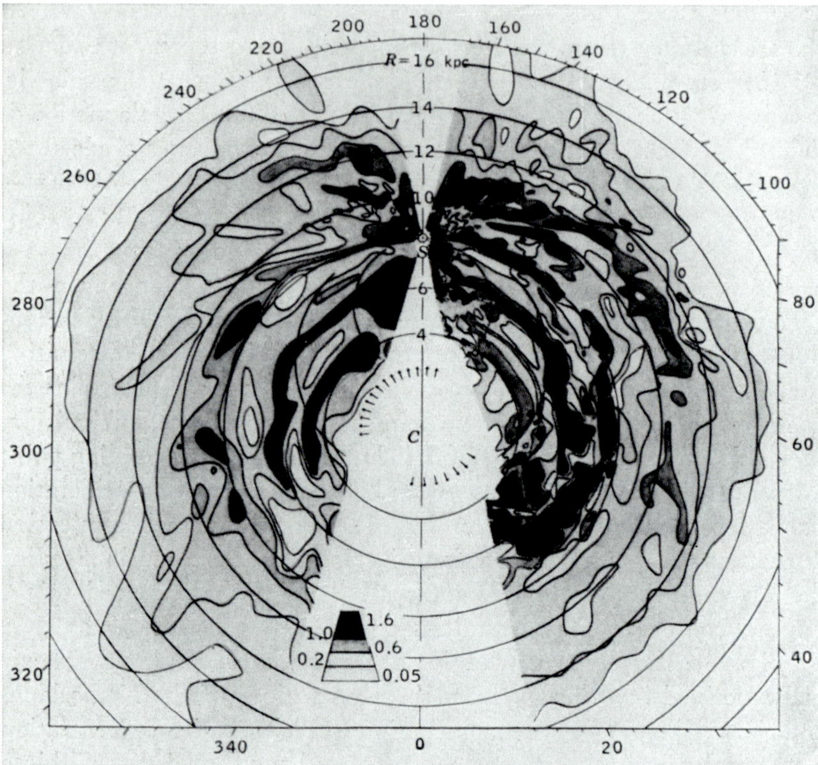

Fig. 24.2. Distribution of neutral hydrogen in the Galactic plane. Each point has been assigned the maximum density which would be seen in projection against the plane. Contours have then been drawn in accordance with the density scale (lower left) which gives the number of atoms per cubic centimetre. The Galactic longitude l^{II} is indicated at the outer edge

by a positive ion (proton). Only in rare cases does recombination occur directly into the ground-state; in most cases, cascade transitions occur through several energy-levels with the emission of less energetic photons $h\nu$. As an estimate, we can say that, for every absorbed Lyman photon $h\nu > 13.5$ eV or $\lambda < 912$ Å, about one Hα photon is emitted. If we may now assume that the nebula absorbs practically all the Lyman radiation from the star, then following Zanstra we may work back from the observed Hα emission of the nebula to infer the

Lyman emission from the star. If we compare this with its visual radiation, then from Planck's formula we can estimate the temperature of the star (or we may use the more exact theory mentioned in Chapter 18). We obtain values for the O and B stars that lie about in the range of the spectroscopically determined temperatures.

We may look at the same process from another standpoint: The number of recombinations per cm^3 is proportional to the number of electrons per cm^3 (N_e) times the number of ions per cm^3 making captures. However, since each hydrogen atom on ionization yields one electron, the latter number is also about equal to N_e. The Hα brightness at any particular place in the nebula must therefore be proportional to the so-called

$$\text{emission measure EM} = \int N_e^2 \, dr \qquad (24.7)$$

defined by B. Strömgren. The integration is along the sightline, generally with r measured in parsecs. For diffuse nebulae, EM is of the order of some thousands or more; for weaker HII regions it is a few hundred. If we estimate the electron density N_e, taking the depth and width of the nebula to be about equal, we realize that in the HII regions the electron density is of the same order as the density of neutral atoms in HI regions (Fig. 24.2), that is $N_e \approx 10$ electrons/cm^3. In the large diffuse nebulae, like the Orion nebula, on the other hand, values $N_e \approx 10^4$ electrons/cm^3 are attained; so these are to be regarded as condensations in the interstellar gas.

In interstellar space, if an electron in its encounter with a proton is not captured but only deflected, then a free-free continuum is produced (Fig. 17.1). This is too weak to be observed in the optical range but, in the range of cm and dm radio wavelengths, we observe it as thermal free-free radiation from diffuse nebulae. As we see at once, its intensity is again proportional to the emission measure EM. However, while optical observations in Hα are strongly obstructed by interstellar absorption (see below), especially near the galactic plane, radio waves pass through dark interstellar clouds without being absorbed. Taking this into account, the agreement between the emissivity as measured by optical astronomy and by radio astronomy is everywhere satisfactory.

Besides the free-free radiation of the HII regions, in the region of centimetre waves radio astronomers have recently observed transitions of hydrogen atoms between neighbouring energy levels with quantum-numbers $n \approx 100$. Further, in the decimetre region in 1966 they discovered the analogous lines of helium with $n = 160 \rightarrow 159 \rightarrow 158$ and $157 \rightarrow 156$. This opens up the prospect of determining in cosmic objects the abundance ratio H:He, which would be supremely important for cosmology and which hitherto no one could think of doing.

We also observe the free-free radiation of ionized hydrogen in centimetre and decimetre waves in the whole extent of the galactic plane. We recognize it from the fact that, according to the theory, for emission from an optically thin layer the intensity I_ν (per unit frequency interval) is independent of the frequency. If we analyse the dependence of the measured intensity $I_\nu(l^{\text{II}})$ in regard to its dependence upon the galactic longitude l^{II}, we can calculate the density of electrons, or of ionized hydrogen, N_e (taking account of its inhomogeneity) as a function of distance R from the galactic centre. In Fig. 24.3 we take the results got by G. Westerhout and compare them with the mean density distribution of neutral hydrogen (from Fig. 24.2). According to this, the degree of ionization of the hydrogen has a pronounced maximum at $R \approx 3.5$ kpc, immediately outside the expanding spiral arm at $R = 3$ pc.

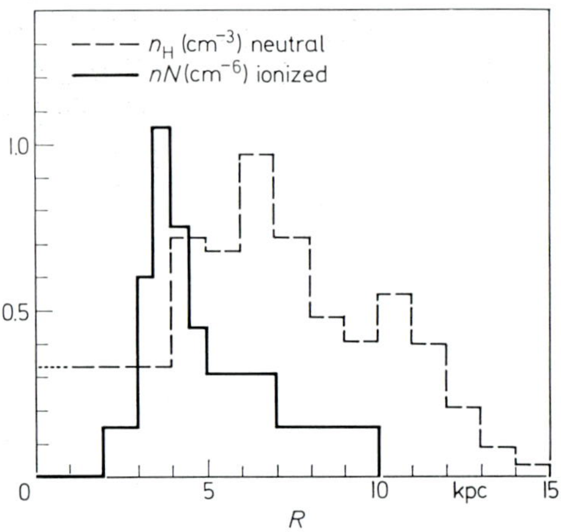

Fig. 24.3. *Density distribution* of ionized and neutral hydrogen in the galactic system. For the neutral hydrogen the mean number of atoms per cm³ is plotted directly, whereas for the ionized hydrogen it is only possible to determine the emission measure (equation 24.7) and hence the product nN, where n is the average space density and N is the average particle density within an H II region; N is about 5 to 10 particles per cm³

In the Galaxy there are altogether about 6×10^7 solar masses of ionized hydrogen and 1.4×10^9 solar masses of neutral hydrogen out of a total mass of about 10^{11} solar masses. The mean density of the hydrogen is about 0.6 atom/cm^3 or 1.0×10^{-24} g/cm^3. In our system, the 2 percent or so of interstellar matter plays no part having dynamical significance. However, we must beware of too hastily taking over this conclusion for other galaxies.

Fig. 24.4. *Orion Nebula,* a diffuse or galactic nebula, and its emission spectrum

The optical spectra of gaseous nebulae (Fig. 24.4) arise in conditions so far removed from thermodynamic equilibrium that a theoretical treatment of individual elementary processes becomes necessary. If r is the radius of the nebula and R that of the exciting star, the radiation from the star comes into operation only with a dilution factor $W = R^2/4\,r^2$. For example, with orders of magnitude $R \approx 1$ solar radius and $r \approx 1$ pc we should have $W \approx 10^{-16}$. Consequently in the atoms and ions only the ground-states and long-lived metastable states (having extremely small probabilities of transition to lower energies) are appreciably occupied. According to I. S. Bowen (1928) the following processes therefore contribute to the excitation of the luminosity of the nebula:—

1. As we have seen, *hydrogen* and *helium*, the two most abundant elements, are ionized by the dilute stellar radiation in the Lyman continuum. Recombination of ions and electrons then takes place into all possible quantum states. From these, for the most part in cascade-transitions, the electrons finally fall back into the ground state. Thus we obtain the whole spectrum of H, HeI, and possibly HeII, as recombination-radiation.

2. *Permitted transitions* of many ions, such as OIII, NIII, . . . are excited by fluorescence. For example, in the recombination of He^+ (often as the final stage of a cascade process) the resonance line $1^2S - 2^2P$ λ 303.78 Å is emitted. As it happens, this can be absorbed by the OIII ion in its ground state so that the 3d $^3P_2$ term is excited. From this, a whole set of OIII lines are emitted, that are observed in the ultraviolet. However, from the term-scheme we know further that the last of such a cascade of transitions is often given by the resonance line OIII λ 374.44 Å. Again thanks to a fortuitous coincidence of term differences, this can excite a particular term of the NIII spectrum, which then emits directly observable lines in the photographic region, and so on.

3. *Forbidden transitions.* A great stir was caused in 1927 by I. S. Bowen's discovery that the lines λ 4958.91 and λ 5006.84 Å, which are strong lines in all nebular spectra and which astronomers had for a long time attributed to a mysterious element "nebulium", could be interpreted as "forbidden transitions" in the OIII spectrum. These originate in transitions from a deep-lying metastable (long-lived) term in the ground term of the ion. While the usual permitted lines (dipole radiation) have transition probabilities of the order 10^8 sec^{-1}, these are, for example, only 0.0071 or 0.021 sec^{-1} for the [OIII] nebular lines just mentioned*. In this case, we have to deal with magnetic dipole radiation (the analogue to that of a frame antenna), in other cases with electric quadrupole radiation of the ion. The excitation of the metastable level results from collisions with electrons that have been produced by the photo-ionization of H and He. In such inelastic collisions the electrons obviously give up energy. Consequently the electron temperature is lower than we should at first have expected from the temperature of the star; it is usually about 8000° to 10 000° K.

---

* Forbidden transitions are denoted by enclosing the symbol of the spectrum in square brackets.

In regard to the physics of their luminosity, but *not* to their cosmic status, the *planetary nebulae* (so-called because of their appearance) are related to the diffuse gaseous nebulae (Fig. 24.5). The Zanstra method already explained gives for the central stars temperatures of 30 000° up to 150 000° K. The luminous envelopes, whose apparent sizes in the case of the nearest such objects are some minutes of arc, have radii of the order $\sim 10^4$ AU, electron density (e. g. from the emission measure) $\sim 10^3$ to $\sim 10^4\,\mathrm{cm}^{-3}$ and an electron temperature $\sim 10^{4\circ}$ K. The observed splitting of the lines indicates an expansion of the envelope — the front part approaches us and the back part recedes — of about 20 km/sec. Thus a planetary nebula is continually replenished from its central star; it is not made of interstellar matter. R. Minkowski and others have found a strong concentration of planetary nebulae towards the galactic centre; thus they belong to the (older) disk-population of the Galaxy. Their absolute magnitudes (there being not much doubt as to their distances) show that in the HR diagram they represent the transition from hot bright stars to white dwarfs.

After this digression we return to further possible ways of studying interstellar matter. The discovery of the interstellar calcium lines CaII λ 3933/3968 Å by J. Hartmann in 1904 gave the very first indication of the existence of interstellar atoms or ions. In fact, it was seen that in the double star δ Orionis the H and K lines did not share in the effects of orbital motion; so one spoke of "stationary lines". In the course of time interstellar lines have been discovered that come from the following ions, atoms and molecules:

$$\text{NaI, KI, CaI, CaII, TiII, FeI; CH, CN, CH}^+ . \qquad (24.8)$$

By far the most intense are the H + K lines of CaII and the D lines of NaI. We have to do throughout with transitions from the ground state, in confirmation of our earlier considerations. In 1926, A. S. Eddington solved in its essentials the problem of the ionization of interstellar matter at great departures from thermal equilibrium. At such low pressures, ionization occurs obviously only through photoionization from the ground state, e. g. in the case of neutral calcium having $\chi_0 = 6.09$ eV through radiation with $\lambda < 2040$ Å. On the other hand, recombination occurs wholly through 2-body encounters between an ion and an electron. The degree of ionization is then such that the processes of ionization and of recombination balance each other.

If we now compare the interstellar lines, say, of CaII with CaI or NaI, the intensity-ratio is, surprisingly, not very different from that in about an F star. How can we account for this? In interstellar gas,

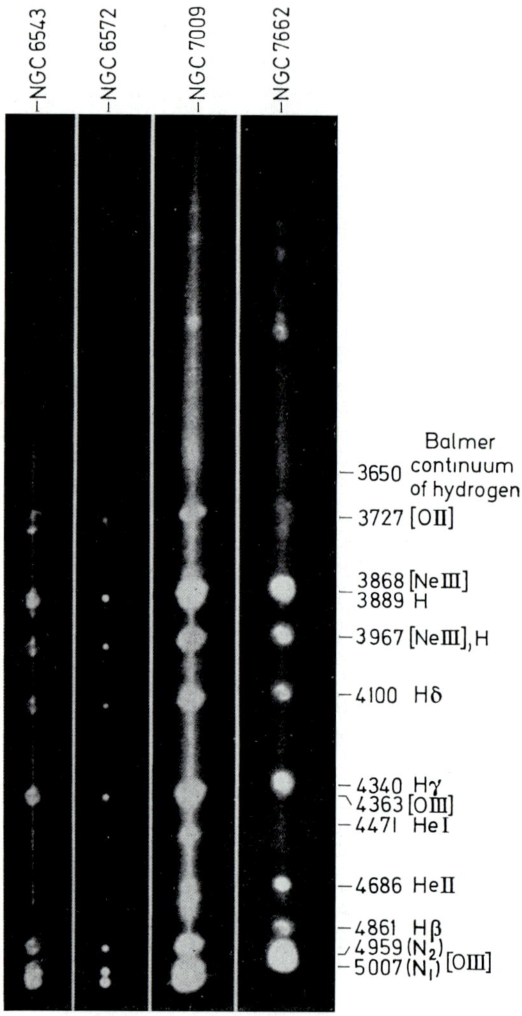

Fig. 24.5. *Spectra of planetary nebulae,* taken with a slitless spectrograph, with the result that the distribution over the envelope of the light from various lines can be readily seen. In the case of NGC 6543 the continuous spectrum of the central star is also clearly visible

as in the stellar atmosphere, the stated processes must balance one another. However, on the one hand, in interstellar matter the electron density N_e is about 10^{16} times smaller than in the stellar atmospheres (5 compared with 2×10^{16} cm^{-3}) and the number of recombinations per ion per second is correspondingly smaller. On the other hand, we have already estimated that the radiation field in interstellar space is diluted by a factor $W \approx 10^{-16}$ compared with that in a stellar atmosphere, so that the number of ionizing processes per second is reduced by about the same factor. In interstellar space both processes therefore proceed about 10^{16} times more slowly than in a stellar atmosphere, but the degree of ionization remains about the same!

As J. S. Plaskett, O. Struve and others showed, the equivalent widths W_λ of the interstellar lines increase in a fairly regular way with the distance r of the star in whose spectrum they are observed. Conversely, they can therefore be used in order to estimate the distance of suitable objects. Moreover, interstellar lines show doppler displacements which correspond to about half the differential galactic rotation given by (23.21) for the star concerned. This is evidence that the interstellar gas occupies the space between the stars in the galactic plane more or less uniformly. Later observations by Th. Dunham, W. S. Adams and others with better dispersion have shown, however, that the interstellar calcium lines are often composed of several components, which can be associated with the various spiral arms of the Galaxy traversed by the light-path. After allowing for local differences of ionization and excitation, it is found that the abundance-distribution of the chemical elements in the interstellar gas and in diffuse gaseous nebulae is everywhere the same. It is also not very different from that given by our analysis of the spectra of stars of the spiral-arm population and of the disk-population of our Galaxy (Table 19.1).

Spectroscopic observations have established that molecules and molecular ions CH, CN, CH^+ also occur in interstellar gas. Radio-astronomers have recently discovered the absorption lines of the hydroxyl radicle OH in the decimeter-wave region. On the other hand, the possible significance of the hydrogen molecule H_2, whose presence is hitherto verifiable neither optically nor radio-astronomically, is still largely undecided.

It is to be expected that, with the help of space-probes outside the Earth's atmosphere, still further interstellar lines will be discovered in the ultraviolet spectral region ($\sim$1000 to 3000 Å).

The continuous absorption and reddening of starlight by *cosmic dust* discloses quite other aspects of interstellar material. Against the background of bright star-clouds, especially in the southern Milky Way, we can even with the naked eye discern dark clouds such as the well-known "Coal Sack" in the Southern Cross (Crux), the "dark

Fig. 24.6. *The southern Milky Way,* showing the "Coal Sack", a local *dark nebula* in the constellation of the Southern Cross. On the far left, lying in the Milky Way, is α Centauri, the brightest star in the southern sky; the brightest globular cluster, ω Centauri, is seen in the upper left-hand corner about equidistant from the top and left-hand edges

nebula" in Ophiuchus, etc.. E. E. Barnard, F. Ross, M. Wolf and others have obtained very beautiful photographs with relatively small, wide-angle cameras (Fig. 24.6). They show a strong concentration of the dark clouds in the galactic plane. The well-known "splitting" of the Milky Way is obviously caused by an elongated dark cloud. Photographs of distant galaxies give a still clearer picture of the association of dark nebulae with spiral arms. Max Wolf first estimated the distances of certain dark nebulae with the aid of the diagram bearing his name: On a plate with photometric standards one counts the number of stars $A(m)$ in the apparent-magnitude range $m - \frac{1}{2}$ to $m + \frac{1}{2}$ per square degree within the region of the dark clouds and in one or more neighbouring comparison fields. Were all stars to have the same absolute magnitude $\overline{M}$, then a dark cloud that causes an absorption of Δm magnitudes in the range of distance r_1 to r_2, that is to say in the range of reduced (i. e. "absorption free") distance-moduli $(m - \overline{M})_1 = 5 \log r_1/10$ to $(m - \overline{M})_2 = 5 \log r_2/10$, will diminish the star-numbers $A(m)$ in the manner shown schematically in Fig. 24.7. Because

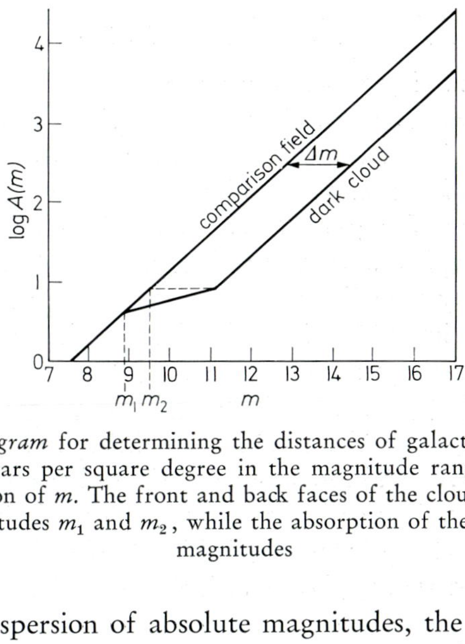

Fig. 24.7. *Wolf diagram* for determining the distances of galactic dark clouds. The number $A(m)$ of stars per square degree in the magnitude range $m - \frac{1}{2}$ to $m + \frac{1}{2}$ is plotted as a function of m. The front and back faces of the cloud correspond to the mean stellar magnitudes m_1 and m_2, while the absorption of the cloud is Δm stellar magnitudes

of the actual dispersion of absolute magnitudes, the accuracy of the method is low, but it suffices to show that many of the conspicuous dark clouds are no more than a few hundred parsecs away. Probably even the great complexes in Taurus and in Ophiuchus are connected with each other past the Sun.

Bright diffuse nebulae with a continuous spectrum, the so-called *reflection nebulae* such as those that envelope the Pleiades, occur where bright stars with temperatures below some 30 000° light up clouds of dust. On the plates we can often see directly the passage from dark clouds to bright nebulae.

In our Milky Way, as also in distant galaxies, the form of the dark nebulae gives the impression that structures of a few parsecs across have been deployed over a distance of a hundred parsecs or more in the direction of the spiral arm.

Although the absorption of starlight in extended, and often not sharply delimited, regions of dark nebulosity is readily noticed, it was not until 1930 that the knowledge of a *general interstellar absorption and reddening* became widespread. Such absorption has a large effect in the photometric measurement of great distances.

If a star of absolute magnitude M is at distance $r=1/p$ parsec, and if its light suffers absorption amounting to γ mag/pc on its way to us, then by generalizing our earlier formula (14.2) the apparent magnitude is given by

$$m-M=5(\log r-1)+\gamma r. \tag{24.9}$$

We call the quantity $m-M-\gamma r$ the *reduced* or *absorption-free* distance modulus. In order to bring out the significance of this important formula, in Fig. 24.8 we have illustrated the relation between the distance modulus $m-M$ and the distance r, or the reduced distance modulus, for $\gamma=0$ (zero absorption) and absorption of $\gamma=0.5$, 1.0, 2 mag/1000 pc. With absorption of 1 to 2 mag/kpc our view is practically cut off, as in a fog, at a few thousand parsecs.

R. Trumpler in 1930 gained the first well-founded ideas of the value of γ for average interstellar absorption in the galactic plane by comparing the ways in which angular diameter and luminosity fall off with distance in the case of open clusters. He was thus able to obtain a direct relation between geometric and photometric measurements of distance. Just as important was Trumpler's discovery that the absorption is always accompanied by a reddening of the starlight. Also more recent spectrophotometric investigations have confirmed that the interstellar reddening (change of colour-index) is always proportional to the interstellar absorption (change of apparent magnitude), which is to say that *the wavelength-dependence of the interstellar absorption coefficient* is everywhere the same.

Away from discernable dark clouds, on average in the galactic plane we can reckon on absorption (visual) of $\gamma\approx 0.3$ mag/kpc; if we do not exclude the dark clouds we find up to 1 to 2 mag/kpc. As to the distribution of absorbing material perpendicular to the galactic plane, i. e. the dependence of γ upon galactic latitude b^{II}, we gain a picture from E. Hubble's discovery of the "zone of avoidance" (1934).

Hubble studied the distribution over the sky of distant galaxies brighter than a given limiting magnitude. In the galactic polar caps their number per square degree $N(m)$ is almost constant. From 30°

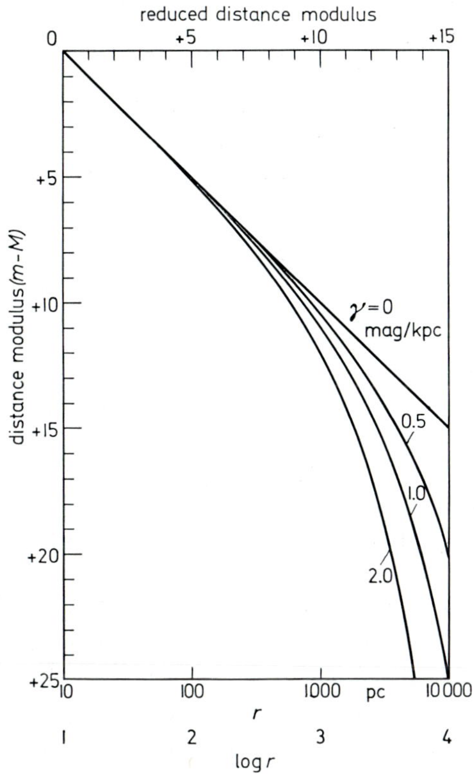

Fig. 24.8. *Relation between distance modulus m—M and distance r* (parsecs) for stars with no interstellar absorption ($\gamma = 0$) compared with that for an (assumed uniform) interstellar absorption of $\gamma = 0.5$, 1 and 2 mag/kpc respectively. The *reduced* distance modulus corresponds to the true distance, that is to $\gamma = 0$

to 40° galactic latitude $N(m)$ falls off more and more steeply towards the galactic equator, in such a way that in the vicinity of this equator there exists a zone almost devoid of galaxies. Hence Hubble concluded that the absorbing material in the Galaxy forms a flat disk, in the middle plane of which we find ourselves with the result that extragalactic objects suffer absorption about $0^{m}\!.35$ cosec b^{II}. Observations of stars in our galactic neighbourhood then show further that the entire half-value thickness of the absorbing layer in this neighbourhood is about 300 pc. As was found later, this corresponds to about the thickness of the hydrogen gas.

Spectrophotometric measurements showed that in the photographic and visual region the dependence of the interstellar absorption or of γ upon the wavelength λ is well represented by the law $\gamma \sim 1/\lambda$. Hence, as well as from the photometry of objects of known colour, we obtain as the relation on the average between, for instance, the weakening of the V-magnitude A_V and the increase of the colour index $B-V$, the so-called colour-excess $E_{B-V} = \Delta(B-V)$

$$A_V = (3.0 \pm 0.2)\, E_{B-V} \tag{24.10}$$

This relation and corresponding ones in other colour-systems are very important because, as already remarked, the space distribution of interstellar dust is so irregular that reckoning with mean values in any single case is highly uncertain. For stars (e.g. bright B stars) whose absorption-free colour index one knows from neighbouring examples, it is much better to measure the colour index and to infer the absorption from the colour excess. Thus it appears that in fact even near the galactic plane there are astonishingly transparent isolated "windows". From measurements of two colour indices, e.g. $U-B$ and $B-V$, we can indeed derive both the absorption-free colour-index and the interstellar absorption.

In 1949 W. A. Hiltner and J. S. Hall made the astonishing observation that the light from distant stars is partly linearly polarized and that the degree of polarization grows about in proportion to the interstellar reddening E_{B-V}. The electric vector of the light-waves (perpendicular to the conventional plane of polarization) vibrates preferentially parallel to the galactic plane.

How do we stand now in regard to the physical nature and origin of interstellar dust? How can we account for the wavelength-dependence of the absorption and the interstellar polarization?

Coarse particles (sand) $\gg \lambda$ absorb and scatter independently of the wavelength λ; according to Rayleigh's law, very fine particles (molecules, in particular) $\ll \lambda$ do so as λ^{-4}. Interstellar grains must therefore have diameters of the order of the wavelength of visible light, ~ 6000 Å or $0.6\,\mu$; their mass is then about 10^{-13} g. According to van de Hulst, the average density of interstellar dust amounts to about 1.3×10^{-26} g/cm³, that is only about 1 percent of all interstellar matter. The interstellar polarization indicates that the particles are anisotropic, very likely needle-shaped and partly aligned. L. Davis and J. Greenstein attribute the orientation to a galactic magnetic field of about 10^{-5} gauss, of which the lines of force are preferentially more or less along the spiral arms. The anistropic structure of the interstellar grains support the view that they should not be regarded as counterparts of interplanetary meteor-dust (powder), but according to J. Oort they should be ascribed to condensation (smoke) from the interstellar gas. Nevertheless, the latter idea has the difficulty that then the reverse process of destruction must also be exhibited, since otherwise we should not understand why at any rate not all the heavy atoms are condensed. Maybe we should picture the particles as loose hydride-flakes.

Even more difficult is the interpretation of the broad interstellar absorption bands discovered by P. W. Merrill in 1934. The strongest is about λ 4430 Å with a half-width of 26 Å. An explanation of this absorption by plasma oscillations in small metallic spheres appears to be possible. Whether it is the correct explanations remains in question.

In conclusion, we try to picture the disposition and significance of interstellar matter in the Galaxy (and similar galaxies) and to mention one or two additional points: The gas and dust form a flat disk, of which the thickness from $r=3$ kpc to 10 kpc is about 200 pc. Within this disk, the gas and dust are concentrated towards the spiral arms; along the arms they form clouds, which are further subdivided in finer structures. In general along the arms the field-lines of the interstellar magnetic field are frozen into the highly conducting plasma. The OB associations of bright blue stars and the younger galactic clusters are mostly embedded in diffuse nebulae, which we must regard as condensations of the interstellar matter. We shall see that these groups of stars have been formed out of the interstellar material. In the evolution of a galaxy, a basic problem is obviously the magneto-hydrodynamic problem of the formation of gas- or plasma-arms in which, amongst other things, the condensation of the dust occurs. Stellar associations, stellar clusters, etc., are later formed in the gas- or dust-arms. In the next chapters we shall support and extend these ideas.

25. Internal Constitution and Energy Generation of Stars

As long ago as 1913 H. N. Russell was convinced as to the significance of his diagram for the study of stellar evolution. However, the interpretation of the diagram, and therewith a theory of stellar evolution based on observation, first became possible in conjunction with the study of the internal constitution of the stars. The older work of J. H. Lane (1870), A. Ritter (1878—89), R. Emden (whose *Gaskugeln* appeared in 1907) and others could be based essentially only upon classical thermodynamics. A. S. Eddington then succeeded in combining their results with the theory of radiative equilibrium and with the Bohr theory of atomic structure, which had meanwhile been developed. His book *The internal constitution of the stars* (Cambridge, 1926) formed the prelude to the whole development of modern astrophysics. We can grasp the basic ideas of Eddington's theory with only a very modest use of mathematics:—

a) *Hydrostatic equilibrium:* In a star, consider a volume-element having base 1 cm² and height dr, where r is the distance from the centre of the star. Its mass $\varrho(r) \times 1 \times \mathrm{d}r$, where $\varrho(r)$ is the density, is under the gravitational attraction of all the mass M_r inside distance r. We have

$$M_r = \int_0^r \varrho(r)\, 4\pi r^2\, \mathrm{d}r \quad \text{or} \quad \mathrm{d}M_r/\mathrm{d}r = 4\pi r^2 \varrho(r). \tag{25.1}$$

According to the newtonian law of attraction, this mass produces at its surface a gravitational acceleration $G\, M_r/r^2$ where G is the gravitation-constant. Therefore the pressure changes over the depth of our volume element by the amount

$$-\mathrm{d}P \quad = \quad \varrho(r)\,\mathrm{d}r \;\cdot\; GM_r/r^2 \tag{25.2}$$

that is $\qquad$ force/cm² = mass/cm² × acceleration

The *hydrostatic equation* for the star is therefore

$$\mathrm{d}P/\mathrm{d}r = -\varrho(r)\, G\, M_r/r^2\,. \tag{25.3}$$

In most stars, P is practically equal to the *gas-pressure* P_{g}; the radiation-pressure P_{r} has also to be taken into account explicitly only in very hot and massive stars, so that $P = P_{\mathrm{g}} + P_{\mathrm{r}}$. Using (25.3) we can easily estimate the pressure at, for example, the centre of the Sun ($P_{\mathrm{c}\odot}$) if on the right-hand side we substitute for $\varrho(r)$ the mean density $\bar{\varrho}_\odot = 1.4$ g cm^{-3} and for M_r half the solar mass $M_\odot = 2 \times 10^{33}$ g. Then

$$P_{\mathrm{c}\odot} \approx \tfrac{1}{2}\, \bar{\varrho}_\odot G\, M_\odot / R_\odot = 1.3 \times 10^{15}\ \mathrm{dyn/cm^2} \text{ or } 1.3 \times 10^{9}\ \mathrm{atm}\,. \tag{25.4}$$

A more exact calculation shows that (25.4) holds good exactly for a homogeneous sphere, and generally, assuming only that $\varrho(r)$ increases monotonically inwards, it gives a least value for P_{c}. More appropriate models give for the Sun a value of the central pressure about a hundred times larger.

b) *Equation of state of the material.* The relation, at each position, between pressure P, density ϱ, and the temperature (absolute) T, which we must now introduce as the third parameter describing the state of the material, is expressed by the *equation of state.* Following Eddington, we start with the equation of state of an ideal gas

$$P_g = \varrho \mathcal{R} T/\mu \tag{25.5}$$

where $\mathcal{R} = 8.314 \times 10^7$ erg/degree/mole is the universal gas-constant and μ is the mean molecular weight. We may employ (25.5) so long as the interaction between neighbouring particles is sufficiently small compared with their thermal (kinetic) energy. On the Earth, we are accustomed to this no longer holding good at densities of about $\varrho \geqq 0.5$ to 1 g/cm³. In stars this limit is raised a great deal by *ionization.* In particular, the commonest elements H and He are completely ionized in all stars below a relatively small depth. A more exact calculation shows that in such a plasma the interaction per particle is smaller than the thermal energy kT, where k is the Boltzmann constant, so long as the number of electrons per cm³ $N_e < 10^8\, T^3$.

The *mean molecular weight* μ for complete ionization is equal to the atomic weight divided by the number of all the particles, i. e. nuclei plus electrons. Thus we have for hydrogen $\mu_H = \frac{1}{2}$, for helium $\mu_{He} = \frac{4}{3}$, and for all heavier elements $\mu \approx 2$. We can readily calculate μ for any given mixture of chemical elements.

Using equation (25.5) we can estimate the mean temperature inside the Sun taking $\overline{P}_g \approx \frac{1}{2} P_{c\odot}$ according to (25.4), $\varrho = \bar{\varrho}_\odot$ and $\mu = \mu_H = 0.5$, obtaining

$$\overline{T}_\odot \approx 6 \times 10^6\ {}^\circ\mathrm{K}. \tag{25.6}$$

At densities $\varrho > 10^2$ to 10^4 g/cm³ (depending on the temperature) the nature of the equation of state for the material in stellar interiors first changes because of the onset of (gas-)*degeneracy.* On this basis, R. H. Fowler was able in 1926 to make comprehensible the enormous densities in the interiors of *white dwarf stars* (and extremely red dwarfs of later types). The Pauli principle provides the physical basis; this requires that — in the first place in an atom — no simple quantum state may be occupied by more than one electron. Then E. Fermi and P. A. M. Dirac applied this principle to the statistics:—

In a degenerate gas *all* electronic quantum states are fully occupied up to a certain maximum energy. Just as in the electron shell of a heavy atom, in a completely degenerate gas the temperature concept becomes meaningless; the pressure P again depends only on the density ϱ, and in fact we have

$$P = 9.91 \times 10^{12} (\varrho/\mu_E)^{\frac{5}{3}}. \tag{25.7}$$

Here μ_E means the mass (in atomic-weight units) which in total belongs to one electron; thus for hydrogen $\mu_E = 1$, for helium $\mu_E = 2$, and so on.

In the case of still higher densities, say $\varrho > 2 \times 10^6$ g/cm³, the maximal energy of the electrons is higher than their rest-energy $m c^2$. In that case we must invoke the theory of special relativity. The maximal energy of the electrons and the pressure again depend only on the density, and we obtain as the equation of state of the *relativistically degenerate gas*

$$P = 1.231 \times 10^{15} (\varrho/\mu_E)^{\frac{4}{3}}. \tag{25.8}$$

c) *Temperature distribution and energy transfer in stellar interiors.* In order to calculate the temperature distribution $T(r)$ in the interior of the star, we must examine the nature of the energy-transport. Inefficient transfer leads to a steep temperature gradient; efficient transfer leads to a low gradient. (Let anyone who questions this, use his fingers to hold, with one end in a flame, first a matchstick and then a metal nail!)

Following K. Schwarzschild's investigations of the solar atmosphere, A. S. Eddington first considered *energy-transfer by radiation*, i. e. radiative equilibrium (see Chapter 18, also in regard to notation).

From the intensity of the radiation I we calculate the flux of radiation πF; we multiply by $\cos\vartheta$ and integrate over all directions. Denoting now the element of optical depth corresponding to mean mass-absorption coefficient $\varkappa$ by $d\tau = -\varkappa\varrho\, dr$ we obtain from (18.4) for the net flux of the total radiation

$$\pi F = -\int_0^\pi \frac{dI}{\varkappa \varrho\, dr} \cos^2\vartheta \, 2\pi \sin\vartheta \, d\vartheta \,. \tag{25.9}$$

In the stellar interior the radiation field is almost isotropic, so that on the right we can take out the integration over ϑ; it is

$$\int_0^\pi \cos^2\vartheta \, 2\pi \sin\vartheta \, d\vartheta = 4\pi/3 \,.$$

Further, we can here at once write for I the value given by the Stefan-Boltzmann law as a function of T, that is

$$\pi I = \sigma T^4. \tag{25.10}$$

We often write the radiation constant σ in the form $\sigma = a\,c/4$, where c is the light-speed; with a so defined, the energy-density of the black-body radiation field is then simply $a\,T^4$ erg cm^{-3}.

Thus we first obtain, when written in full,

$$\pi F = -\frac{4}{3}\pi \frac{d}{\varkappa \varrho\, dr}\left(\frac{a\,c}{4\pi} T^4\right) \tag{25.11}$$

and the total radiative energy that flows outwards per second across the spherical surface of radius r becomes $L_r = 4\pi r^2 \cdot \pi F$ or

$$L_r = -4\pi r^2 \frac{4}{3} a\,c \frac{T^3}{\varkappa \varrho} \frac{dT}{dr}\,. \tag{25.12}$$

At the surface of the star $r = R$ the quantity L_r becomes the observable luminosity L; see also (14.5). At temperatures of about 6×10^6 °K the maximum of the Planck radiation curve $B_\lambda(T)$ is from (11.23) around $\lambda_{\max} = 5$ Å, that is in the X-ray region. Here the absorption coefficient is determined by the bound-free and free-free transitions of the atomic states that have not yet been fully "ionized away". In the mixture of elements such as that in the Sun and in stars of population I and the disk-population, these are high stages of ionization of the more abundant heavy elements like O, Ne, ... In "metal-poor" subdwarfs we must also take account of the contributions made by H and He.

Besides energy-transport by radiation, under the conditions discussed on page 187 in the case of the Sun, for example, a hydrogen- and helium-convection zone may also occur in the stellar interior. Other convection zones may arise also in connexion with the nuclear generation of energy. According to L. Biermann, in all such convection zones (apart from boundary regions) energy-transport by convection (the ascent of hot material and the descent of cooled material) exceeds by far that by radiation. The connexion between the temperature T and the pressure P is then determined by the familiar adiabatic relation

$$T \sim P^{1-1/\gamma} \tag{25.13}$$

where $\gamma = c_p/c_v$ is the ratio of the specific heats at constant pressure and at constant volume. By logarithmic differentiation with respect to r we derive

$$\frac{1}{T}\frac{dT}{dr} = \left(1 - \frac{1}{\gamma}\right)\frac{1}{P}\frac{dP}{dr} \tag{25.14}$$

and thence the temperature-gradient in a convection-zone

$$\frac{dT}{dr} = \left(1 - \frac{1}{\gamma}\right)\frac{T}{P}\frac{dP}{dr}\,. \tag{25.15}$$

As we have already explained in Chapter 20, the results of the calculation of convective energy-transfer are even today still very unsatisfactory in quantitative respects.

As we shall here pursue no further, in a degenerate gas energy transport takes place by *heat-conduction* by free electrons.

d) *Energy generation in stellar interiors by nuclear reactions.* Making use of the knowledge that, since the formation of the Earth some 4.5×10^9 years ago, the luminosity of the Sun has not changed significantly, J. Perrin and A. S. Eddington realized even in 1919/20 that the hitherto considered mechanical or radioactive energy sources did not nearly suffice for supplying the radiation of the Sun. So they came to conjecture that in the interiors of the Sun and stars *nuclear energy* must be generated by the conversion or, as we also say, the "burning" of hydrogen into helium.

Using Einstein's relation (1905)

$$\Delta E = \Delta m\, c^2 \tag{25.16}$$

the total energy ΔE liberated by the union of four hydrogen atoms to give one helium atom, $4\,H^1 \rightarrow He^4$, can easily be calculated from the mass difference Δm between the initial and final states. The mass of four hydrogen atoms amounts to 4×1.008145 units of atomic weight (each equal to 1.660×10^{-24} g), while that of the helium atom is 4.00387 units. The difference corresponds to

$$0.02871 \text{ units of atomic weight} \sim 4.768 \times 10^{-26}\,\text{g} \sim 4.288 \times 10^{-5}\,\text{erg} \sim 26.72\,\text{MeV}. \tag{25.17}$$

If we suppose that, for example, one solar mass of pure hydrogen were converted into helium, we should obtain a quantity of energy of 1.27×10^{52} erg, which could supply the present rate of energy output of the Sun $L_\odot$ for a life-span of 105 billion years (billion $= 10^9$).

The rapid advance of nuclear physics in the 1930s enabled H. Bethe and others in 1938 to work out the nuclear reactions that are possible at temperatures of about 10^6 to 10^8 degrees in solar material and in other mixtures of chemical elements. Experimental investigations, particularly those of W. A. Fowler, of reaction cross-sections at low proton-energies have contributed very essentially to our knowledge of energy generation in stars. We collect the important reactions in accordance with the usual scheme

$$\text{initial nucleus (reacts with \ldots, gives out \ldots) final nucleus.} \tag{25.18}$$

We use the notation

$$\begin{array}{llll} \mathrm{p} & \text{proton} & \alpha & \mathrm{He}^{++}\text{ particle} \\ \beta^+ & \text{positron} & \mathrm{e}^- & \text{electron} \\ \gamma & \text{radiation-quantum} & \nu & \text{neutrino}\,. \end{array} \tag{25.19}$$

By combination with a thermal electron e^-, an emitted positron β^+ immediately produces two γ-quanta.

Because of their minute cross-sections, neutrinos escape from stellar interiors without suffering collisions. We shall therefore write on the right-hand side following the equations of each reaction the "usable" energy liberated and in brackets the energy escaping in the form of neutrino-radiation. The slowest reaction, which determines the time required for the reaction-chain, is printed in heavy type.

We begin with the proton-proton or pp chain:

$$\mathbf{H^1\,(p,\,\boldsymbol{\beta}^+\boldsymbol{\nu})\,D^2} \rightarrow \mathrm{D^2\,(p,\gamma)\,He^3} \begin{array}{l} \nearrow \mathrm{He^3\,(He^3, 2\,p)\,He^4} \quad \ldots \\ \qquad\qquad\qquad\quad \nearrow \quad \ldots \\ \searrow \mathrm{He^3\,(\alpha,\gamma)\,Be^7} \searrow \quad \ldots \end{array}$$

$$\left.\begin{array}{ll} \ldots\ldots\ldots\ldots\ldots\ldots\ldots\ldots & 26.21 + (0.51)\ \mathrm{MeV} \\ \ldots\ \mathrm{Be^7\,(e^-,\nu)\,Li^7 \rightarrow Li^7\,(p,\,2\,\alpha\,\gamma)\,2\,He^4} & 25.92 + (0.80)\ \mathrm{MeV} \\ \ldots\ \mathrm{Be^7\,(p,\gamma)\,Be^8 \rightarrow Be^8 + \beta^+ + \nu \rightarrow 2\,He^4} & 19.5\ \ + (7.2)\ \ \mathrm{MeV}. \end{array}\right\} \tag{25.20}$$

Which one of the branchings predominates depends upon the temperature and the composition of the material.

A second possibility for the conversion of hydrogen into helium is offered by a reaction-cycle which H. Bethe and C. F. v. Weizsäcker had studied earlier (1938). In it the elements C, N, O take part and essentially determine the speed of the process, but at the end of the process they are quantitatively reinstated.

CNO cycle

$$\begin{array}{lcc} & \nearrow \mathrm{N^{15}\,(p,\alpha)\,C^{12}} & \mathrm{N^{15}\,(p,\gamma)\,O^{16}} \\ & \downarrow & \downarrow \\ \mathrm{O^{15}\,(\beta^+\nu)\,N^{15}} \nearrow & \mathrm{C^{12}\,(p,\gamma)\,N^{13}} & \mathrm{O^{16}\,(p,\gamma)\,F^{17}} \\ \uparrow & \downarrow & \downarrow \\ \mathbf{N^{14}\,(p,\gamma)\,O^{15}} \nwarrow & \mathrm{N^{13}\,(\beta^+\nu)\,C^{13}} & \mathrm{F^{17}(\beta^+\nu)\,O^{17}} \\ & \downarrow & \downarrow \\ & \searrow \mathrm{C^{13}\,(p,\gamma)\,N^{14}} & \mathrm{O^{17}\,(p,\alpha)\,N^{14}} \\ & \text{main cycle} & \text{auxiliary cycle} \\ & 25.03 + (1.69)\ \mathrm{MeV} & 24.74 + (1.98)\ \mathrm{MeV} \end{array} \tag{25.21/22}$$

The energy generation comes principally from the cycle written on the left. In the Sun, for example, the cycle on the right is performed about 2200 times less often.

As we must remark here, in a steady state the relative abundances of the participating isotopes are determined as functions of temperature and pressure as in a state of "radioactive equilibrium".

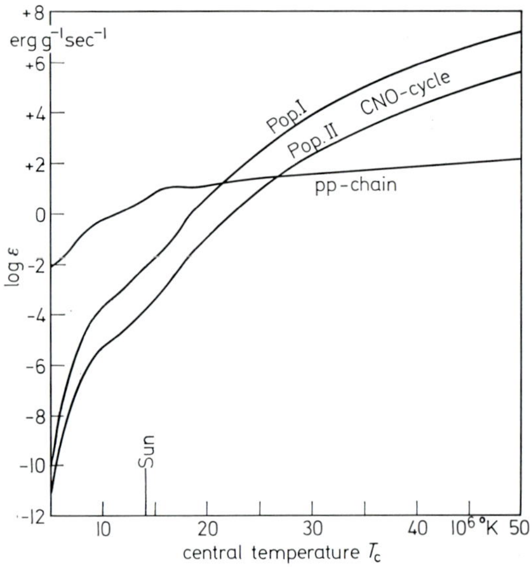

Fig. 25.1. Rate of energy generation ε in stars of various central temperatures T_c, for a central density $\varrho_c = 100$ g cm^{-3}. The pp-chain dominates in cooler stars such as the Sun, and the CNO-cycle dominates in the hotter stars. (After W. A. Fowler, 1959)

Again following W. A. Fowler, in Fig. 25.1 the mean energy-generation in erg g^{-1} sec^{-1} by "hydrogen-burning" is shown on a logarithmic scale for stars with central temperatures T_c from 5 to 50×10^6 °K and, somewhat arbitrarily, a central density $\varrho_c = 100$ g cm^{-3}. The age of the stars is assumed to be 4.5×10^9 years, so that the curves apply to the older population I and disk-stars, and to younger population II stars. The composition by mass assumed for the former is

$$\begin{array}{llll} \text{H} & \text{C} & \text{N} & \text{O} \\ 50 & 0.3 & 0.1 & 1.2 \text{ percent (the remainder is He);} \end{array} \quad (25.23)$$

for population II the abundance of CNO is reduced by a factor 25. The most important result is: Cool stars with central temperatures up to $T_c = 21 \times 10^6$ °K for population I or 27×10^6 °K for population II (in particular also for the Sun with $T_c \approx 13 \times 10^6$ °K) produce their energy by the pp-chain; hotter stars (in the upper part of the main sequence) produce it by the CNO cycle.

In the interior of a star, suppose that a considerable part of the hydrogen has been used up and that, as a result of contraction, the temperature has risen to more than 10^8 °K. As E. J. Öpik and E. E. Salpeter remarked in 1951/52, *helium-burning* then sets in first to give C^{12} in accordance with the 3α-process:

This begins with the slightly endothermic combination of two helium nuclei

$$He^4 + He^4 + 95 \text{ keV} = Be^8 + \gamma . \qquad (25.24)$$

The Be^8 is present with a certainly minute concentration and is then further built up by the reaction

$$Be^8 (\alpha, \gamma)\, C^{12} \text{ with an energy-release of 7.3 MeV per } C^{12} \text{ nucleus.} \qquad (25.25)$$

Reckoned per He-atom, this burning yields only 2.46 MeV, that is about 10 percent of the energy released in forming the atom. Starting from here, the following nuclei, having mass-numbers that are multiples of four, can be built up by further (α, γ)-reactions with comparable energy-production:—

$$C^{12} (\alpha, \gamma)\, O^{16} \quad 7.15 \text{ MeV};$$
$$O^{16} (\alpha, \gamma)\, Ne^{20} \quad 4.75 \text{ MeV}; \quad Ne^{20} (\alpha, \gamma)\, Mg^{24} \quad 9.31 \text{ MeV}. \qquad (25.26)$$

All these reactions are interesting, not only on account of the energy generated, but also in regard to the production of heavy elements. In particular, Ne^{20} can lead to the reaction-sequence

$$Ne^{20} (p, \gamma)\, Na^{21} \rightarrow Na^{21} (\beta^+ \nu)\, Ne^{21} \rightarrow Ne^{21} (\alpha, n)\, Mg^{24}. \qquad (25.27)$$

The last reaction appears suitable for providing neutrons (n) for the synthesis of heavy elements, which, because of the strong coulomb fields, can obviously not be achieved under any circumstances by charged particles.

The application to the theory of the internal constitution of the stars is basically entirely simple: Let ε be the energy generated per second per gramme of stellar matter by all the nuclear reactions proceeding at the values of the temperature T and density ϱ concerned. Therefore in a spherical shell r to $r + dr$ energy $\varrho\, \varepsilon\, 4\, \pi\, r^2\, dr$ erg/sec is generated, and the energy-flux L_r [see (25.2)] increases in accordance with the equation

$$dL_r/dr = 4\pi r^2 \varrho \varepsilon. \tag{25.28}$$

e) *Summary: the fundamental equations of the theory of the internal constitution of the stars and general consequences.* For a better grasp, we bring together the equations of the theory of the internal constitution of the stars from sections a) to d); today their solution is normally got with the use of a large electronic computer. To these are added the equation of state [section b)] and the (in practice very complicated) equations that relate the parameters of the material, ε and $\varkappa$ or γ, to two of the state-parameters P, T, ϱ. As is important for what follows, all these relations depend essentially upon the chemical composition of the matter.

Hydrostatic equilibrium under the influence of self-gravitation (25.3)	$dP/dr = -\varrho G M_r/r^2$
and (25.1)	$dM_r/dr = 4\pi r^2 \varrho$
energy-generation (25.28)	$dL_r/dr = 4\pi r^2 \varrho \varepsilon$
energy-transfer: radiation (25.12)	$\frac{dT}{dr} = -\frac{3}{4ac}\frac{\varkappa\varrho}{T^3}\frac{L_r}{4\pi r^2}$
energy-transfer: convection adiabatic (25.15)	$\frac{dT}{dr} = \left(1 - \frac{1}{\gamma}\right)\frac{T}{P}\frac{dP}{dr}$.

(25.29 a—e)

Finally, our problem is fully determined by the boundary conditions:

α) In the centre of the star, we must obviously have

$$\text{for } r = 0: \quad M_r = 0 \text{ and } L_r = 0 \tag{25.30}$$

β) At the surface of the star, the equations for the stellar interior must in principle pass over into those of theory of stellar atmospheres already discussed. So long as we are interested only in the internal structure, we can use the above equations up to $r = R$, say, where $T \to 0$. We can make some general consequences of the theory intuitively clear; obviously we can also deduce them by formal calculations.

Consider a mass of gas M furnished with given energy-sources L. This system, which at first forms an indefinite picture as regards its arrangement in space, we now consolidate in imagination into a star with mass M and luminosity L. Provided a stable configuration is indeed possible, this will adjust itself to a definite radius R. On the other hand, L is related to the radius R and the effective temperature T_e by

$$L = 4\pi R^2 \sigma T_e^4 \tag{25.31}$$

$$\text{luminosity} = \text{stellar surface} \times \text{total flux.}$$

So the effective temperature of our star is fixed. Thus for stars having the same structure and composition (which we must not overlook), so-called homologous stars, a unique relation between mass M, luminosity L, radius R (or effective temperature T_e) is satisfied

$$\varphi(M, L, T_e) = 0. \tag{25.32}$$

A. S. Eddington discovered such a relation in 1924. According to his calculations, the dependence of the function φ on T_e is so weak that he was able to speak simply of the *mass-luminosity relation*. At first, the agreement with observation seemed to be really good; later many exceptions were found, but in the light of the general theory these are not at all unexpected.

Further, we take into account the fact that, from the theory of nuclear energy-generation, ε is determined as a function of the state-parameters (for example, T and ϱ). This produces a further relation between the three quantities M, L, T_e. For stars in a steady state, having the same structure, we have therefore an equation of the form

$$\Phi(L, T_e) = 0. \tag{25.33}$$

This implies that in the HR-diagram, or in the colour-magnitude diagram, these stars lie on a definite line. This assertion is often called the Russell-Vogt theorem. In fact, we shall become acquainted with a line of this kind in the colour-magnitude diagram which forms the so-called *zero-age main sequence*. On the other hand, the presence of red giants and of supergiants then shows that already at least one further parameter comes into play. As we shall see, this is the chemical composition of the star, which changes with the age.

26. Colour-Magnitude Diagrams of Galactic and Globular Clusters. Stellar Evolution and Abundances of the Elements

Our current ideas about the origin, evolution and fate of stars originate in the study of colour-magnitude diagrams of stellar clusters. Here we have groups of equi-distant stars. From the magnitudes and colours measured photoelectrically with high accuracy, the subtraction of a common distance modulus and a single correction for interstellar absorption and reddening leads to the values of the (most commonly employed)

absolute magnitude M_V

and the true

colour index $B - V$.

As we saw, the colour-magnitude diagram is basically equivalent to the HR diagram. However, while for very faint stars, down to about 21^m, it is still possible accurately to measure colour indices, it is no longer possible to photograph spectra that can be classified.

To fix the ideas we once again briefly summarize the important properties of both sorts of cluster:—

1. Galactic or open clusters. Their distances are determined by comparison of proper motions μ and radial velocities V (see (23.7)), or by using the method of spectroscopic parallaxes, or by photometric comparison of stars in the lower part of the main sequence with corresponding stars of a standard cluster (usually the Hyades) or with stars of our neighbourhood. Galactic clusters (Fig. 23.5) contain from a few dozen to several hundred stars, their diameters are of the order of 1.5 to 20 pc; they are always found (Fig. 23.7) in or near the spiral arms of the Galaxy and, like other objects belonging to population I, they move in nearly circular paths about the galactic centre. We know some 400 galactic clusters; taking account of regions of the Galaxy that are very remote or obscured by dark clouds, we estimate the total number to be about 20 000. Closely related to the galactic clusters are the looser moving clusters, as well as the OB- and T-associations mentioned in Chapter 23.

Loose clusters, in which the star-density is not much greater than in the region round about, will disperse even on kinematic grounds after about one galactic revolution ($\sim 2.5 \times 10^8$ years). Also more compact clusters will be gradually dissipated by the gravitational fields of gas-and star-clouds. All the same, detailed calculations allow the Pleiades, for example, which is still a fairly compact cluster, a life of some 10^9 years.

2. *Globular clusters.* In our galactic system we know something over one hundred globular clusters. We have already discussed at length the photometric determination of their distances with the help of cluster-variables or RR Lyrae stars. A typical globular cluster contains within a region of about 40 pc diameter several hundred

thousand stars, so the mean star-density is about ten times larger than in galactic clusters. Towards the centre of the cluster the star-density increases so strongly that the night sky must be really bright! The absolute magnitude of a globular cluster is around -8^M. Analogously to known results in the kinetic theory of gases, we can estimate the total mass of a globular cluster from the dispersion in the radial velocities of its stars. In this way we get values of a few times $10^5 M_\odot$. Since the globular clusters travel round the galactic centre in elongated elliptic paths, they pass through the galactic plane about every 10^8 years. Because of their compact structure, the resulting "jolt" has no significant effect.

After these preliminary remarks, we turn to the colour-magnitude diagrams, first of galactic and then of globular clusters.

3. *Colour-magnitude diagrams of galactic clusters.* We mention briefly the pioneering work of R. Trumpler in the 1930s on the HR diagrams etc. of galactic clusters. In the following we rely upon the more recent investigations made photoelectrically, or at anyrate using

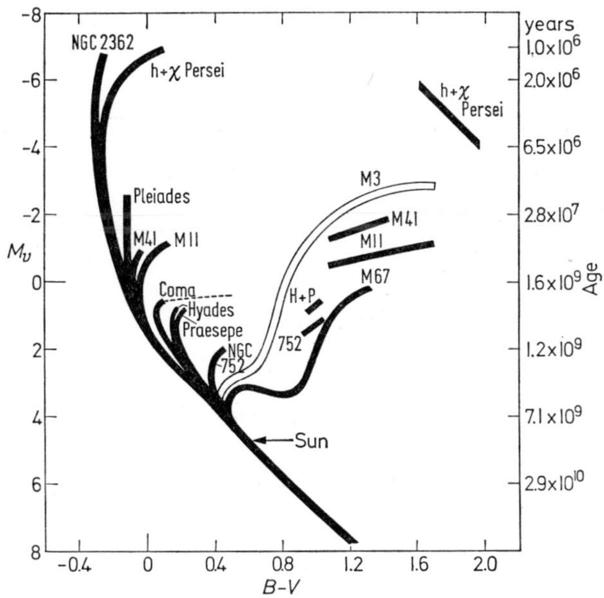

Fig. 26.1. *Colour-magnitude diagrams of galactic cluters* (after A. R. Sandage, 1956): visual absolute magnitude M_V plotted against colour index $B-V$. The age of a cluster may be read off from the vertical scale on the right by noting the position of the "knee", i. e. the point where the main sequence turns off to the right. The globular cluster M 3, included for comparison, is approximately as old as the oldest galactic clusters (e. g. M 67)

photoelectric luminosity-scales, by H. L. Johnson, W. W. Morgan, A. R. Sandage, M. Walker and others. In deciding if any particular star is a member of a nearby cluster, we are guided by proper-motions and maybe radial velocities; progress is here naturally not so rapid.

Following A. R. Sandage (1956) in Fig. 26.1 we show a collection of the colour-magnitude diagrams (M_V versus $B-V$) of the galactic clusters NGC 2362, h + χ Persei, Pleiades, M 41, M 11, Coma Berenices cluster, Hyades, Praesepe, NGC 752, M 67 and, for later comparison, the globular cluster M 3. The lower parts of the main sequences (up to about the Sun G 2) can be brought into coincidence without any forcing. On the other hand, higher up sooner (in h + χ Persei at the O and B stars) or later (in Praesepe at about the A stars) the sequence turns off to the right. At about the absolute magnitude of the turn-off (the so-called "knee"), we find on the right-hand side at large (positive) $B-V$ some red giant stars. In M 67 and similar clusters the passage from the main-sequence to the red-giant branch takes place continuously. For comparison with the rather schematic presentation in Fig. 26.1, in Fig. 26.2, 26.3 we show the actual observational material for Praesepe (H. L. Johnson 1952) and M 67 (Murray,

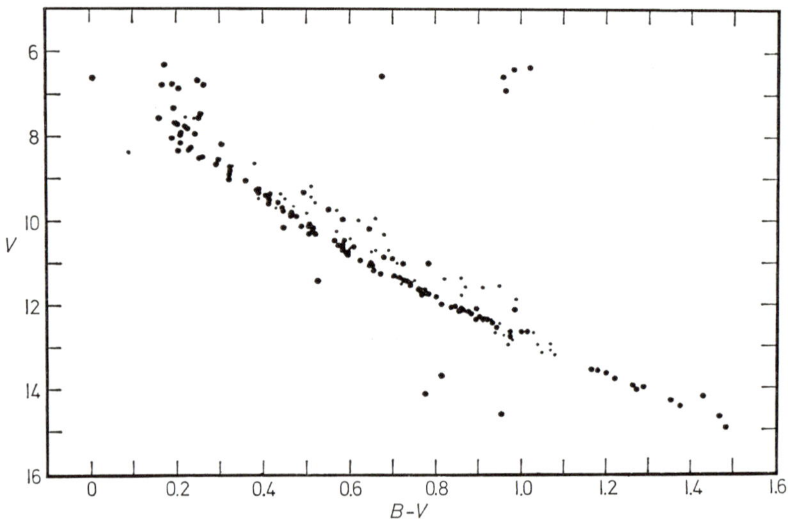

Fig. 26.2. *Colour-magnitude diagram of Praesepe:* apparent magnitude V plotted against B—V. Distance modulus for this cluster is $6^{m}.2 \pm 0.1$. The stars lying about 1 magnitude above the main sequence are most probably binaries. The intrinsic scatter of magnitudes on the main sequence is less than $\pm 0^{m}.03$

Corben and Allchorn 1965). We shall call attention to further details in connexion with the theory of stellar evolution.

4. *Colour-magnitude diagrams of globular clusters.* The structure of the colour-magnitude diagrams of globular clusters remained unclear until 1952 when, under the leadership of W. Baade of the Mt. Wilson and Palomar Observatories, a group of young astronomers A. R. Sandage, H. C. Arp, W. A. Baum and others set about determining the main sequence — which even in favourable cases begins somewhere in the region 19^m to 21^m. In this way, a comparison with stars of our neighbourhood and with those of galactic clusters was possible for the first time. Fig. 26.4 (after A. R. Sandage 1953) shows for example the colour-magnitude diagram of Messier 3. To the *main sequence* from the faintest magnitude $m_{pv} = 21.5^m$ to 19^m, is joined on the upper

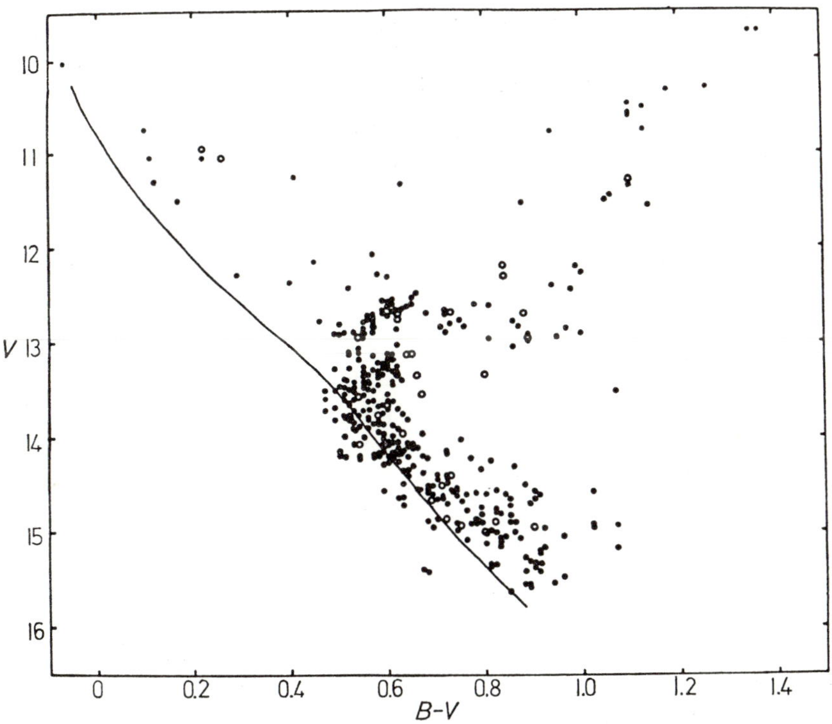

Fig. 26.3. *Colour-magnitude diagram of the old galactic cluster M 67* (Royal Greenwich Observatory, 1964). Only stars with full-weight proper motions have been included. The apparent distance modulus is $9^m.76$. The solid line shows the standard (zero-age) main sequence. The "blue stragglers" at the upper left all belong to the central region of the cluster ($r < 6'.7$)

right-hand side a branch with *red giants* and then *super giants.* From these latter there runs lower down to the left the *horizontal branch* extending to blue stars to the left of the main sequence (for, say, our

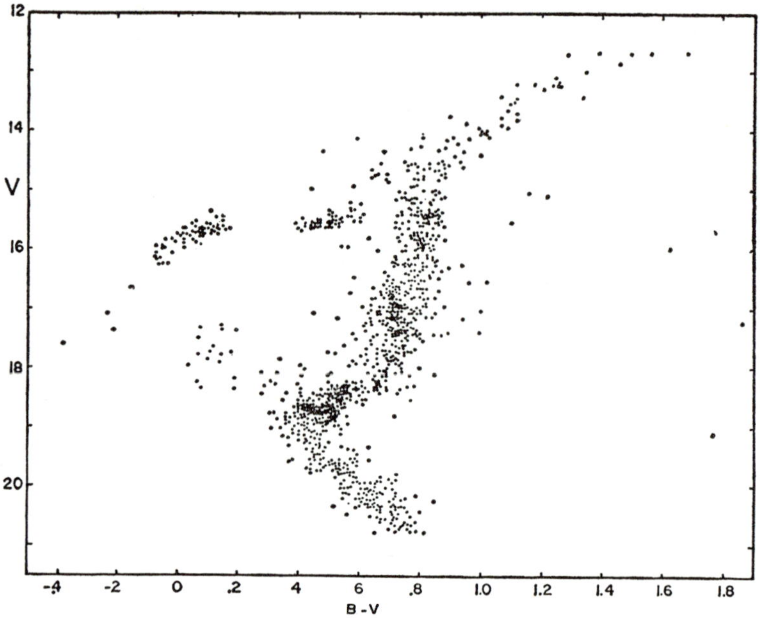

Fig. 26.4. *Colour magnitude diagram of the globular cluster M 3.* Distance modulus is $15^{\mathrm{m}}.7$. The cluster variables occur in the gap in the horizontal branch at $V \approx 15^{\mathrm{m}}.7$, $0.2 < B-V < 0.4$

neighbourhood). In the interval of colour-indices $B-V=0.2$ to 0.4, there is a well-defined gap in the horizontal branch in which the *cluster-variables* are found (not plotted in Fig. 26.4). Also the colour magnitude diagrams of other globular clusters show that *all* stars in this interval are variable. In Fig. 26.1 we have shown the colour-magnitude diagram of M 3 for comparison with galactic clusters (the horizontal branch being omitted). Here again we shall discuss further details in connexion with theoretical results.

5. *Nuclear evolution of the stars. Interpretation of the colour-magnitude diagrams of galactic and globular clusters. Formation of the chemical elements.* After these preliminaries we shall first consider, as an example, the energy-balance of the Sun. Its central temperature (calculated according to Eddington's theory) $T_c = 13 \times 10^6$ °K has obviously adjusted itself (see Fig. 25.1) so that the pp-process is responsible for the energy-generation.

Provisionally treating the Sun's material as homogeneous, its mass $\mathfrak{M}_\odot = 1.983 \times 10^{33}$ g consists of about 60 percent hydrogen. The complete conversion of this into helium by the pp-process (25.20) would produce 0.71×10^{52} erg. At its present luminosity $L_\odot = 3.84 \times 10^{33}$ erg/sec, the Sun would therefore consume 10 percent of its hydrogen — which would mean a noticeable change in its properties — in 6×10^9 years. Since the time when the Earth acquired its solid crust 4.5×10^9 years ago, the Sun can in fact have scarcely changed.

How is it then with the energy-balance of other main-sequence stars? Their central temperatures T_c rise from lower values at the cool end of the main sequence up to about 35×10^6 °K in B0 stars etc. Somewhat above the Sun, therefore, the CNO cycle takes over the energy-generation (Fig. 25.1), without any essential change in its efficiency (equation (25.20) (25.21)). From known values of the masses $\mathfrak{M}/\mathfrak{M}_\odot$ and luminosities $L/L_\odot$ (i. e. the energy-production) of main-sequence stars we now easily calculate the time in which 10 percent of the hydrogen is consumed, which we call for short the evolution-time t_E (Table 26.1)

$$t_E = 6 \times 10^9 \, \frac{\mathfrak{M}/\mathfrak{M}_\odot}{L/L_\odot} \text{ years.} \tag{26.1}$$

Table 26.1. *Stars of the main sequence and their evolution-times*

Spectral type	Effective temperature T_e	Mass $\mathfrak{M}/\mathfrak{M}_\odot$	Luminosity $L/L_\odot$	Evolution-time t_E in years
O 7.5	38 000°	25	80 000	2×10^6
B 0	33 000°	16	10 000	1×10^7
B 5	17 000°	6	600	6×10^7
A 0	9 500°	3	60	3×10^8
F 0	6 900°	1.5	6	1.5×10^9
G 0	5 800°	1	1	6×10^9
K 0	4 800°	0.8	0.4	12×10^9

Since their time of formation which (to anticipate later results) we cannot put earlier than $\sim 10 \times 10^9$ years ago, main sequence stars below G0 have used up only a small fraction of their hydrogen. On the other hand, hot stars of early spectral types "burn" away their hydrogen so quickly that they can have come into existence only a relatively short time ago, of the order t_E. The age of the O and B stars is indeed appreciably shorter than the time of revolution of the

Galaxy in our neighbourhood ($\sim 2.5 \times 10^8$ years); such stars must therefore have been formed in their present surroundings. Before we investigate further the origin of the stars, we first consider their evolution away from the main sequence.

The course of this evolution depends crucially upon whether the material, that is changed by nuclear processes in the stellar interior, mixes with the rest of the material, or whether it remains in position or remains within the relevant convection zone if one is present. F. Hoyle and M. Schwarzschild were probably the first to show in 1955 that only the latter view leads to an acceptable theory of stellar evolution and also can be made dynamically plausible.

We can illustrate what happens in detail by the example of a star of 5 solar masses (R. Kippenhahn and others 1965). Let this begin its evolution as a fully-mixed main sequence star of type about B5 belonging to population I, i. e. with chemical composition (by mass)

$$\text{hydrogen } X: \text{ helium } Y: \text{ heavy elements } Z = 0.602 : 0.354 : 0.044. \tag{26.2}$$

We show its further evolutionary track in Fig. 26.5 in a theoretical colour-magnitude diagram taking the effective temperature T_e as abscissa and the luminosity $L/L_\odot$ as ordinate and using logarithmic

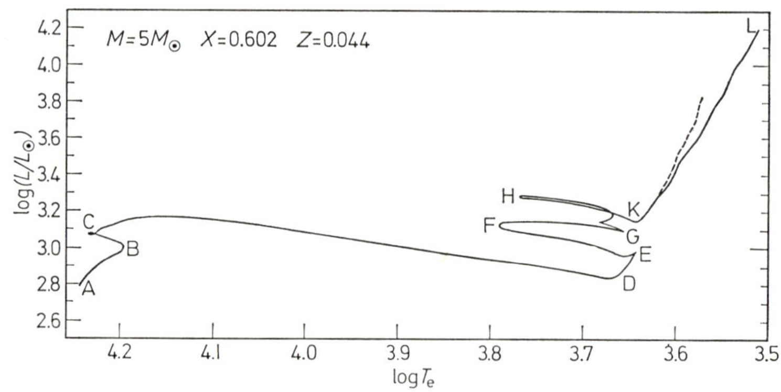

Fig. 26.5. *Evolutionary track* of a star of 5 solar masses in the theoretical colour-magnitude diagram

scales. In Fig. 26.6 are shown the changes in the stellar interior following the temporal sequence $A \to B \to \ldots K \to L$ using a time-scale which is different in different parts. Instead of the distance r from the centre of the star, $\mathfrak{M}_r/\mathfrak{M}$ from (25.1) is plotted in order to show what fraction of the stellar mass is in action at the moment. The homogeneous starting-model has effective temperature $T_e = 17\,500$, radius $2.58\,R_\odot$ and absolute bolometric magnitude $M_{bol} = -2.24$. At the

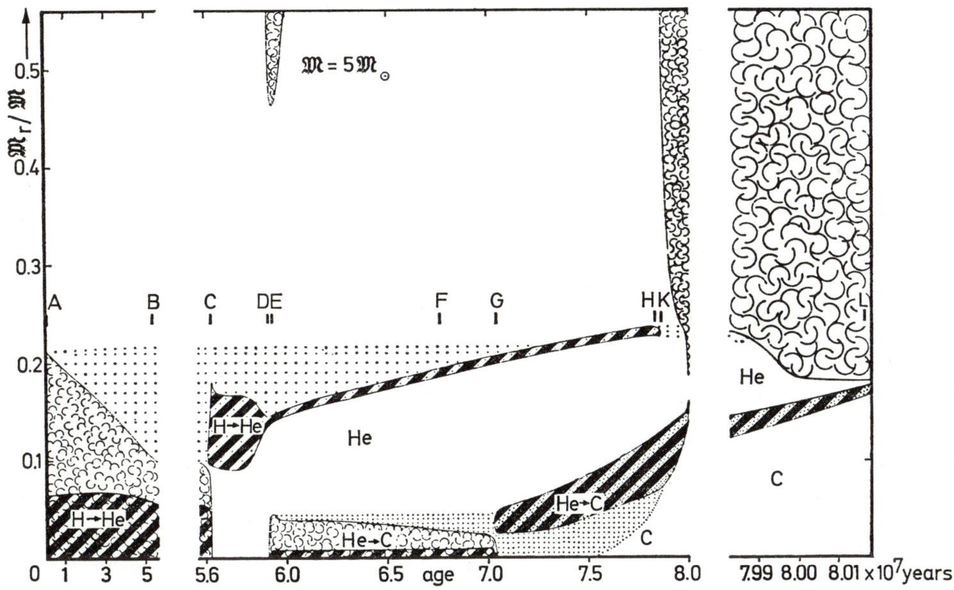

Fig. 26.6. *Time variation of conditions in a stellar interior.* The age given on the abscissa refers to the time which has elapsed since the star left the main sequence. Letters *A* to *L* refer to the corresponding letters in Fig. 26.5 which label the evolutionary track. The ordinate $\mathfrak{M}_r/\mathfrak{M}$ is the fraction of the mass inside a radial distance *r*. The "cloudy" areas denote convective zones; hatched areas denote regions where nuclear energy generation ε exceeds 10^3 erg g^{-1} s^{-1}; dotted areas denote regions where the H (or He) content decreases towards the centre. (After R. Kippenhahn, H.-C. Thomas and A. Weigert, 1965)

centre the temperature is $T_c = 26.4 \times 10^6$ °K and the pressure is given by $\log P_c = 16.74$ or $P_c = 5.5 \times 10^{10}$ atmospheres. Close to the centre we have a hydrogen-burning region where hydrogen is burnt to give helium by the operation of the CNO cycle. To this hydrogen-burning region is joined a convection zone within which the reaction products become mixed. This phase of the evolution (A→B→C) lasts 5.6×10^7 years corresponding approximately to our estimated evolution-time t_E (Table 26.1). When the core is burnt out, for a short time (C→D→E $= 0.3 \times 10^7$ years) there is set up a hydrogen-burning zone in the form of a shell. Then at E there arises at the centre a $3\,He^4 \rightarrow C^{12}$ burning region in which at central temperatures now from

$T_c \sim 130$ to 180×10^6 °K the 3α process ((25.24), (25.25)) takes over the energy-production. When the helium-core is also burnt out, a helium-burning zone in the form of a shell is set up. However, from E to K a thin hydrogen-burning shell which is steadily progressing outwards also continues to contribute appreciably to the energy-generation. From K the star rapidly assumes the status of a red supergiant. All the later stages of evolution during which the star with $T_e \approx 6500°$ several times traverses the region of cepheid-instability are run through relatively quickly. To the original 5.6×10^7 years spent in the immediate vicinity of the main sequence (A $\rightarrow$ C) there has to be added only another 2.4×10^7 years before arriving at the last red-giant state that can be worked out.

The progression of nuclear evolution — in common with stars of other masses — we can summarize in the following way:

An originally homogeneous star forms at its centre a hydrogen-burning region in which the energy generation proceeds, in the case of the higher central temperatures (large stellar masses) by means of the CNO cycle, in the case of lower central temperatures (small stellar masses) according to the pp-process. In the T_e, L or colour-magnitude-diagram the still homogeneous star is located upon a line which we term the standard main sequence, or less accurately the zero-age main sequence (ZAMS). We shall return to the question of its empirical determination. The stars remain close to the main sequence until an appreciable fraction of the hydrogen is burnt up, i. e. for a time-interval $\approx t_E$ (Table 26.1). In the T_e, L-diagram the evolutionary track then first moves upwards and to the right (as M. Schönberg and S. Chandrasekhar had already calculated in 1942) i. e. into the region of red giants and supergiants.

We can now interpret the colour-magnitude diagrams of galactic clusters (Fig. 26.1) as an *age-sequence:* h $+ \chi$ Persei with its extremely luminous blue supergiants, which burn away their hydrogen so extravagantly, is a quite young stellar cluster. The turn-off from the main sequence, the so-called "knee", at $M_V \approx -6$ indicates (Table 26.1 or the right-hand scale in Fig. 26.1) an age of a few million years. The few red supergiants to the right of the upper end of the main sequence are separated from the latter by the so-called "Hertzsprung gap" which has been known empirically for a long time. Becoming less marked, the gap extends down to about F0 III stars. The gap is easily explained by the fact that, for example in Fig. 26.5 (i. e. for a star of 5 $M_\odot$) the portion of track C → D last for only 0.3×10^7 years, compared with 2.1×10^7 years for the ensuing red-giant stage or 5.6×10^7 years for the hydrogen-burning phase on the main sequence.

The colour-magnitude diagrams of, for example, the Pleiades ... Praesepe ... down to M 67, in which the main sequence turns off to the giant branch further and further down, indicate ever greater ages. The colour-magnitude array of the galactic cluster NGC 188 may lie even a little below that of M 67, but the corrections for interstellar absorption and reddening present difficulties for this object. Without question, there is no diagram that turns off significantly lower down, i. e. there exists a *maximum age* for galactic clusters which, according to the best model-calculations at the present time is about 10×10^9 years. This is evidently also the *age of the galactic disk.*

The aggregate of colour-magnitude diagrams of galactic clusters in Fig. 26.1 up to fairly bright M_V possesses a well-defined envelope, from which the nuclear evolution towards the right begins. This is obviously the standard main sequence, or zero-age main sequence, which has already been demanded by the theory. In its upper part it lies a little below the main sequence of luminosity-class V (Fig. 15.4) and merges into this at the G-stars (Table 26.2).

Table 26.2. *Zero-age main sequence* (H. L. Johnson, 1964)

$B-V$	M_v
−0.20	−1.10
−0.10	+0.50
0.00	+1.50
+0.10	+2.00
+0.20	+2.45
+0.30	+2.95
+0.40	+3.56
+0.50	+4.23
+0.60	+4.79
+0.70	+5.38
+0.80	+5.88
+0.90	+6.32
+1.00	+6.78
+1.10	+7.20
+1.20	+7.66
+1.30	+8.11

We can best interpret the familiar colour-magnitude diagram of the field-stars in our neighbourhood as that of a mixture of stars from the remnants of many associations and clusters that have become dispersed in the course of time. The calculated evolution-times make it at once understandable that the main sequence conforms closely to the standard main sequence. The (at first sight not easily understandable) bunching together of the yellow and red giants in Russell's giant branch, according to A. R. Sandage, is ascribed to the fact that in this region the evolutionary tracks of the more massive and luminous stars from left to right and those of the less massive and fainter stars from the lower main sequence towards the upper right run together as in a funnel. Particularly in regard to their masses, the giant stars in our neighbourhood may not be treated as a homogenous group.

We turn now to the colour-magnitude diagrams of the globular clusters (Figs. 26.1, 26.4), which actually formed the starting point for the modern theory of stellar evolution. They very much resemble the diagrams of the older galactic clusters with the difference that their giant branch lies some 3 magnitudes brighter. According to the model-calculations of F. Hoyle and M. Schwarzschild (1955), this is a consequence of the fact that, as members of the extreme halo-population II, the globular clusters have *very low metal content* (of the order of one hundred times smaller than the solar value). This is shown by the spectra of their red giants. The opacity and the energy-generation are considerably different, as a result. The turn-off of the colour-magnitude diagrams at the same place as that of NGC 188 shows that *all* globular clusters (in the halo) have an age of 10 to at most 15×10^9 years, only a little more than that of the oldest galactic clusters (in the disk). Thus the halo must have been formed during a relatively short time at the outset of galactic evolution.

In the colour-magnitude diagrams of globular clusters the *horizontal branch* (Fig. 26.4) runs back from the red-giant branch (upper right) towards the lower left as far as to the A and B stars of absolute magnitude $M_V \approx 1$. Within the horizontal branch there exists the gap occupied by the pulsating cluster-variables or RR Lyrae stars. At the present time, we still cannot say how the turn-round of the evolution in the region of the red giants proceeds. Spectroscopic observation may tell against the several-times conjectured catastrophic evolution, e. g. the star HD 161 817 in the horizontal branch (somewhat left of the RR Lyrae gap) in the chemical composition of its atmosphere (Table 19.1) does not differ in the slightest from the subdwarfs of the population II main sequence.

Further evolution probably leads to the domain of planetary nebulae and Wolf-Rayet stars, and thence, obviously with continual loss of mass, to the white dwarf stars, the "cemetery of the colour-magnitude-diagram".

The formation of the heavy elements must have proceeded along with the evolution of the Galaxy (halo → disk → spiral arms). In the halo we find stars having metal-abundance* ranging from 1/200 of

---

* We mean the abundances of all the spectroscopically detectable elements from C to Ba. On account of their small abundances, the lightest elements Li, Be, B and the heaviest elements are not detectable. As to helium, see below.

the "normal" value up to that of population I stars and disk-stars. On the other hand, the metal-abundances of the latter do not differ among themselves up to the accuracy of current measurements (about a factor 2). In the spiral arms and in the disk there are no metal-poor stars. The quantitative analysis of stellar spectra (Table 19.1) has shown further that the *relative* abundances of the heavy elements (C to Ba) are the same in the metal-poor and in the metal-rich stars, while the ratio of *all* the heavy elements to hydrogen increases by a factor up to 200.

Unfortunately we do not know how helium behaves, since only the stars of the population II main sequence below F5 to G0 have remained unchanged since the initial state of our Galaxy. We can expect no helium lines in their spectra on account of their low temperatures. Also the investigation of the internal constitution of these stars has not yet reached such a degree of perfection as would permit the determination of the helium-abundance (Y) with any certainty. Thus we are not yet in a position to say whether the raw material of our Galaxy consisted of almost pure hydrogen (with less than $^1/_{200}$ of the present heavy elements) or whether from the start a considerable proportion of helium was mixed with it. In the first case, in the subdwarfs, for example, the helium abundance ought to be reduced in proportion to the metal-abundance; in the other case, the ratio H:He would be about the same in all stars *.

We are able to shed light upon the helium problem from another angle. Since the predominant part of the mass of our Galaxy consists of the disk- and spiral arm-population I, it therefore consists of some 35 percent helium. We can easily reckon how much energy was produced by the formation of this helium from hydrogen. We can then compare this energy with that which the Galaxy *would have* radiated at its present absolute luminosity during its past lifetime ($\sim 10 \times 10^9$ years). We find, if the helium now in existence has been formed from hydrogen during the evolution of the Galaxy, then in the past the Galaxy must have been on the average some 15 times brighter than it is at present, or for, say, an interval of 10^9 years it must have been

* The theory of an origin of all the elements in a cosmic "big bang" (Chapter 22), which was widely accepted for some time, must now be given up on account of difficulties having to do with nuclear physics and astronomy (metal-poor subdwarfs). Whether the theory may be retained in a restricted form for the case of helium is at present probably not yet settled.

about 150 times brighter, and so on. Later on we shall have to weigh up this important evidence along with other observations on the evolution of galaxies.

On the other hand we consider the origin of the *heavy elements* from the standpoint of nuclear physics, in connection with the fundamental work of E. M. and G. R. Burbidge, W. A. Fowler, F. Hoyle (B^2FH, as previously noted) and others since 1957. That is to say, we start with the measured abundances of the chemical elements and their isotopes in the Sun, the meteorites and the Earth's crust, and ask, What nuclear processes could have led to this abundance-distribution? At what temperatures and densities did they operate? In what astronomical bodies could such conditions be realized?

In connection with nuclear energy-generation we have already discussed

a) Conversion of hydrogen into helium at about 1 to 5×10^7 °K.

b) Formation of heavy nuclei from helium. At about 1 to 2.5 $\times 10^8$ °K C^{12}, O^{16}, Ne^{20} are formed. In this connection, further reactions, in particular (25.27), produce neutrons which take part in further syntheses.

Other processes lead to:—

c) Synthesis from α particles of still heavier nuclei having mass-numbers which are multiples of 4, up to Ca^{40}, at $T \approx 10^9$ °K; the "α-process".

d) The "e-process" generates the elements of the iron group V, Cr, Mn, Fe, Co, Ni in thermal equilibrium at about 4×10^9 °K and at a proton-neutron ratio of about 300.

e) The "s-process" consists in neutron-capture (by light nuclei or Fe-elements) which proceeds slowly in comparison with the associated β-decay processes. The s-process produces, for example, Sr, Zr, Ba ... Pb ...

f) As "r-process" we denote the corresponding neutron-captures that proceed rapidly in comparison with the associated β-decays. This process yields many of the really heavy elements, in particular the radioactive elements U^{235} and U^{238} at the expense of the iron-group.

g) The "p-process" produces the rare light isotopes of the heavy elements in a hydrogen-rich medium (protons) at about 2.5×10^9 °K.

Then the next question is, How is the formation of the heavy elements, or the increase of their abundance relative to hydrogen from the old halo population, through the disk-population, to the youngest population I of the spiral arms, bound up with the evolution of the Galaxy?

In their work of around 1957, E. M. and G. R. Burbidge, W. A. Fowler and F. Hoyle have tried to find a place for the s-processes (which are unavoidable on the grounds of nuclear physics) in the later stages of stellar evolution, and for the r-neutron-processes in supernovae explosions, about which admittedly we still know little. The following picture then presents itself: The Galaxy would have begun as an almost spherical cloud, corresponding roughly to the halo, composed of almost pure hydrogen. Here the first stars would have been formed, of which the cooler subdwarfs still survive at the present time. The massive stars evolved quickly, and heavy elements arose partly in connection with energy-generating nuclear processes, partly (by neutron processes) in supernovae-outbreaks. In the latter, and in the final contraction of fully evolved stars towards white dwarfs, material rich in heavy elements would be dispersed again into the interstellar medium. It would provide the raw material for forming a second generation of metal-rich stars, and so on. Since the metal abundances of the disk-population (e. g. the Sun) and of the youngest spiral-arm population I (e. g. τ Sco, 10 Lac; Table 19.1) do not markedly differ from each other, the formation of the heavy elements had almost come to an end when the galactic disk was formed out of a large part of the mass of the original halo. The processes of star-construction and star-destruction must therefore have proceeded much more rapidly at the low density of the halo than after the formation of the denser disk. This presents a considerable difficulty for the theory described. Consequently, the great uniformity of the abundance-ratios for all the heavy elements is difficult to understand, since they must have arisen through very varied processes acting in varied cosmic circumstances.

6. *The initial contracting phase of stellar evolution. Star formation.* We turn back once more to the starting point of our deliberations, and ask, *How do the stars reach the main sequence?* and then, *How and from what did the stars arise?*

The close association in space of young, absolutely bright O and B stars with gas- and dust-clouds in the spiral arms of our Galaxy and the Andromeda galaxy strongly suggests that quite generally stars are formed in and from cosmic clouds of diffuse matter. The only available energy-source they have at first, that is until some nuclear reaction gets under way, is gravitational energy (contraction-energy).

It is fitting to recall here that as early as 1846, shortly after the discovery of the law of conservation of energy, J. R. Mayer raised the question of the origin of the radiative energy emitted by, say, the Sun. He considered the fact that a mass of meteorites m falling into the Sun would give up energy of amount, according to (6.28),

$$m\,G\,\mathfrak{M}/R \tag{26.3}$$

as heat energy, where G is the gravitation constant; $\mathfrak{M}$, R denote the mass and radius of the Sun. Since the mass of infalling meteorites is in fact very small, H. von Helmholtz 1854 and Lord Kelvin 1861 were able to show that the contraction of the Sun itself would be a more significant source of gravitational energy.

As we see immediately from (26.3), the energy released in the formation of a gas-sphere of radius R from initially dispersed material, or in a considerable contraction of a star of radius R, is always of the order of magnitude

$$E_{\text{contr.}} \approx G\,\mathfrak{M}^2/R. \tag{26.4}$$

In the case of the Sun, for example, this is about 3.8×10^{48} erg; the content of the Sun in thermal and ionization energy is of the same order. The energy $E_{\text{contr.}}$ could supply the emission from the Sun at its present luminosity $L_\odot = 3.84 \times 10^{33}$ erg/sec for only

$$E_{\text{contr.}}/L_\odot \approx 30 \text{ million years.} \tag{26.5}$$

What then is the track in the colour-magnitude diagram of a star that is formed by the contraction of initially widely dispersed material? In 1961, C. Hayashi showed that, as soon as the hydrogen is partly ionized, the star has an extended convective envelope. In the colour-magnitude diagram it moves steeply downwards, with slightly rising temperature, until it reaches the almost horizontal evolutionary track for radiative equilibrium, previously computed by L. G. Henyey and others. It then approaches the standard main sequence, moving along this track towards the left. According to I. Iben's more precisely calculated models, a star of 1 solar mass, for example, contracts at $T_e \approx 4000°$ in about 10 million years. Then its luminosity increases again by a factor about 2 and after a further 50 million years, or so, it reaches the standard main sequence. The fact that nuclear reactions already make some contribution to the energy on the way explains why this time somewhat exceeds that in (26.5).

Towards the upper part of the main sequence the contraction time corresponding to (26.5) becomes steadily smaller; a B0 star requires only about 6×10^4 years. On the other hand, stars of smaller mass form themselves more slowly; an M star of 0.5 $M_\odot$ requires about 1.5×10^8 years. Contracting masses below a certain value do not at any stage reach the temperature needed for setting nuclear processes in operation. S. S. Kumar has shown (1963) that this limiting mass for population I material is 0.07 $M_\odot$, and for metal-poor population II material it is 0.09 $M_\odot$. Smaller masses form cool, wholly degenerate bodies that are naturally called *black dwarfs*. That bodies such as the dark companion of Barnard's star (page 132) with mass about 0.0015 $M_\odot$ or the large planets, like Jupiter with about 0.001 $M_\odot$ were formed as black dwarfs appears to be quite possible.

In young galactic clusters, which we recognize by their bright blue stars, M. Walker in fact found to the right of the lower main sequence stars of mean and of late spectral types, which by their T Tauri-variability, Hα emission lines, in some cases fast rotation, etc., are indicated to be young stars in the course of formation. For example,

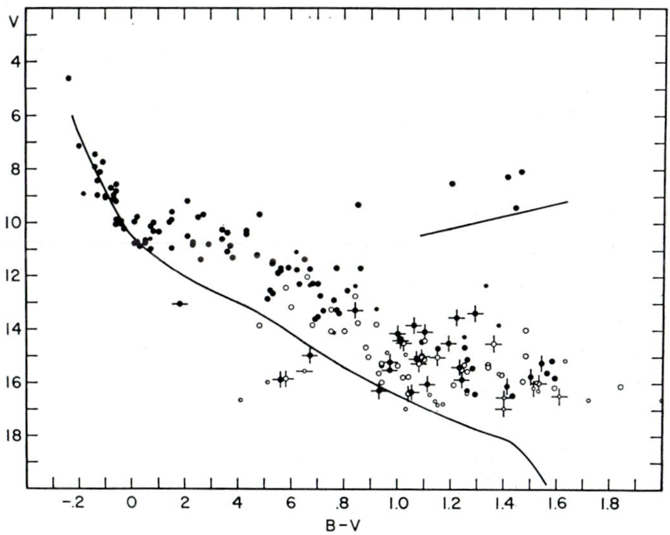

Fig. 26.7. *Colour-magnitude diagram* of the very young galactic cluster NGC 2264 (after M. WALKER, 1956).

● photoelectric measurements | variable stars

○ photographic measurements — stars with Hα emission

The curves indicate the standard (zero-age) main sequence and the giant branch, corrected for uniform interstellar reddening of the cluster. The apparent distance modulus is $9^{\mathrm{m}}.7$, and the distance 800 pc. The lifetime of the cluster (about) 3×10^6 years) has been insufficient to enable stars to reach the main sequence below about spectral type A0

Fig. 26.7 shows following M. Walker (1956) the colour-magnitude diagram of the cluster NGC 2264; according to the evidence of the brightest stars, this has an age of only 3×10^6 years. In agreement with the theory stars below spectral type A0 ($< 3\ M_\odot$) have therefore not yet reached the standard main sequence.

Dynamical considerations, into which we cannot enter here, indicate that only relatively large masses of gas, say 10 to $10^3\ M_\odot$, can become unstable and condense into stars. In fact as long ago as 1947, V. A. Ambartsumian had extracted from the observations the important inference that, at anyrate for the most part, stars are formed in times of the order of 10^7 years in groups with masses of about $10^3\ M_\odot$. These are the OB associations with bright blue stars and the T associations with cooler stars, notably T Tauri variables of small absolute luminosity, the two sorts often occurring together. An OB association, as in Orion or Monoceros, with its enormous bodies of gas, which are ionized by the short-wavelength radiation from the O and B stars immersed within them, first attracts the attention of the observer as an HII region that shines brightly in Hα light. The expansion of the gas heated by the stars that have already been formed is the probable reason why further stars move away from the centre with speeds of the order of 10 km/sec. Thus the associations are *not permanent.* The expansion-age got by extrapolating the stellar motions backwards in general agrees with the evolution-age of the brightest stars. For the Orion nebula and the Trapezium stars that excite its radiation (and certain similar systems) we find an age of only about 1.5 to 3×10^4 years. In contrast, the Orion cluster itself must be about one hundred times older.

In a number of cases, individual stars travel away from their OB association with much larger speeds up to 200 km/sec. A. Blaauw sees such a "runaway star" as the surviving component of a fast-revolving double star whose primary component had been suddenly dispersed in space as the result of an explosion (possibly as a class II supernova).

7. *Pathological stellar evolution.* Quite briefly and without any pretence of completeness, we now mention certain "pathological" kinds of star. Their spectral analysis is still only starting, and for the most part their evolution is still not understood.

The *helium stars* and *the carbon stars,* in whose spectra He or C is anomalously strong while the behaviour of the heavy elements is about normal, suggest the notion that here "burnt" material from the interior has somehow made its way to the surface. Helium stars occur in population II (perhaps at the left-hand end of the horizontal branch) and apparently also in the population I main sequence. Carbon stars occur in the giant branch in the T_e, L-diagram in about luminosity-class III.

The *Ap stars* with their enormous magnetic fields and the *metallic-line stars* occupy in the colour-magnitude diagram a strip of the main sequence from cooler B stars as far as F-type stars. Here again we do not at all understand the spectra.

Looked at from the standpoint of the theory of the internal constitution, the crucial difficulty of the problems raised lies in the computation of the convective zones. The hydrodynamics of convective energy-transport is still only making a beginning. Therefore we still cannot say for certain whether one or several convective zones in a star may establish a connection between the nuclear-burning zone and the stellar atmosphere.

8. *Stellar statistics and stellar evolution. Rates of formation of stars.* By statistical studies of stars with known parallaxes, during the 1920s J. Kapteyn, P. J. van Rhijn and others derived the luminosity function for stars in our neighbourhood (Chapter 23). Here we restrict ourselves to stars still on the main sequence and in Fig. 26.8 we plot the luminosity function $\Phi(M_V)$ = number of main-sequence stars with absolute magnitudes $M_V - \frac{1}{4}$ to $M_r + \frac{1}{4}$ per cubic parsec in the solar neighbourhood *. The fact that from $M_V \approx 3.5$ this number falls off rapidly towards brighter magnitudes was related by E. E. Salpeter (1955) to the fact that stars fainter than $3\overset{M}{.}5$ have been accumulating since the formation of the Galaxy $T_0 \approx 10^{10}$ years ago, without significant change. Brighter stars, on the other hand, after the lapse of about the evolution time t_E (Table 26.1) from the epoch of their formation, move away from the main sequence and ultimately, after a total time which is certainly small compared with T_0, they become white dwarfs. Thus we can readily calculate the *zero-age luminosity function* $\Psi(M_V)$ giving the number of stars *that have been formed* since the start of the Galaxy in the magnitude interval $M_V \pm \frac{1}{4}$, per cubic parsec (external conditions being assumed to remain the same). For brighter stars we have

$$\Psi(M_V) = \Phi(M_V) \cdot \frac{T_0}{t_E(M_V)} \; ; \tag{26.6}$$

* $\Phi(M_V)$ is often defined instead for the magnitude-interval $M_V \pm \frac{1}{2}$.

for fainter stars $\Psi(M_v)$ goes over continuously into $\Phi(M_v)$. In Fig. 26.8 we have included the zero-age luminosity function $\Psi(M_v)$ following A. R. Sandage, who has extended the calculations made by E. E. Salpeter.

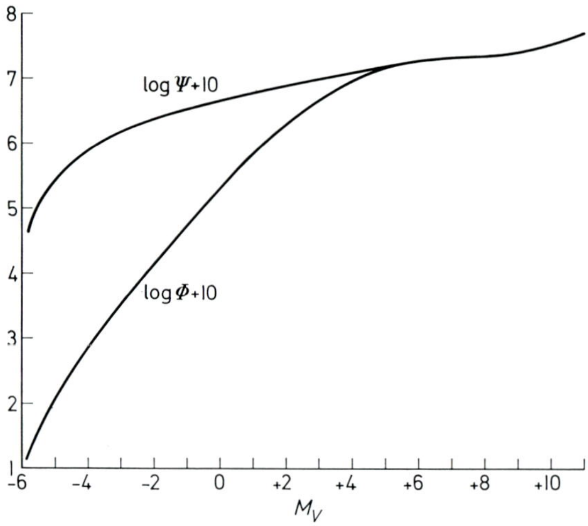

Fig. 26.8. *Luminosity function* $\Phi(M_V)$ and *zero age luminosity function* $\Psi(M_V)$ of main sequence stars in the solar neighbourhood. The functions Φ and Ψ give respectively the present number of stars per cubic parsec in the magnitude range $M_V - \frac{1}{4}$ to $M_V + \frac{1}{4}$, and the number of stars born in this range since the formation of the Galaxy

If our ideas are correct, then the luminosity function of young galactic clusters must correspond to the zero-age luminosity function Ψ, and not to the Φ of our neighbourhood. In Fig. 26.9 M. Walker has compared the luminosity functions of three very young clusters, which he has studied, with the calculated function Ψ, derived from the luminosity function Φ for our neighbourhood. The differences between the total stellar densities in clusters and in our neighbourhood is taken into account by appropriate displacements of the logarithmic ordinate-scale. The shape of the curves shows remarkable agreement. This supports the view that the resolution of an original body of gas into stars proceeds throughout according to the same initial luminosity function $\Psi(M_v)$.

The difference $\Psi(M_v) - \Phi(M_v)$, summed over all M_V, represents those stars that have evolved off the main sequence at any time since the origin of the Galaxy. By far the majority of these stars must now be white dwarfs. Actually the space-density of white dwarfs calculated from Fig. 26.8 agrees well with the observed value, certainly as well as could be expected in view of the uncertainty of the data.

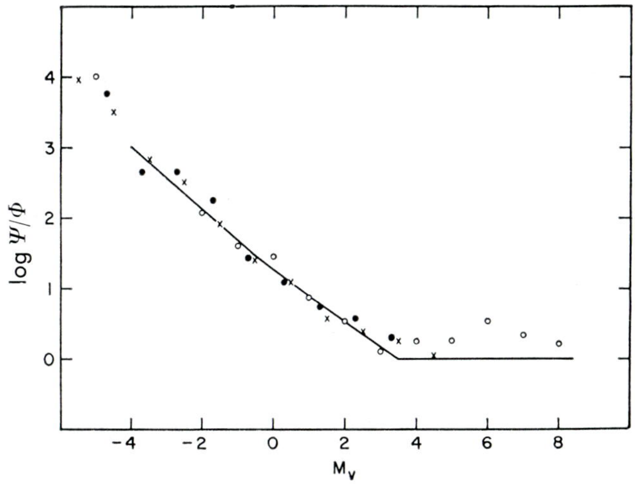

Fig. 26.9 *Comparison of observed luminosity functions of young galactic clusters* NGC 6530 (dots), NGC 2264 (circles), and the Orion Nebula cluster (crosses) with E. E. Salpeter's zero age luminosity function. The ordinate is the logarithm of the ratio of the observed luminosity function of the cluster (points) and of the zero age luminosity function $\Psi(M_V)$ (line), to the observed luminosity function $\Phi(M_V)$ for main sequence stars in the vicinity of the Sun. The luminosity functions of the clusters have been adjusted for the star density in each cluster relative to that of the solar neighbourhood. (After M. Walker, 1957)

As remarked, the foregoing calculations involve the unsatisfactory assumption that star formation has proceeded in the same way everywhere and at all times. Consequently, M. Schmidt and S. v. Hoerner have attempted to work out the formation and evolution of stars within the setting of a suitable model of the Galaxy, which was however again treated as unchanging with time. But since, as we found empirically, the formation of heavy elements and probably an accompanying phase of intensive star-formation was associated in time with the contraction of the galactic disk, these more complicated calculations may not at present take us much further than the simple estimates made by Sandage and Salpeter.

The latter have also studied the luminosity function (including giant stars) of the globular cluster M 3, in order to obtain insight into the evolution of the stellar population II. Apart from a maximum, or peak, at $M_V \approx 0$, which is populated by cluster-variables, the luminosity function of the globular cluster differs little from that of our neighbourhood. Therefore Sandage filled in the region fainter than $M_V = +6$, where the stars are too faint to be observed, in accordance with the van Rhijn luminosity function. Then we have for the entire cluster:

	Number	Mass		
Luminous stars	588 000	1.75×10^5	$2.45 \times 10^5\,M_\odot$	(26.7)
White dwarfs	48 500	0.70×10^5		

While half the luminosity of the cluster comes from stars brighter than $M_V = -0.14$, half the mass is reached only at $M_V = +11.28$. The overall ratio of mass/luminosity is $M/L \approx 0.8$ in solar units in exellent agreement with other studies (including dynamical studies) of M 92.

Thus the evolution of the metal-poor halo-population II differs from that of the metal-rich disk-population essentially by the formation of the horizontal branch in the colour-magnitude diagram, while only small differences subsist between the luminosity functions of the two old stellar populations.

27. Galaxies

The advance into cosmic space beyond the confines of the Milky Way into the realm of distant galaxies — or extragalactic nebulae, as they used to be called — and the beginnings from that of a *cosmology* founded upon observation must for all time belong to the greatest achievements of our century.

We mentioned earlier (page 221) the classic catalogue of Messier (M) 1784 and also, arising out of the work of W. and J. Herschel the *New General Catalogue* (NGC) of 1890 by J. L. E. Dreyer and his *Index Catalogue* (I.C.) of 1895 and 1910, the notations of all of which are still in use today. Apart from the Magellanic Clouds in the southern sky, the brightest galaxy, the Andromeda nebula (Fig. 27.1) — which S. Marius had already observed in 1612 — has, for example, the catalogue numbers M 31 and NGC 224. In modern times we mention the *Reference catalogue of bright galaxies* of G. and A. de Vaucouleurs (1964) (which grew out of the *Shapley-Ames catalogue* of 1932) which covers the whole sky and lists 2599 galaxies brighter than 14^m. Unsurpassed illustrative material is given in *The Hubble atlas of galaxies* by A. Sandage, Mt. Wilson and Palomar Observatories, 1961.

Fig. 27.1. *Andromeda galaxy* M 31 (NGC 224) and its physical companions, the elliptical galaxies M 32 (NGC 221) and NGC 205 (lower left). M 31 has an inclination of 11°.7, and a distance of 680 kpc. Mount Wilson and Palomar Observatories, 48″ Schmidt camera

In the 1920s the cosmic status of the "spiral nebulae" was the subject of strenuous debate amongst astronomers. Here we can only just mention the important contributions made by H. Shapley, H. D. Curtis, K. Lundmark and others.

In 1924 E. Hubble succeeded in partially resolving the outer parts of the Andromeda galaxy M 31 and of some other galaxies (to use the modern terminology) into stars, and in identifying various objects with known absolute magnitudes, which could be used as a basis for photometric distance-determination:

a) *Cepheids.* The light-curves of these (Fig. 27.2), with periods of from 10 to 48 days, were first interpreted naturally in terms of the "general" period-luminosity relation. About 1952 W. Baade and

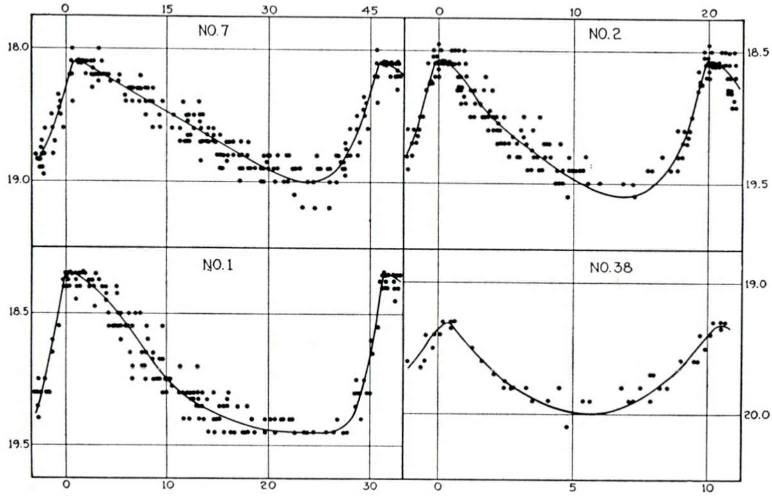

Fig. 27.2. *Light curves of four Cepheids* in the Andromeda galaxy M 31, after E. Hubble, 1929. The abcissae given the time in days, and the ordinates the photographic magnitude

others realized as a result of discrepancies regarding the absolute magnitude of red giants stars, that the "classical" cepheids of population I, in particular, therefore, the long-period cepheids of the Andromeda galaxy, are about $1^{\mathrm{m}}.5$ *brighter* than the corresponding cepheids of population II, to which the W Virginis and RR Lyrae variables of the globular clusters belong. This then required an increase of extragalactic distances by a factor about 2.

b) *Novae.* Hubble was able to detect and make use of novae, whose light-curves fully resemble those of galactic novae. The very much brighter S Andromedae with $m_{\mathrm{v.\,max}} \approx 8$ observed by Hartwig in 1855 was later recognized to be a supernova.

c) *The brightest non-variable stars.* Again certain corrections were subsequently needed, after a fraction of the "brightest stars" proved to be groups of stars or HII regions. Here we cannot enter into the questions (very difficult in their details) of the exact determination of the absolute magnitude of various sorts of objects nor of the equally important improvement of the magnitude-scale for fainter stars.

At present, the most probable value for the distance of the Andromeda galaxy is 680 kpc or 2.2 million light years.

W. Baade first succeeded in 1944 in resolving the central part of the Andromeda galaxy and of the neighbouring smaller galaxies M 32 and NGC 205 (Fig. 27.1) by using the utmost refinement of photographic technique. Here the brightest stars are red giants with $M_V = -3$, while the brighter blue stars of the spiral-arm population I are absent.

As a result of E. Hubble's investigations it was definitely established that galaxies like M 31 and others are widely similar to our Milky Way Galaxy. What can we say then about the distribution of the galaxies in the heavens?

As we saw, the almost galaxy-free zone around the galactic equator between about $b = \pm 20°$, the so-called "zone of avoidance" is produced by a thin layer of absorbing material in the equatorial plane of the Milky Way.

Fairly commonly we observe *groups* of two, three ... galaxies that obviously belong together physically. The Milky Way along with the Andromeda galaxy M 31, its two (physical) companions (Fig. 27.1) and about 20 other galaxies form the so-called *local-group*. Again in higher galactic latitudes, where galactic obscuring clouds are absent, the galaxies show a non-uniform distribution in the sky. As Max Wolf had already noted in a number of cases, they form *clusters of galaxies* like, for example, the Coma cluster (in Coma Berenices) at about 10^8 pc = 100 Mpc (megaparsec) distance which includes several thousand galaxies in a region of about 3 Mpc diameter.

Now consider more closely the manifold forms of galaxies. E. Hubble was able to arrange these in a scheme of classification (Fig. 27.3) which, with improvements (already incorporated in the figure) also forms the basis for the *Atlas of galaxies* and at the same time is more exactly specified by the *Atlas*. As in the case of the Harvard sequence of spectral types, it is clear in advance that such a purely descriptive scheme need by no means represent an evolutionary sequence.

The elliptical galaxies E0 to E7 have rotationally-symmetric form with indications of some further structure. The observed (apparent) ellipticity is naturally that determined by the projection of the (true) spheroid on the sky. The statistics of the apparent ellipticities show that the true ellipticities of the E-galaxies are fairly uniformly distributed. The surface brightness decreases uniformly from the centre towards the outside.

From the ellipticals there is a steady transition to the *spiral galaxies*. These are all more strongly flattened. However, here we have a fork in the sequence.

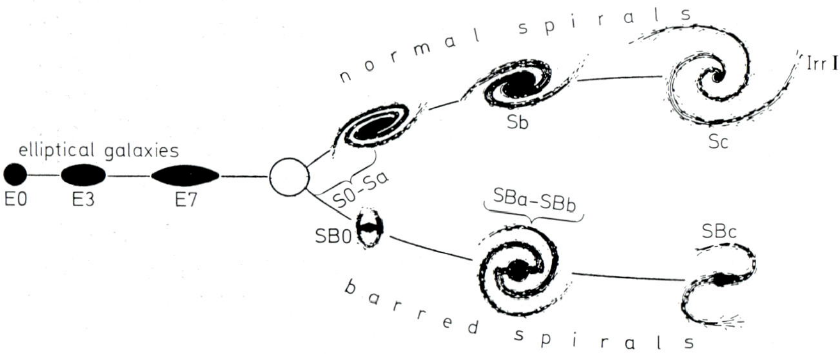

Fig. 27.3. *Classification of galaxies*

The *normal spirals* (S) have a nucleus from which the spiral arms spring more or less symmetrically. In the case of the *barred spirals* (SB) a straight "bar" comes out of the nucleus, and to its ends the spiral arms are attached almost perpendicularly.

Between the spiral arms in normal, as well as in barred, spirals there are great numbers of stars so that in photometric recordings the arms do not stand out at all prominently.

The sequence of types S 0 . . . Sc or SB 0 . . . SB c is characterized by the property that the *central region* — also called the nucleus or the lens of the spiral galaxy — becomes relatively smaller, while the windings of the *spiral arms* become more open, as we proceed along the sequence. As an example, M 31 and the Milky Way Galaxy both belong to type S b.

The Hubble Atlas further divides the S and SB spirals into two subclasses according as the arms spring directly out of the nucleus, or out of the ends of the bar (suffix s), or come tangentially out of an inner ring (suffix r).

To the Sc galaxies (like M 33 in the local group) are joined continously the irregular galaxies Irr I (not shown in Fig. 27.3). These relatively rare systems show in the first place no rotational symmetry and no well-defined spiral arms, etc. The best-known examples are our neighbours in the local group, the Large Magellanic Cloud (LMC;

Fig. 27.4) and the Small Magellanic Cloud (SMC). These are situated in the southern heavens at distances of about 55 and 63 kpc. More exact investigation has shown that the irregular luminosity distribu-

Fig. 27.4. *Large Magellanic Cloud* (LMC). The numerous H II regions stand out clearly in this Hα photograph

tion of the Irr I-systems is only brought out by the bright blue stars and neighbouring gas nebulae that feature strongly in the blue plates. In agreement with radio-astronomical 21-cm measurements of the location and velocity distribution of the hydrogen, the substrate of fainter red stars shows that the main body of, for instance, the Magellanic Clouds exhibits a much more regular, flattened form and considerable rotation.

The Irr II systems, which only superficially resemble the Irr I, such as M 82 have recently been recognized as galaxies in the nucleus of which an explosion of unimaginable violence is occurring (see page 308).

As must be stressed once again, in the Hubble sequence the galaxies are classified only according to their form. Considered physically, as B. Lindblad has emphasized, we clearly have an ordering according to increasing angular momentum (flattening). The study of the distances of the galaxies has shown that, for any one Hubble type, their absolute luminosities and diameters are spread over a wide range. In this sense, we speak of *giant* and *dwarf* systems. Anticipating what will be said later, there is a steady transition from the giant ellipticals, which are rich in stars and have $M \approx -19.5$, through objects like the Andromeda companion NGC 205 (Fig. 27.1) with $M \approx -15$ to the dwarf ellipticals, which are poor in stars and have $M \approx -10$. In an analogous way, compared with the Irr I giant systems like the Large Magellanic Cloud LMC, there are systems some 3^M fainter, like the irregular dwarf galaxy IC 1613, which certainly belongs to our local group. Doubtless for dynamical reasons, by contrast there exist no dwarf spirals.

In the *spectra of galaxies,* in the first place the Fraunhofer absorption lines establish that the galaxies are composed mainly of *stars.* Emission lines of HII regions are observed in the spiral arms of types S also SB b and c, and even more markedly in Irr I systems. Some quite different origin must obviously be found for the very broad emission lines in the nuclei of the so-called Seyfert galaxies and of the Irr II systems like M 82 (see page 308) as well as those of many giant ellipticals.

The doppler shifts of the absorption and emission lines (the latter are better for measuring) give information about the radial velocity and the rotation of a galaxy. For example, Fig. 27.5 shows the measurements made by H. W. Babcock (1939) for the Andromeda galaxy M 31. Later observations by N. U. Mayall (1950) on 32 emission objects, as well as radio measurements of the 21-cm line by the Leiden group, confirm these as a whole, but they show not inconsiderable quantitative differences. A decrease of the rotational velocity corresponding to the transition to Kepler motion in the outer parts of galaxies, can be observed in only a few favourable cases.

From Fig. 27.5 we find for M 31 a radial velocity of approximately -300 km/sec. Since the galactic coordinates of the Andromeda galaxy are $l^{\mathrm{II}} = 121°$, $b^{\mathrm{II}} = -21°$, this is for the most part the reflection of the rotational motion of our own Galaxy.

As already remarked, the rotational velocity must first increase inwards from the outside. After passing through a flat maximum, it then decreases inwards. Secondary maxima and minima in Fig. 27.5 at $\pm 3'$ and $\pm 9'$ are not confirmed by the more recent measurements of G. Münch (1964). Right at the centre, the Andromeda galaxy is known to have a nucleus of $2''.5 \times 1''.5$ (which looks almost starlike on under-exposed plates). In 1960 A. Lallemand, M. Duchesne and M. F. Walker, using an electronic image converter, were able to measure its rotation. They found (Fig. 27.6) a further sharp maximum in the rotational velocity of 87 km/sec at only 2.″2 from the centre.

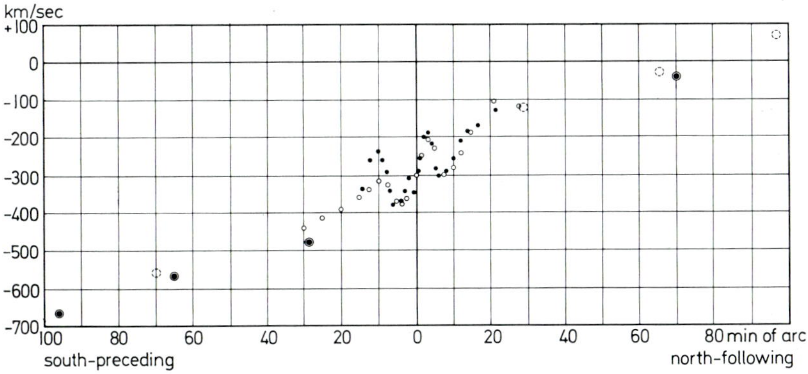

Fig. 27.5. *Radial velocity measurements in the Andromeda galaxy,* after H. W. Babcock, 1939. M 31 is approaching us with a velocity of —300 km/s; when this has been subtracted from the velocity of each point, the remainder is the rotation velocity, which is seen to be the same on either side of the centre (some reflected observed points are shown as broken circles). At a distance of 680 kpc, two points separated by 10 kpc, about the Sun's distance from the galactic centre, would subtend an angle of 50.′5

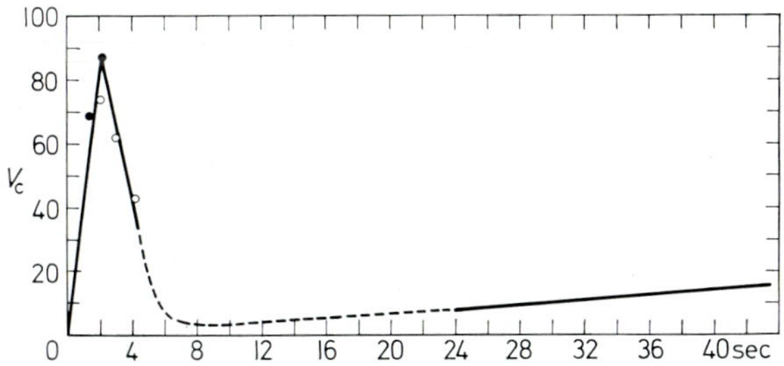

Fig. 27.6. Rotation of the *innermost* part of the nucleus of the Andromeda galaxy M 31. The abcissa is marked in *seconds* of arc. The solid line for $r > 24''$ has been taken from the measurements of H. W. Babcock

Taking careful account of the inclination of the galaxy of 11.°7, of non-radial velocity components, etc., and assuming circular orbits, it was sought to compute the mass and mass-distribution of the galaxy from the rotational velocities. As was to be expected, the total mass of M 31 of about $3 \times 10^{11} M_\odot$ agrees closely with that of our Galaxy.

Of great interest are the properties of the innermost nucleus (which is not observed in our Galaxy). In a region of only 7.4 pc are crowded together 1.3×10^7 solar masses (Period of rotation $\sim 5 \times 10^5$ years); the mass-density comes to $1.5 \times 10^3\, M_\odot/\mathrm{pc}^3$ or about 10^4 times that in our neighbourhood. The mass-luminosity ratio of 3.6 (in solar units) shows, however, that the object is composed of normal stars.

For the classification of the spectra of galaxies we must generally be content with a treatment of the nucleus (if this is predominant), otherwise of the whole system, with a dispersion of only 100 to 400 Å/mm. It aims to give information about the stellar populations of the system and to relate their classification to the Hubble sequence.

Since the spectrum of a galaxy consists of the superposition of many stellar spectra (composite spectrum), other things being equal, in the short-wavelength region hot blue stars will dominate the spectrum, and in the long-wavelength region cooler red stars will do so. W. W. Morgan and N. U. Mayall (1957) restricted themselves principally to the interval $\lambda\, 3850-4100$ Å:—

A-*systems* have broad Balmer lines; in the range $\lambda\, 3850-4100$ the spectrum corresponds to spectral type A, near $\lambda\, 4340$ to that of a F8 star. A typical representative is the Irr I galaxy NGC 4449 which is similar to the Magellanic clouds. The Hubble types Irr I, Sc, SBc all come in here.

F-*systems* correspond in the violet to spectral type F, and near $\lambda\, 4340$ to G. A typical example is the Sc galaxy M 33 = NGC 598; the Sb spirals also occur here.

K-*systems* produce a spectrum that can be interpreted as the superposition of the spectra of normal (i. e. not metal-poor) G8- to early M-giants (CN criterion) with fainter F8- to G5-stars. Earlier spectral types make no appreciable contribution. The prototype is the Andromeda galaxy M 31 = NGC 224. Here one can also see that the spectra of the central region and of the disk are not significantly different. Besides the large Sb and Sa spirals, the corresponding barred spirals, giant ellipticals like the well-known radio-galaxy M 87 = NGC 4486 or NGC 4636 as well as "dust-free" Sb and Sa systems are all included. Between the main types A – F – K one can insert the intermediate types AF and FG.

To summarize once again, Fig. 27.7 shows which regions in the HR diagram according to Morgan and Mayall make essential contributions to the spectrum in the case of A systems and in the case of K systems.

Closely linked with the spectral type is the *colour index,* for example $C = m_{pg} - m_{pv}$, of the galaxies. Since spirals seen edge-on show absorption and reddening by their own interstellar material,

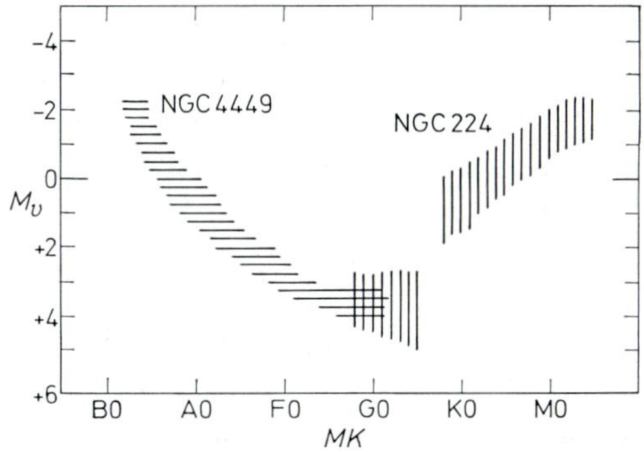

Fig. 27.7. *Hertzsprung-Russell diagram* showing the regions principally contributing to the light of (a) A-System NGC 4449 (Irr I) (b) K-system NGC 224 = M 31 (Sb). As in the case of the Milky Way, the cooler parts of the main sequence do not contribute significantly

on the side presented to us, we must correct for these (as well as for galactic absorption and reddening). E. Holmberg (1964) has made the so-derived true colour indices C^* the starting point of a comprehensive study, which should give us a quantitative survey of the whole realm of galaxies. Table 27.1 gives for the sequence of Hubble-types first once more the spectral types according to Morgan and Mayall, then the colour index $m_{pg} - m_{pv}$, corrected for absorption in the galaxy itself and in our Galaxy, and the absolute visual magnitude. (The determination of distances or distance-moduli are later to be discussed more fully.) Then follows the mass determined from the rotation and the ratio of mass to visual luminosity (referred to the Sun) and the mean mass-density for the volume within a suitably defined bounding surface. From the 21-cm measurements we can also infer the mass of atomic hydrogen in a galaxy; we have written down the fraction of the total mass which it contributes. According to our present views, this part of the mass is available for forming further stars. We must not hide the fact that all the data in Table 27.1 are subject to considerable uncertainty and large (physical) dispersion. The columns Irr I and E hold good essentially for giant galaxies; the luminosities and masses of dwarf galaxies are much smaller. Finally, we have left aside "compact" galaxies and "radio" galaxies (Chapter 28).

Table 27.1. *Average properties of galaxies of the various Hubble types according to E. Holmberg 1964. Normal and barred spirals are taken together; E refers to giant ellipticals (dwarf ellipticals have $M_{\rm phot} \approx -10$)*

Hubble type	Irr I	Sc	Sb	Sa	E, So
Spectrum	A	AF	F FG	K	
True colour-index $C_* = m_{\rm pg} - m_{\rm pv}$	0.12	0.22	0.41	0.53	0.77
True abs. vis-magnitude M_*	-17.3	-19.8	-21.0	-20.6	-19.6
Mass ($M_\odot = 1$)	1.0×10^9	1.6×10^{10}	1.3×10^{11}	1.6×10^{11}	2.0×10^{11}
Mass/Visual luminosity ($\odot = 1$)	0.9	1.4	3.6	6.6	~ 22
Mass-density in $M_\odot/{\rm pc}^3$	~ 0.003	0.013	0.025	0.08	0.16
H percent of mass	37	20	3	1.3	$\lesssim 0.2$

Here we must return to the difficult problem of the direct photometric measurement of the distances and absolute magnitudes of the galaxies. The employment of single stars as distance-indicators is essentially limited to the local group and certain neighbouring clusters. Here we see that the absolute magnitudes cover the enormous range of $M_{\rm pg} = -20$ to -10, or four powers of 10, from giant systems like our own and the Andromeda galaxy down to extreme dwarf systems like the Draco system (E2) or IC 1613 (Irr I). The dwarf systems, indeed, make scarcely any contribution in luminosity or mass, but as regards numbers they form at least 50 percent of the population of cosmic space.

For the photometric determination of *greater* distances, we are thrown back upon statistical methods which depend upon total luminosities of galaxies. Here we make use of the luminosity-function of galaxies, which expresses the fraction of galaxies in, say, the range $M_{\rm pg} \pm \frac{1}{2}$ as a function of the absolute magnitude $M_{\rm pg}$. In Fig. 27.8 we show the luminosity function (in contrast to Fig. 26.8 the ordinate is not logarithmic) as it was derived, first by Hubble (1936, the interest of this curve being largely historical), by E. Holmberg (1950) from neighbouring galaxies, then by F. Zwicky (1957) from clusters of galaxies, the number of dwarf galaxies fainter than $M_{\text{phot}} \approx -12$ must be rather uncertain.

After these, necessarily extensive, preliminaries and by appealing to further observations, we now seek to gain insight into the composition of various stellar populations (see page 213 and Chapter 26) and also their occurrence in our Milky Way Galaxy and in distant galaxies of various types. Starting from the ideas developed in Chapter 26 concerning the evolution of stars and stellar clusters, it is natural to proceed to consider problems of the *evolution of galaxies*. Therefore we recall in the first place the investigations of W. Baade and others during the years 1944—60 and distinguish (see Table page 342) first in our Galaxy and then also in distant galaxies:

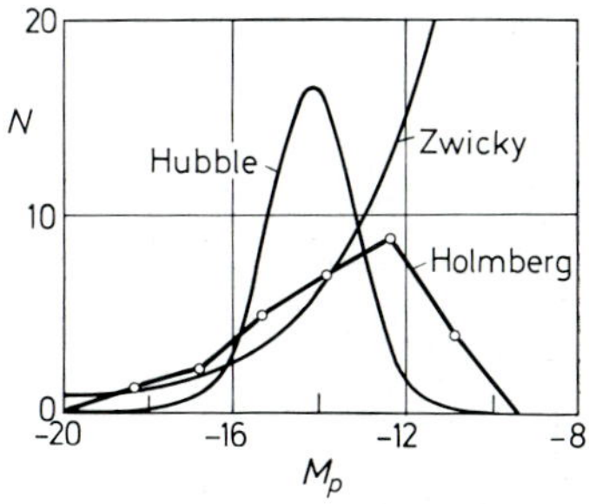

Fig. 27.8. *Luminosity function of galaxies* according to E. Hubble, 1936, E. Holmberg, 1950 and F. Zwicky, 1957

Extreme population I. To this belong all "young" formations in the spiral arms: interstellar gas, young galactic clusters with their O and B stars on the one hand, T Tauri and flare-stars on the other hand. The gas is concentrated into a disk of only about 200 pc half-value thickness; the stars that are manifestly formed from this material are still more strongly concentrated. Accordingly the velocity-components perpendicular to the galactic plane are small, $W \approx 8$ km/sec. The concentration towards the galactic centre is decidedly weak. The luminosity function corresponds to Salpeter's $\Psi(M_v)$ (Figs. 26.8, 26.9). In remote systems the blue stars at the upper end of the main-sequence with $M_{pg} = -7$ to -8 are the first features of population I to be identified. The A0 supergiants of the Magellanic Clouds reach -10^M. If red (super) giants are present, as for example in h and χ Persei, these differ little in absolute magnitude from the brightest blue stars. The spectroscopically determined abundance of heavy elements (relative to hydrogen) in the gas and in the stars of early spectral types is everywhere "normal". By way of the *older population* I with A-stars, N. Roman's "strong-line" and then the "weak-line" stars (interpreted as showing greater or lesser metal-abundance) we arrive at the important disk population.

Disk-population. The stars in the galactic disk between the spiral arms are still everywhere strongly concentrated towards the galactic plane (half-value thickness ~500 pc; W≈ 18 km/sec), but in the disk they are also concentrated towards the galactic centre. Besides the main body of stars in our neighbourhood, which like the Sun and stars of extreme population I move in nearly circular orbits around the galactic centre, the disk population includes the *planetary nebulae* (the concentration of which towards the galactic centre was found by R. Minkowski) and the ordinary *novae*, and among *variables* the RR Lyrae stars with periods $<0\overset{d}{.}4$. The chemical composition in the stars and in the planetary nebulae can be studied by quite different methods. Within the present accuracy of quantitative spectroscopic analysis (~factor 2) a difference from that of the extreme population I is *not* to be inferred*. The HR diagram of the disk population obviously agrees largely with that of the old galactic clusters like M 67 or NGC 188. It contains no bright blue stars since the main sequence turns off to the right at about $+3\overset{M}{.}5$ or about F5. Instead, the brightest stars are red giants of absolute magnitude $M\approx 0$ with colour index $\sim +1.3$. That these have "normal" metal abundance is shown by the obvious strength of the CN bands; so we speak of "CN giants"**. The mean colour index of the disk population, of which the luminosity function is the van Rhijn function $\Phi(M)$, is determined by the red giants and its value (C_{int}) is $+0\overset{m}{.}85$.

Intermediate population II, which joins on to the disk population, includes the high velocity stars of spectral types F to M, and the long period variables with $P<250^{d}$. The essential difference from the disk population lies in the weaker concentration towards the galactic plane and correspondingly greater *W*-velocity components, but more especially in the fact that here we now find stars whose metal abundances are reduced by factors from 1.5 to 20.

Extreme halo population II, into which the intermediate population II merges continuously, includes the majority of globular clusters. The colour-magnitude diagrams of these differ from those of the old galactic clusters like M 67 (Figs. 26.1, 26.3, 26.4) in that the giant branch extends to absolute magnitude -3^{M}, as contrasted with 0^{M}

* As regards the small differences between "strong-" and "weak"-line stars found by N. Roman (by visual estimation) the situation is still not completely clear.

** The abundance of such diatomic molecules is proportional to the square of the abundance of the heavy elements.

for M 67, and it lies somewhat more to the left (blue). The mean colour index of such a system is therefore $+0^{m}.56$, as against $+0^{m}.85$ for old galactic clusters. The brightest stars are red giants with -3^{M}, which are distinguished by weak CN bands. All these differences are related to the fact that in the halo population II the metal abundances are reduced by factors of from 10 to 200, as is confirmed by (unfortunately still not numerous) spectral analyses. Since in the F to M stars of the intermediate population II, and still more of the extreme halo population II, the numerous metal lines in the ultraviolet are weakened, in a two-colour diagram in which $U-B$ is plotted against $B-V$ these stars lie above the normal stars. Thus they can be recognized even beyond the reach of spectroscopic observation. The difference $\delta(U-B)$ of the colour indices of these stars from those of normal main-sequence stars of population I we call the ultraviolet *colour-excess*, cf. Fig. 23.10. Next to the brightest red giants with -3^{M}, the most characteristic stars are the cluster variables or RR Lyrae stars with periods $>0^{d}.4$, which contrast with the more infrequent type II cepheids or W Virginis stars having periods of about 14—20 days. However, the halo population II occurs not only in the globular clusters but, loosely distributed, it populates the whole halo of the Milky Way which forms an almost spherical subsystem of the Galaxy with a radius of 15 to 20 kpc. We at once recognize these high velocity stars by their large radial velocities and especially by their large W velocity components perpendicular to the galactic plane. We call the main-sequence stars of this population *subdwarfs*. For if we try to fit them to the usual spectral classification, then because of their weak metal-lines we should assign them to classes that are too "early" for their temperatures. In an HR diagram they would then lie to the left of the normal main sequence, i. e. below it. On the other hand, if we — more reasonably — plot the bolometric luminosity against the effective temperature of the stars, then the subdwarfs and stars with normal metal abundance fall on almost the same main sequence.

From the colour-magnitude diagrams we were already able in Chapter 26 to infer that the halo population II forms the oldest, the disk population a slightly younger, and the extreme (spiral-arm) population I the youngest subsystem of our Galaxy. The younger subsystems manifestly show an increasingly marked concentration towards the galactic plane and — as a dynamical corollary — smaller average velocity components W perpendicular to the plane. On the other hand, the concentration towards the galactic centre

	Age	Brightest stars	Metal abundance
Halo population II	10—15 $\times 10^9$ years	Red giants $-3^M.5$; CN weak	$\sim 1/_{200}$ to $1/_5$
Disk population	mainly about 10×10^9 years	Red giants $0^M.0$; CN strong	Normal (as in the Sun)
Spiral-arm population I	10^5 to 10^{10} years	Blue O and B stars -7^M to -8^M; a few fainter red giants	

decreases. At the same time, the metal abundance increases, although by no means uniformly but in such a way that among high velocity stars of the intermediate population II almost normal abundances occur and that from the disk population to the extreme population I there is no further increase. Also the kind of stellar cluster varies according to age: Typical galactic clusters exist only in the galactic plane, typical globular clusters only in the halo. However, N. U. Mayall and W. W. Morgan noticed that there are some globular clusters whose integrated spectra and colour indices show that their stellar population corresponds to that of the metal-rich old galactic clusters like M 67; at the same time they show a considerable concentration towards the galactic plane, corresponding largely to that of the intermediate population II. Also among the typical globular clusters in the halo there is incidentally a transition from extremely metal-poor clusters like M 92 (estimated at 1/200 normal) to clusters that are much less poor in metals.

As we outlined in Chapter 26 (section 5, page 273), E. M. and G. R. Burbidge, W. A. Fowler and F. Hoyle (1957) have proposed the following picture of the evolution of the Galaxy, in which we can fix the times of the most important occurrences by appeal to the colour-magnitude diagrams of clusters, etc.: Some 10 to at most 15×10^9 years ago there came into existence an almost spherical body of hydrogen, the halo. Whether it already contained a significant proportion of helium and perhaps traces of metals (certainly less than 1/200 present amounts), we do not know. Evidently in rapid succession, globular clusters and stars were formed that travelled round the galactic centre in highly elliptic paths. In connexion with the further evolution and death of these stars, as white dwarfs, heavy elements were also formed that, having become mixed with the interstellar gas, were available for the formation of further generations of stars. As the galactic disk was formed, the formation of heavy elements almost came to a halt, since the chemical composition of the youngest stars is scarcely different from that of the older stars of the disk. If the disk came into existence as a result of collapse from the halo, this occurrence — according to simple estimates from celestial mechanics — could have lasted only a few hundred million years, i. e. a time of the order of the period of rotation of the Galaxy. The majority of the disk stars would thus have an age of about 10×10^9 years.

Then spiral arms were formed in the disk, first from gas and dust. Ever since — right up to the present time — stellar clusters, associations, etc., of population I have been formed from these.

We entertain no illusions as to the hypothetical character of this picture; in the next chapter we shall return to it from the viewpoint of radio astronomy and high-energy astronomy.

Here we first study the stellar populations in galaxies of various Hubble types and of various sizes (masses); we distinguish between them only as indicated in the table alongside (page 294).

We seek also to gain for ourselves a picture of the time-dependences involved. We begin at the left-hand end of the Hubble sequence with the dwarf ellipticals. In a masterpiece of astronomical observation, W. Baade and H. Swope have studied most accurately the *Draco system,* a dwarf galaxy belonging to the local system with 48′ diameter, which is only about ten times larger than a globular cluster. Its colour-magnitude diagram (Fig. 27.9) as regards location and extent of the giant branch, in the structure of the horizontal branch with numerous RR-Lyrae stars, and indeed in every respect, resembles that of the metal-poor globular clusters of our Galaxy. Also the luminosity function with a sharp peak at the luminosity of the RR Lyrae stars exactly matches that of M 3 which is taken for comparison. Since moreover gas and dust are completely absent, we must conclude that the Draco galaxy, like the globular cluster M 92 or the subdwarfs of our halo, has so to say come to a standstill in the first stages of its evolution as a pure population II system. Also, because of the similarity of the colour-magnitude diagrams, its absolute age must largely correspond to that of the old globular clusters ($\approx 10 \times 10^9$ years).

Giant ellipticals present a different picture. In NGC 205 (Fig. 27.1), a medium-sized system, the brightest stars are indeed metal-poor red giants of population II of $-3^{\mathrm{M}}\!.5$. However, the mean

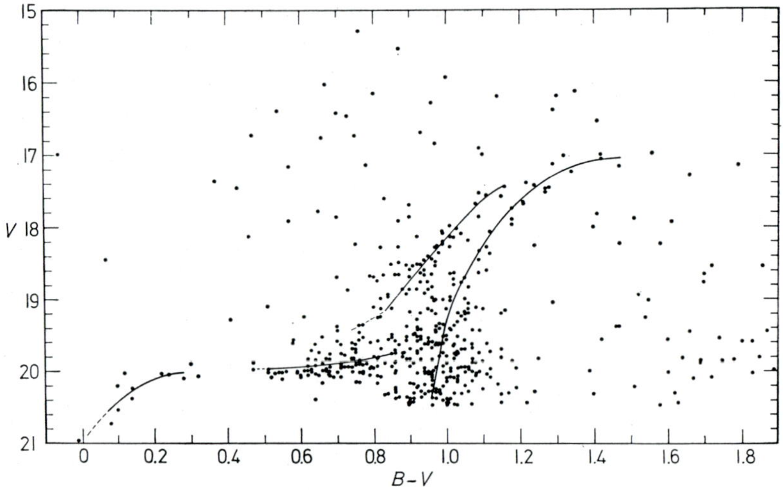

Fig. 27.9. *Colour-magnitude diagram of Draco System.* This dwarf galaxy is at a distance of 99 kpc and has a diamter of 1.4 kpc. The system is very similar to a metal-poor globular cluster, but is about ten times larger. (After W. Baade and H. Swope, 1961)

colour index and the integrated spectrum, i. e. the predominant part of the light, correspond to the metal-rich old disk population. Admittedly, even the brightest stars of this population cannot be distinguished as individuals. In some larger elliptical galaxies there are observed also shreds of gas- and dust-clouds in which, as could be expected, bright blue stars and gaseous nebulae are being formed.

As regards spiral galaxies we can be quite brief, since they largely resemble our own Sb Galaxy. The Andromeda galaxy, for instance, has a halo with numerous globular clusters, the brightest of which with $-9^{\mathrm{M}}\!.4$ exactly matches our ω Centauri with $-9^{\mathrm{M}}\!.8$. W. Baade was able to distinguish seven spiral arms in the disk; all contain gas and dust with blue stars and their surrounding gaseous nebulae embedded in them. The main part of the light of the disk between the arms, as the spectrum and colour indices confirm, comes here also from metal-rich stars of the (old) disk population, while the brightest red giants belong to the metal-poor population II. Thus, and in regard also to its origin, we must see M 31 as a true twin brother (though somewhat smaller) of our Galaxy. We may extend this conclusion also to the Sb and similar spirals of other clusters, which in all respects resemble M 31 but which we cannot study in as much detail.

From the Sc galaxies there is a continuous transition to the Irr I galaxies at the right-hand end of the Hubble sequence. We have already remarked that the "irregular" appearance, for example, of the Magellanic Clouds only comes from the fact that the gas and the bright blue stars that occur in it are situated in irregularly distributed OB associations of about 20 to 100 pc in diameter. The absolute magnitudes of their young O stars range up to -7^M, that of the A0 supergiants even as far as -10^M. Besides there are similar but much larger systems of diameter about 500 pc, and in other galaxies up to 3000 pc, which we can describe as *super-associations.* Possibly the formation of stars from the gas is a hierarchical process. All this concerns extreme population I.

The population I features are, however, embedded in a much more regularly distributed substratum of fainter stars. The rotation of the whole system is shown by the radial velocities of stars and by radio measurements of the 21-cm line. The mean colour index shows that this is a genuine disk population. There are, however, a great number of globular clusters with typical cluster variables and with colour indices $C = +0.58$ which exactly match the old halo population II of our Galaxy. Thus there can be no doubt that in the Magellanic Clouds, as in our Galaxy and the rest of the local group, star-formation set in about 10×10^9 years ago. Further, we can scarcely do other than carry this conclusion over to other clusters of galaxies whose members, so far as we can determine, do not differ in the slightest from those of the local group. Thus *in all galaxies the formation of stars began about ten billion years ago.*

On the other hand, that the Hubble sequence is not an evolutionary sequence follows from the great differences between masses (Table 27.1). Out of an irregular galaxy of $10^9 M_\odot$ one cannot produce an elliptical galaxy of $2 \times 10^{11} M_\odot$. The galaxies must have acquired their present state in their "early years". On the other side, as a result of comprehensive observations, W. Baade has verified that there exist no "young E galaxies", which would at once attract attention by the presence of blue stars and gaseous nebulae. The differences between galaxies of various types in population-content and in form is therefore connected with the fact that in the different systems star-formation proceeded at very different rates. This *cannot* depend upon the total mass or the mass-density of the systems, since there exist giant and dwarf systems among the ellipticals and among the Irr I galaxies. As already remarked, what distinguishes these types can really be only their *angular momentum.* Thus the rate of star-formation in a system appears to be closely bound up with its angular momentum. The physical mechanism of this connexion is at present just as obscure as the origin of the characteristic configurations in the Hubble sequence.

28. Radio Emission from Galaxies: Exploding Galaxies

Besides the well-known "optical window" from the ozone limit at about 3000 Å to the infrared at about 22 μ the Earth's atmosphere has a second range of transparency from $\lambda \approx 1$ millimetre to as far as the onset of ionospheric reflexion at $\lambda \approx 30$ metres. Since K. G. Jansky's discovery (1931) radio astronomy has exploited this spectral range.

In Chapter 9 (pp. 91—92) we have described radio telescopes and radio interferometers, for the precise measurement of the position and intensity-distribution of cosmic radio sources. By combining accurately calibrated measurements at various wavelengths λ, or frequencies $\nu = c/\lambda$, we obtain *radio spectra.* In recent times it has also become possible to measure the *polarization* of radiation in radio-frequencies from cosmic sources. In Chapter 20 (pp. 185—194) we became acquainted with the radio emission from the Sun. This is composed of the thermal radiation of the quiet Sun produced by free-free transitions in the plasma of the corona at 1 to 2 million degrees, and also of the non-thermal radiation of the disturbed Sun, which, along with other phenomena of solar activity, is produced by plasma oscillations, synchrotron radiation (see below) etc.

As regards the radio radiation of the Galaxy, we have already to some extent discussed the thermal part in Chapter 24 (pp. 232 ff. and 237—238). We have to do, on the one hand, with the line-emission of neutral hydrogen (HI regions) at $\lambda = 21.105$ cm or $\nu = 1420.40$ MHz and of the OH radical $\nu = 1665.40$ and 1667.36 MHz as well as 1612.20 and 1720.55 MHz and, on the other hand, with the continuum of free-free radiation of ionized hydrogen, etc., especially in the HII regions and gaseous nebulae near the galactic plane and in planetary nebulae. We recognize such thermal continua by the fact that in the case of emission by an optically thin layer the intensity, or the flux per unit frequency interval S_ν, is nearly independent of ν. The radiation-temperature T_ν according to the Rayleigh-Jeans law* behaves approximately as $T_\nu \sim \nu^{-2}$. In contrast to this, the radio emission of the Galaxy coming from all over the sky, as well as that of the strong radio-sources, shows a spectrum expressed approximately by

$$I_\nu \sim \nu^{-0.7} \quad \text{or} \quad T_\nu \sim \nu^{-2.7}. \tag{28.1}$$

Also taking account of absorption and self-absorption, this cannot be ascribed to thermal emission. Again, the radiation-temperatures of more than 10^5 °K measured at long wavelengths can scarcely be interpreted as thermal. So H. Alfvén and N. Herlofson 1950 appealed to the mechanism of *synchrotron radiation,* or magneto-bremsstrahlung, for the explanation of the non-thermal radio continua. These notions were then further developed by Shklovsky, Ginzburg, Oort and others. It has been known to physicists that relativistic electrons (i. e. electrons moving with nearly the speed of light, whose energy E appreciably exceeds their rest-energy $m_0 c^2 = 0.511$ MeV), that describe a circular path in the magnetic field of a synchrotron, emit intense continuum radiation in their direction of motion, the spectrum of which extends into the far ultraviolet. This continuum differs from that of free-free radiation or bremsstrahlung in that the acceleration of the electrons results, not from atomic electric fields, but from a macroscopic magnetic field H.

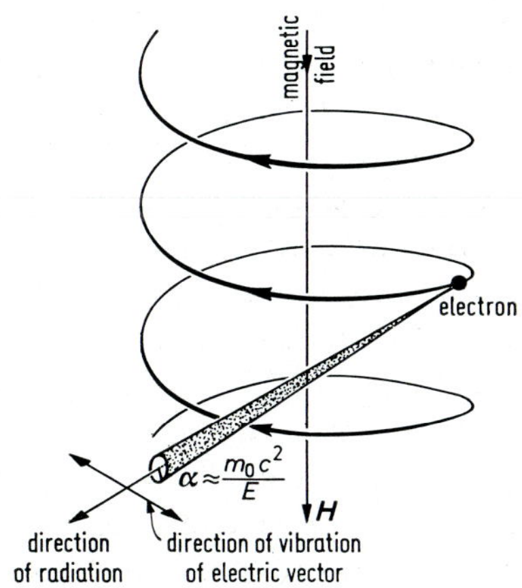

Fig. 28.1. *Synchrotron radiation* of a relativistic electron in a magnetic field H

V. V. Vladimirsky and J. Schwinger developed the theory of synchrotron radiation in 1948—49. It rests upon the following consideration. According to the laws of relativistic kinematics, an electron circulating with almost the speed of light emits radiation within a narrow cone of angle $\alpha \approx m_0 c^2/E$ (Fig. 28.1). Like a lighthouse-beam, this sweeps over the observer in rapid succession so that, taking account of the relativistic doppler effect, he receives a succes-

* Equation (11.25): $S_\nu = 2\,\nu^2 k T_\nu / c^2$.

sion of radiation-flashes, each of duration Δt, say. The spectral resolution or, mathematically speaking, the Fourier analysis of the radiation gives a continuous spectrum, whose maximum lies at the circular frequency $\sim 1/\Delta t$. More exact calculation gives for the frequency of the maximum ν_m [Hz] the formula

$$\nu_m = 4.6 \times 10^{-6} H_\perp E_{eV}^2 \tag{28.2}$$

where $H_\perp$ is the component of the magnetic field (in gauss) perpendicular to the direction of motion of the electron and E_{eV} is its energy in electron volts. If we work with a mean galactic magnetic field of $H_\perp \approx 5 \times 10^{-6}$ gauss, the following electron energies correspond to the various frequencies ν_m or wavelengths λ_m shown:

$$\begin{array}{llllll} \lambda_m = & 3000\,\text{Å} & 30\,\mu & 3\,\text{mm} & 30\,\text{cm} & 30\,\text{m} \\ \nu_m = & 10^{15} & 10^{13} & 10^{11} & 10^{9} & 10^{7}\,\text{Hz} \\ E = & 6.6 \times 10^{12} & 6.6 \times 10^{11} & 6.6 \times 10^{10} & 6.6 \times 10^{9} & 6.6 \times 10^{8}\,\text{eV}. \end{array} \tag{28.3}$$

Even for the production of radiation in the radio-frequency range, and certainly in the visible, electrons must be available with energies in the region of those of cosmic rays. If we can represent the energy-distribution of the electrons by a power law

$$N(E)\,dE = \text{const}\, E^{-\gamma}\, dE \tag{28.4}$$

then the intensity of the synchrotron radiation is given by

$$I_\nu \sim H^{(\gamma+1)/2} \nu^{-(\gamma-1)/2}. \tag{28.5}$$

In 1965 J. L'Heureux and P. Meyer measured the energy-distribution of the electrons in primary cosmic rays in the range 0.5 to 3×10^9 eV (see (28.3)) and they obtained $\gamma = 1.6$. This would give $I_\nu \sim \nu^{-0.3}$ which could still be compatible with the radio-astronomical measurements (28.1). As to the production of the synchrotron electrons in the Galaxy, either they could be *primarily* accelerated along with the heavy ions of cosmic rays, or they could arise *secondarily* from the heavy component through proton-proton collisions. In the first case at high energies about equal numbers of positive and negative electrons must be produced, while in the second case positive electrons would be somewhat favoured.

In 1939 when G. Reber investigated the distribution of radio-radiation over the sky at $\nu = 167$ MHz or $\lambda = 1.8$ m, making use of his radio telescope of modest angular resolving power, he noted its concentration towards the galactic plane and towards the galactic centre. According to equation (28.5) the distribution of galactic radio-emission over the sky gives information about the density of relativistic electrons and their distribution in the galaxy. This is of great significance since the original distribution of directions of the charged particles arriving at the Earth cannot be inferred, because of their complicated deflexions by the terrestrial and interplanetary magnetic fields.

Radio astronomy and also, as we shall see, astronomy with X-rays and γ-rays, presents us with the possibility of learning something about the occurrence of highly energetic electrons in cosmic systems. An obstacle is that we still know little about the magnetic field H that also appears in (28.5). At present we can still not determine how the

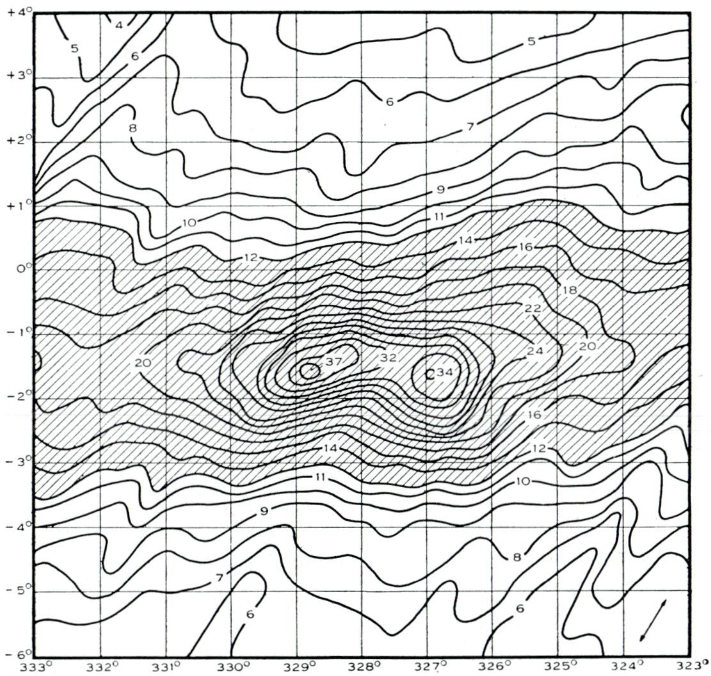

Fig. 28.2. *Survey of the neighbourhood of the galactic centre* at λ 3.5 metres by E. R. Hill, O. B. Slee and B. Y. Mills, 1958. The galactic coordinates are l^{I}, b^{I} (see equation 23.11). The numbers on the isophotes denote radiation temperatures in units of 10^3 °K. The mean direction of the scan, made with the Mills cross, is shown by the arrow in the lower right hand corner

galactic emissivity in radio-frequencies derived from the observations is shared between the factor $\sim H^{1.7}$ and the factor N_e, the density of relativistic electrons in the relevant energy-range.

In addition to the non-thermal radiation from the galactic disk (Fig. 28.2) (which joins up with the thermal radiation of ionized hydrogen at decimetre and centimetre wavelengths) one observes, also in the metre-wavelength region, another component which is distributed relatively uniformly over the whole sky. Following J. E. Baldwin, this has been ascribed to the galactic halo in which there ought to be present (dynamically independent of the stars):

a) relativistic electrons, b) a magnetic field, c) a plasma, in which the magnetic field is "suspended". In order to explain the smallness of its concentration towards the galactic plane, we should ascribe to the plasma either macroscopic velocities (like those of high velocity stars) of some hundreds of km per second, or corresponding thermal velocities, i. e. a temperature of from 10^5 to 10^6 °K.

More recent measurements with ever greater angular resolving power have traced an ever greater proportion of the "halo-radiation" to galactic and, still more, extragalactic radio sources (see below) that are densely distributed over the sky. By now it is questionable as to whether between the many sources anything at all is left over for the galactic radio-halo!

We turn now to the (discrete) radio sources whose investigation in the radio and optical ranges in recent times has led to such exciting advances in astronomy, cosmogony and cosmology.

In 1946 J. S. Hey and his collaborators discovered the first radio-source Cygnus A because of its intensity fluctuations. Simultaneous observations at stations far removed from each other showed later on that these fluctuations are to be interpreted as scintillation produced in the ionosphere, chiefly in the F2-layer at a height of about 200 km. Like optical scintillation which occurs only for stars but not for planets (which have a larger angular diameter), radio scintillation occurs only in the case of sources of sufficiently small angular diameter. For the so-called quasi-stellar radio sources (see below), some of which have angular diameters less than a second of arc, another sort of scintillation has recently been discovered, which is produced in the interplanetary plasma. With the help of giant radio telescopes and radio interferometers (Chapter 9, page 91), M. Ryle at the Mullard Radio Astronomy Observatory of the Cavendish Laboratory in Cambridge (England), A. C. B. Lovell at the Nuffield Radio Astronomy Laboratories at Jodrell Bank (England), B. Y. Mills and the late J. L. Pawsey at the Radiophysics Laboratory of C.S.I.R.O. in Sydney (Australia), their collaborators, and others have discovered more and more radio sources and measured their position, angular diameter or intensity distribution, and radio spectrum.

Once radio-determinations of position had attained sufficient accuracy, the difficult problem of the optical identification of cosmic radio sources could be tackled. We divide up our review according to the type of source and restrict ourselves to some characteristic examples.

A. Non-Thermal Galactic Radio-Sources

1. *Type I supernova remnants.* In 1949 J. G. Bolton and his coworkers succeeded in identifying the radio source Taurus A, which they had discovered, with the Crab nebula (M 1 = NGC 1952; $m_{pg} = 9.0$). W. Baade and R. Minkowski had in 1942 made an intensive study of this highly interesting object. It consists of an inner, almost amorphous region $3\overset{\prime}{.}2 \times 5\overset{\prime}{.}9$ with a continuous spectrum, and an envelope whose bizarre filaments (the "legs" of the Crab) radiate mainly in Hα. As one can infer from records in east Asia, the Crab nebula is the remains of a supernova of type I which in 1054 AD attained a maximum brightness $m_v = -5$ or $M_v = -18$. Confirmation of the date of the outburst and also the parallax are given by comparing the radial velocity of expansion of ~ 1100 km/sec with the corresponding proper motion, which has also been measured. I. S. Shklovsky in 1953 advanced the clever idea that we should interpret the radio radiation and optical continuum of the Crab nebula as synchrotron radiation. The following arguments supported this hypothesis: For thermal radiation the electron temperature T_e would have to be greater than the highest radio radiation temperature T_ν, that is greater than 10^9 °K, while in all gaseous nebulae T_e never exceeds 10^4 °K. Also the amorphous nucleus shows only a continuum in the optical range and no lines. A telling argument in favour of the synchrotron hypothesis came, moreover, from the verification of the polarization of the continuum (Fig. 28.1) which it demands. For the optical spectral range, this was obtained by Dombrowsky and Vashakidze 1953/54, and later with better resolution and accuracy by Oort, Walraven and Baade. In spite of difficulties caused by the Faraday effect, polarization measurements in the centimetre and decimetre range were also secured later on. A penetrating discussion of the lifetime of relativistic electrons in the Crab nebula led Oort to the view that even still, almost a thousand years after the supernova outburst, such electrons are being continually produced.

The remnants of the type I supernova of 1572 (Tycho Brahe) and that of 1604 (Kepler) could be observed radio-astronomically and optically; Cassiopeia A, the strongest radio source in the northern sky may also be the remnant of a supernova.

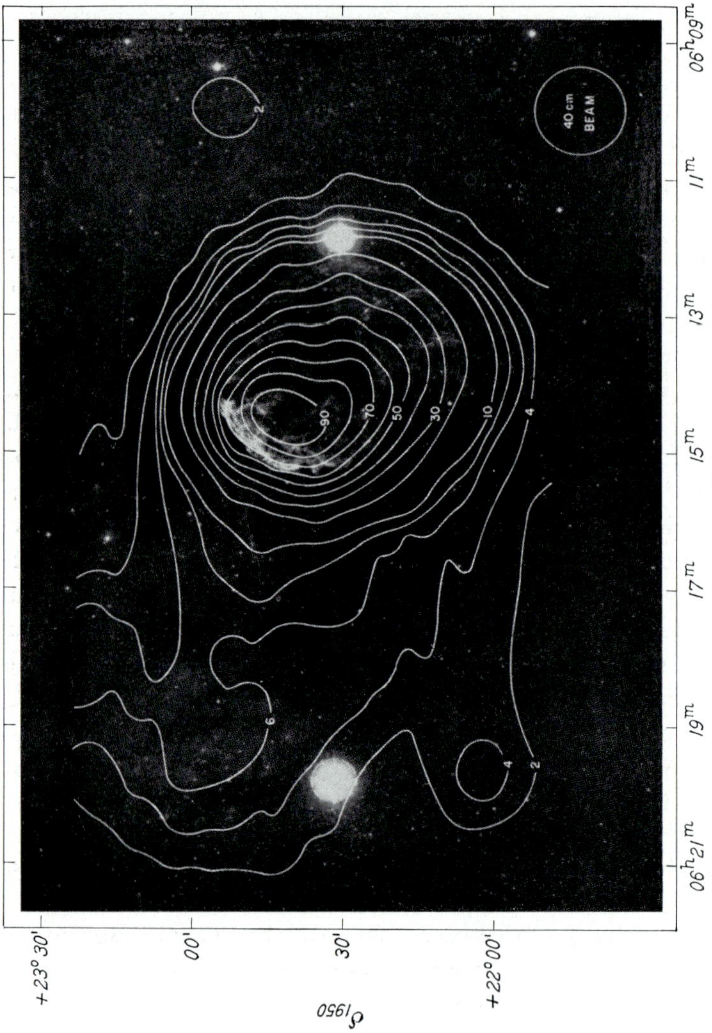

Fig. 28.3. *Distribution of radio brightness* at λ 40 cm, in IC 443, a supernova remnant. The lines of equal radiation temperature are marked in units of 0.95 °K. The radio map is shown superposed on a photograph taken from the Mount Wilson and Palomar Sky Survey

Then pictures taken in Hα-light show a large number of extended ring-shaped or circular nebulae that also emit radio-frequency radiation, like the well-known Cygnus-loop, the Veil nebula or IC 443 (Fig. 28.3). Finally, the north galactic radio "spur", which has been a puzzle for a long time, extending from near the galactic centre almost to the galactic north pole, as well as certain other similar structures, appears to be part of a gigantic ring round the sky. All such radio sources might likewise be connected with old supernovae.

2. *Flare-stars.* It has been known since the early days of radio-astronomy that flares or eruptions on the Sun are associated with extraordinarily intense non-thermal radio emission. The physical mechanisms are extremely complex; apparently both plasma oscillations and the synchrotron process play essential parts. In the 1950s it was discovered that a whole series of red dwarf stars exhibit frequent flares, which resemble spectroscopically the specially strong "white" flares on the Sun. For example, in the case of the dM5.5 star Luyten 726-8B ($m_v = 13.5$, $M_v = 16.6$) since 1900 astronomers have observed 14 flares each giving an increase in luminosity of the whole star by more than 1^m. All these flare-stars, which are similar to T Tauri stars and RW Aurigae stars, are evidently young stars, still in process of being formed. It was natural to ask whether such stars, in spite of their remoteness, also produce detectable *radio* flares. A joint undertaking, ably organized by A. C. B. Lovell, of Jodrell Bank and Sydney, with simultaneous optical observations at the Smithsonian Astrophysical Observatory and several observatories in the USSR, brought the confirmation that in fact *stellar radio flares do exist.* Moreover, in these the ratio of radio radiation to radiation in the optical continuum exceeds the usual value for large solar eruptions 100-fold to 1000-fold. This seems perhaps not so astonishing if we remember that in red dwarf stars the hydrogen convection-zone (and very likely also the magnetic field) is considerably more strongly developed than in the Sun.

B. Extragalactic Radio Sources

1. *Normal galaxies.* In 1950 R. Hanbury Brown and C. Hazard were able to show that the Andromeda galaxy M 31 radiates in radio-frequencies with about the intensity to be expected by reason of its similarity to our Galaxy. More recent observations on M 31 and other normal spirals show that a considerable part of the radio emission can be ascribed to the nucleus and the disk. In M 31 and certain others, but not all spiral galaxies, a radio halo appears to be present. However, on account of its small surface brightness, it stands out feebly from the fluctuating background, so that its true extent, and maybe even its existence, is beset by some uncertainty.

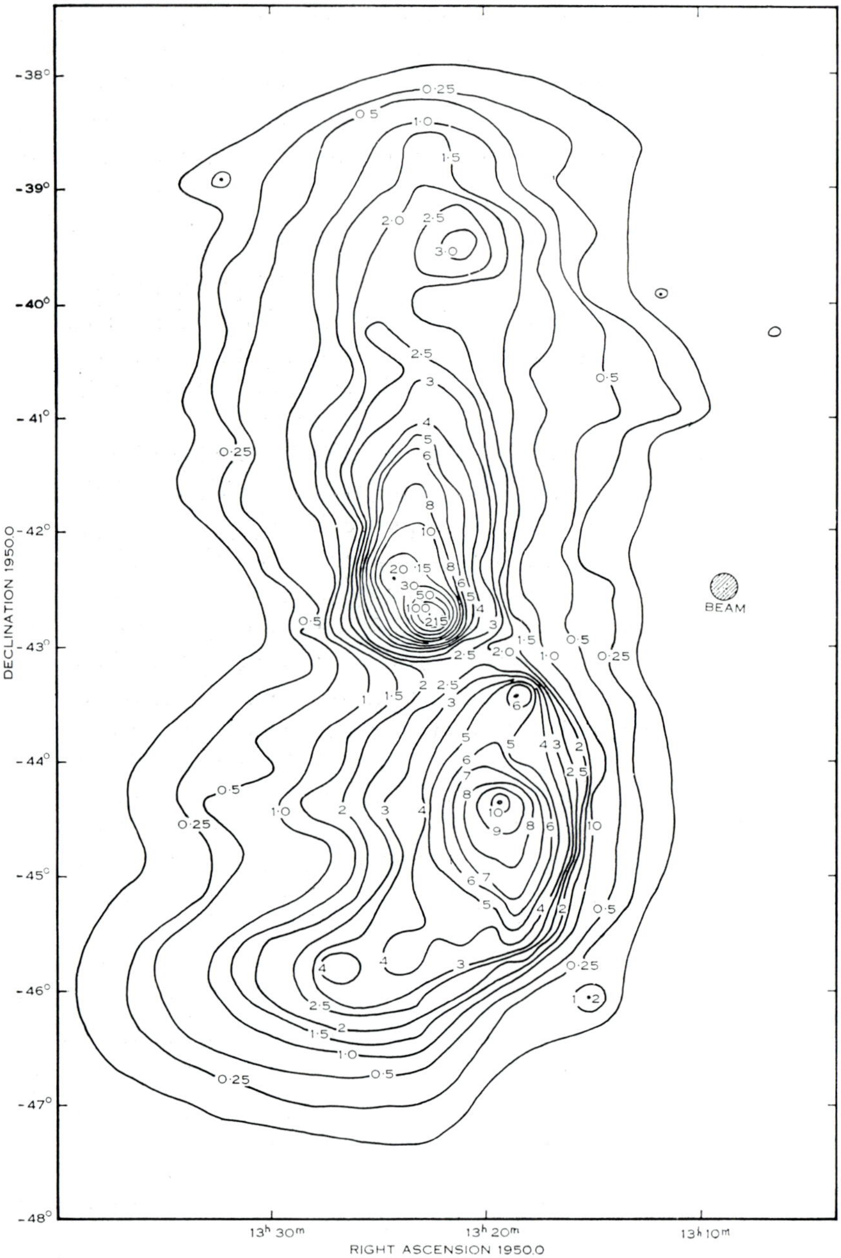
DECLINATION 1950.0
RIGHT ASCENSION 1950.0
BEAM
13h 30m
13h 20m
13h 10m
-38°
-39°
-40°
-41°
-42°
-43°
-44°
-45°
-46°
-47°
-48°

Fig. 28.4 a

2. *Radio galaxies.* By 1954 the measurements of the positions of several of the stronger radio sources had attained such accuracy that W. Baade and R. Minkowski succeeded in identifying them with optical objects. In particular, the second strongest radio source in the northern sky, Cygnus A, could be associated with a remarkably faint object of photographic magnitude $17^{m}_{.}9$. Besides a weak continuum and the Hα line, it shows forbidden lines of [OI and III], [NII],

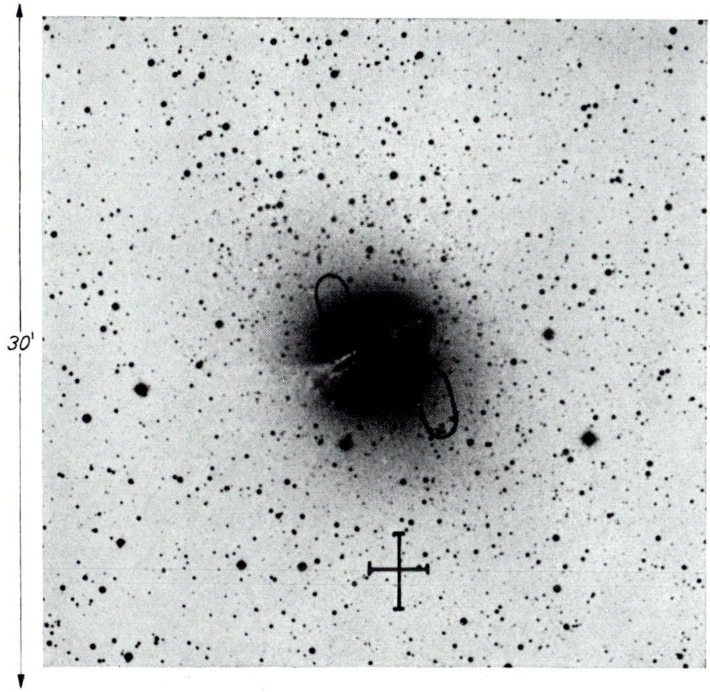

Fig. 28.4 b

Fig. 28.4. *Radio source Centaurus A* (NGC 5128). a) Isophotes of the radio continuum at 1420 MHz (λ 21 cm) by B. F. C. Cooper, R. M. Price and D. J. Cole, 1965. Radiation temperatures are given in °K. The circle (right) of 14′ diameter denotes the resolving power (beamwidth at half intensity) of the aerial. The corresponding 960 MHz isophotes are shown (schematic only). b) The E0 galaxy NGC 5128 (RA $13^{h}\,22^{m}\,31^{s}_{.}6$, $\delta -42°\,45'_{.}4$, 1950) has been identified with the unresolved central source in a).

According to interferometer measurements by P. Maltby (1961) this galaxy itself consists of two sources, which are almost symmetrically placed with respect to the remarkable absorbing lane which surrounds the galaxy. There is no *optical* indication of either the narrow (younger) or wider (older) pair of radio sources. The sole indication of the explosion responsible for flinging the radio sources out of the core is a pair of weak extensions just discernable in Fig. 28.4 b, and approximately along the line joining the two inner radio sources. This line is itself nearly parallel to the rotation axis of the galaxy

[NeIII and V] . . ., with a redshift (Chapter 29) corresponding to 16 830 km/sec. Thus we have to do with an extragalactic object at a distance about 170 Mpc. The optical picture shows two "nuclei" about 2″ apart. Baade and Minkowski first interpreted this as two colliding galaxies. In the collision the stars would be only a little affected; the gas, on the other hand, would be swept out of both galaxies and excited into a state of radio emission. Subsequent investigation has not been able to endorse this picture. Measurements of the radio intensity-distribution of Cygnus A made with the great Jodrell Bank interferometer-system showed that the greater part of the radio emission comes from *two components* separated by about $100'' \approx 80$ kpc along the axis of the optical system. Each is of about 20 kpc diameter and the two are connected together across the "optical" galaxy by a weakly emitting bridge. The components must certainly include a magnetic field and synchrotron electrons. Whether less energetic particles are also present we cannot tell. Optically, *nothing* is detectable that has to do with the radio components. Many strong radio sources show a similar structure with two enormous "balloons", or "plasmons", approximately symmetrically placed with respect to a galaxy along its axis of rotation.

The radio galaxy Centaurus A = NGC 5128 (Fig. 28.4) shows in the first place in the optical picture a rather turbulent band of dark material perpendicular to the axis. In exposures with high contrast, we can trace faint appendages along the rotation axis out to about 40 kpc in both directions. The radio isophotes show two "balloons" at about 3° separation from each other and another pair at about 6′. In both systems the radio radiation is polarized. It is natural to ascribe the occurrence of such double radio sources to gigantic explosions in the interiors of the galaxies concerned, in which plasma with magnetic field and synchrotron electrons is driven out approximately symmetrically towards both sides. Since, for example, in Centaurus A the separation of the "balloons" from the galaxy is about 400 000 light-years and 13 000 light-years, respectively, at anyrate two explosions must have occurred in this galaxy *at least* those numbers of years in the past, since the speed of propagation must be less than c. The energy radiated during the intervals might well come only from the explosions themselves.

3. *M 82, an exploding galaxy.* In 1963, C. R. Lynds and A. R. Sandage succeeded in observing optically an explosion, of such a character as that just described, in the central part of the galaxy M 82.

It is natural to suppose that we are here witnessing the genesis of a radio source similar to Cygnus A or Hercules A.

Going by its membership of the M 81 group, the galaxy M 82 is at a distance of about 3 Mpc; since 1961 it has been known to be a radio source. Superficially it has the appearance of an irregular galaxy. Actually we have to do with a flattened galaxy out of whose nucleus, as spectra and photographs in Hα show, enormous masses of hydrogen are shooting out in both directions along the axis. The speed of each mass is proportional to its distance from the nucleus. We can trace these *filaments* on both sides out to a distance of about 4000 pc from the nucleus of the galaxy; there the speed is about 2700 km/sec. Hence we easily calculate that the explosion took place 1.5 million years ago. From the intensity of Hα we can infer the mass of the ejected filaments to be $\lesssim 5.6 \times 10^6$ solar masses, and the mean density

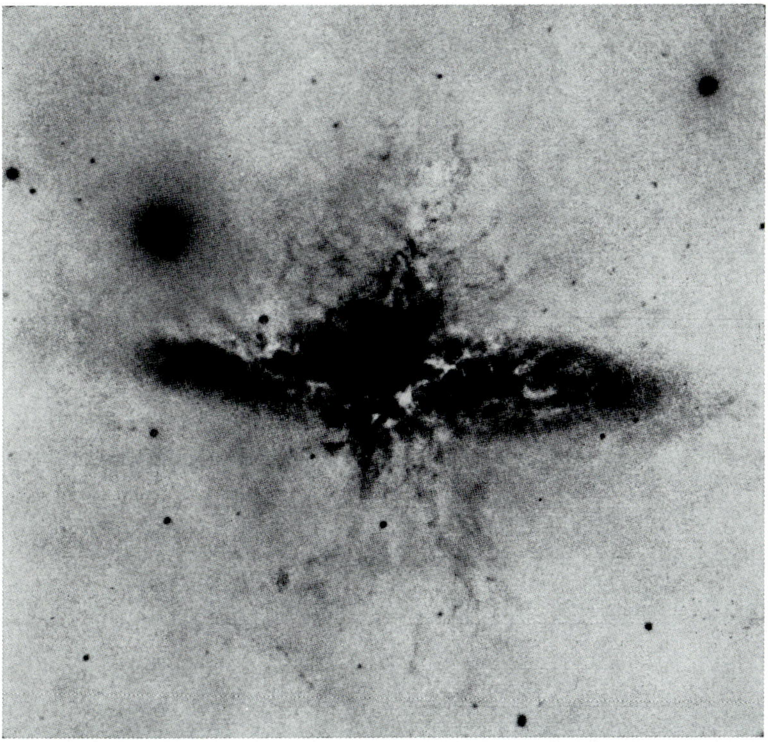

Fig. 28.5. *Radio source M 82.* Hα photograph (negative) taken by A. R. Sandage (1964) with Mount Wilson and Palomar Observatories' 200-inch telescope. The hydrogen filaments were flung out to about 4000 pc. on either side of the disk of the galaxy by an explosion which occurred about 1.5 million years ago

to be about 10 protons/cm^3. The total kinetic energy of the moving bodies of gas comes to $\gtrsim 2.4 \times 10^{55}$ erg.

For comparison we can easily verify that 5.6×10^6 Suns in a time interval of 1.5 million years would radiate about 4 percent of this amount of energy.

The synchrotron emission of M 82 integrated from the region of radio waves into the optical region (see Fig. 28.5) for the 1.5 million years since the explosion (assuming the emission to remain constant in time) amounts to 9×10^{55} erg. This amount of energy must then have been stored up in synchrotron electrons with energies up to about 5×10^{12} eV. According to Lynds and Sandage, this appears to be possible, if the magnetic field is less than 2×10^{-6} gauss. The total amount of energy liberated in the explosion can be only tentatively estimated at 10^{56} to 10^{58} erg. The present emission in the radio-frequency range (from 10^7 to 10^{11} Hz; see (28.3)) for radio-galaxies like Cygnus A is about 10^5 times stronger than that from M 82 (5.1×10^{44} as compared with 4.2×10^{39} erg sec^{-1}). Since the age of these sources is comparable with that of M 82, here total energies estimated at 10^{62} erg have to be produced. Before we put forward suggestions about their origin, we shall report briefly upon another type of source.

4. *Quasi-stellar radio sources (QSS) and quasi-stellar galaxies (QSG).* Measurements with the radio interferometer at Jodrell Bank using baselines up to 132 km, and the analysis with respect to time of the occultation of radio-sources by the Moon, gave a resolving-power better than 1″. It then appeared that a number of not particularly weak 3C sources (third Cambridge catalogue of discrete radio-sources) have diameters of this order. They must therefore possess very high radio surface-brightness. It was possible to identify them with star-like images in plates taken with the 48-inch Schmidt-camera and the 200-inch Hale telescope at Palomar. Finally in 1962/63 M. Schmidt was able to show from the spectra that one has here to do with very compact blue galaxies, whose spectra consist of a structureless continuum and strong emission lines, similar to those of already known radio-galaxies. The enormous redshifts (see Chapter 30) showed that one was dealing with very remote galaxies, which in the optical region out-shine normal spiral galaxies 50- to 100-fold. The designation *quasar* or QSS = quasi-stellar radio source was invented for them. Their radio power (integrated, say, from 10^7 to 10^{11} Hz) corresponds to about that of the strong radio galaxies like Cygnus A. There

appears to be a more or less continuous transition to these (possibly an age-sequence); some quasars also have the typical double structure.

Then in 1965 A. Sandage showed that there exists a very extensive family of quasi-stellar galaxies (QSG) which differ from the QSS by the absence of strong radio emission but which optically are little different. Their relatively large brightness in the ultraviolet makes it possible to distinguish them from stars in a two-colour diagram with $U-B$ plotted against $B-V$, in spite of their star-like images. In magnitudes fainter than $m_{pg} \approx 16$ many of the blue objects in high galactic latitudes (where galactic absorption is weak) are such *blue compact galaxies.* Also the surveys of blue stars in high galactic latitudes by G. Haro in Tonantzintla and by M. Humason and F. Zwicky at Palomar had often lit upon such objects. Their spectra generally resemble those of quasars. Application of the methods of stellar statistics (see Chapter 23, page 216), which must here be generalized to be compatible with relativistic cosmology (Chapter 30), first led to the realization that the QSS are considerably rarer in cosmic space than the QSG and obviously also than normal galaxies.

Must we therefore draw the conclusion that the QSS represent a relatively short-lived transition stage in the evolution of galaxies? Or should we, more fundamentally and contrary to what the present state of the rest of the galaxies seems to teach us (Chapter 27), suppose that quasi-stellar radio sources are continually being formed anew even at the present time?

Let us summarize our knowledge of extragalactic radio sources. In the nuclei of galaxies (dwarf galaxies are probably to be excepted) there occur — more often than once in the same galaxy under certain circumstances — stupendous explosions in which great quantities of synchrotron electrons up to about 5×10^{12} eV, and certainly also heavy particles with comparable energies, are generated. The optical spectra of all "excited" galaxies show emission-lines with doppler effects of many hundreds to thousands of km/sec. In the so-called Seyfert galaxies (only some of which emit detectable radio radiation) one notices only the strongly broadened emission lines in the nucleus. In the giant elliptical M 87 = NGC 4486, the radio source Virgo A, one observes a jet shot out from the centre, that emits bluish polarized light, i. e. optical synchrotron radiation. In the more strongly rotating galaxies two "plasmons" are ejected approximately symmetrically along the axis. In these, synchrotron electrons plus magnetic field (that are obviously jointly derived from the galaxy concerned) produce the radio-radiation, while nothing is to be seen of them in optical frequencies. The energies that can be released in such galactic explosions we may estimate at about 10^{62} erg (and some authors go so far as 10^{64} erg). This corresponds to the nuclear energy $0.007\, mc^2$ of about 8×10^9 solar masses of hydrogen. As the most extreme possibility, we note that 10^{62} erg is equal to the total relativistic rest-energy mc^2 of 6×10^7 solar masses. Since at present we scarcely see how it could be possible for nuclear energy alone to be liberated in the manner required, the possibility has been discussed that a mass of from about 10^7 to $10^9\, M_\odot$ collapses to extreme densities. In such a *gravitational collapse* a considerable fraction of the rest energy mc^2 would be released first as lost potential energy. This raises difficult problems of general relativity theory which we cannot pursue here.

The discovery of galactic explosions with a huge production of energetic particles opens up a line of thought that might lead to a substantial revision of the ideas developed in the preceding chapter concerning the origin of the heavy elements and the evolution of galaxies.

V. A. Ambartsumian (see also page 213) has made the suggestion that as the initial state for the evolution of a galaxy, like our Milky Way Galaxy, we should not consider the condensation of a protogalaxy or of the halo from intergalactic material. On the contrary, we should start from a *compact galaxy.* We may characterize this in the earliest stage as the nucleus of a galaxy without any other attachments. The importance of such structures has recently been recognized by F. Zwicky and A. Sandage and others. Ambartsumian considers that the halo of our Galaxy might have been produced by a mighty explosion, like that of radio double-sources, from a compact primitive galaxy. One would seek also to relate the formation of the heavy elements to this occurrence.

We regard the following considerations as relevant to this concept:

a) The mean value of the z-component of the galactic angular momentum per unit mass of the stars of the halo population II is sursprisingly small. The mean value of the V-velocity-components of globular clusters taken as $+83$ km/sec multiplied by about 10 kpc corresponds to about the mean angular momentum per unit mass for the whole galactic disk. Since, after the condensation of its stars, the halo could no longer exchange angular momentum, this result seemed to fit the idea that the halo and disk came from the same initial body of matter. However, it has recently been shown that for extreme subdwarfs $\overline{V} = +67$ km/sec and, moreover, for metal-poor RR Lyrae variables $\overline{V} = 30$ km/sec only. The angular momentum per unit mass resulting from these values is 2 or 3 times smaller than the mean for the Galaxy. This seems to be immediately understandable if the halo stars come from the nucleus; on the other hand it seems to be scarcely compatible with the idea of condensation from intergalactic material.

b) The extremely rapid formation of stars in the halo ($<2\times10^8$ years) according to B^2FH (E. M. and G. R. Burbidge, W. A. Fowler and F. Hoyle 1957) is difficult to reconcile with the view supported by other observations that star formation is favoured by high densities (about $\sim\varrho^2$).

c) The astonishing uniformity of the relative abundances of the heavy elements from the oldest subdwarfs to the youngest stars of the spiral-arm population I is difficult to reconcile with the view of B^2FH that these elements must have been formed by quite different processes in quite different cosmic structures. In Ambartsumian's view, on the other hand, all the heavy elements were formed in a single "great process".

d) H. Suess (1965) has referred to the fact the abundance-distribution of heavy elements exhibits a series of regularities which are not explained by the B^2FH-theory. They point rather in the direction of the theory advanced in 1949 by M. Goeppert-Mayer and E. Teller of a formation of heavy elements by fusion-processes from a neutron-rich "nuclear fluid".

On the other side, however, Ambartsumian's theory must also explain how it happens that at the beginning of the explosion only hydrogen-rich, i. e. metal-poor, subdwarf-material is driven out, and this is then quickly followed by the rest of the material with large metal abundance of almost the same composition. Since we still know very little about the physics of galactic explosions, it seems too early to pursue further speculations on the subject. Future investigations may make possible a decision between the two views of the "pre-history" of the Galaxy.

29. Cosmic Rays: High Energy Astronomy

To-day one investigates primary cosmic rays, before they have started to interact with the material of the Earth's atmosphere, by use of balloons, rockets and artificial satellites. They include protons, α-particles and heavy nuclei. By means of tracks in photographic emulsions, one can determine their charge-number Z and then one finds the following abundance-distribution compared with ordinary stellar material of population I. The table gives log N referred to hydrogen = 12, where N is the relative number of atoms.

	H	He	Li, Be B	C, N O, F	Heavy nuclei $Z \geqq 10$	$23 \leqq Z \leqq 30$	$Z > 30$
Primary cosmic rays	12.0	11.2	9.4	10.1	9.6	9.0	<6
Stellar material	12.0	11.2	3	9.1	8.0	7.0	4

(29.1)

The abundance of the easily-destroyed nuclei Li, Be, B, which arise as secondaries by spallation of heavier nuclei, by comparison with experiments on large nuclear accelerators, leads to the inference that from their origin right up to their entry into the Earth's atmosphere, the cosmic rays have traversed cosmic material amounting to no more than about 3 g/cm^2.

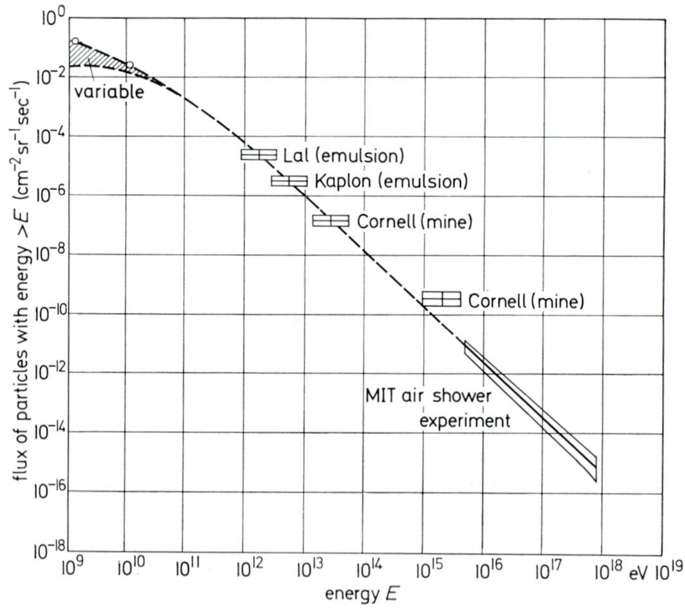

Fig. 29.1. *Integral energy-spectrum of cosmic rays:* number of particles having energy greater than E passing through one square centimetre per unit solid angle per second plotted as a function of E

Neutrino astronomy. Quite fantastic possibilities would be opened up by a detection of cosmic neutrinos *. We have already remarked that several percent of the energy generated inside a star escapes directly into cosmic space as neutrino emission; here we cannot go into other generating processes. The interaction cross-section of the neutrino with respect to every kind of matter allows it to traverse the entire universe without suffering a collision. Thus neutrino astronomy could give us the most direct information, for example, concerning the energy-generating core of the Sun, and on the other hand above all also concerning the neutrino production of the entire cosmos in connection with its evolutionary history and with its interrelations in the large (Chapter 30).

The detection of neutrinos proceeds in a radio-chemical way through their nuclear reactions with Cl^{37} or by detecting their secondary particles using suitable arrays of coincidence counters. R. Davis, Jr. (1964) undertook a first attempt to detect neutrino emission from the Sun. In a 700 metre-deep mine (for protection against secondary particles from cosmic rays) he exposed a tank containing 6.1 tons of C_2Cl_4 for several months in order to accumulate the radioactive Argon-isotope produced by the neutrinos and to detect it radiochemically. He obtained an upper bound of <0.3 neutrino processes per day, being still overwhelmingly affected by disturbing effects. Theoretically, one expects about a fifth part of this upper bound [see equations (25.20)—(25.22)]. Since the receivers employed were evidently still much too small, at the present time an experiment is being prepared using 610 tons of C_2Cl_4 in a 1470 metre deep goldmine.

30. Cosmology

Five years after the measurement of the distances of remote galaxies, E. Hubble in 1929 achieved a second discovery of stupendous significance: *The redshift of the lines in the spectra of distant galaxies increases in proportion to their distance.* We write the redshift

$$z = \frac{\Delta\lambda}{\lambda_0} = \frac{\lambda - \lambda_0}{\lambda_0} \tag{30.1}$$

where λ_0 is the laboratory wavelength and λ the measured wavelength.

* More precisely, we must distinguish between two sorts of neutrinos and the corresponding anti-neutrinos.

If r is the distance and if we interpret the redshift as a doppler shift *, then we have for the speed of recession $v = dr/dt$ of the galaxies the relation

$$v = c\, \Delta\lambda/\lambda = H\, r. \tag{30.2}$$

Apparent exceptions for nearby galaxies — the Andromeda galaxy, for example, is *approaching* us at 300 km/sec — are easily explained as the reflection of the rotation of our Galaxy. Recent measurements of the 21-cm hydrogen line have given the best confirmation of the non-dependence of the effect on wavelength assumed in (30.2).

For the Hubble constant H in 1929 Hubble himself obtained the value 530 km/sec per megaparsec. Following on W. Baade's revision of the cosmic distance-scale (distinguishing between cepheids of populations I and II), in 1968 A. R. Sandage computed the most probable value as

$$H = 75.3 \text{ km/sec per megaparsec.} \tag{30.3}$$

Other recent discussions have given values between 50 and 100 km/sec per megaparsec.

Conversely equation (30.2) is frequently applied, for want of anything better, in order to derive a distance r from the measured redshift in the spectrum of a galaxy that is no longer telescopically resolvable. In the present state of the subject, one must then state what value has been used for H.

One can first interpret the relation (30.2) quite naively by saying that an expansion of the universe from a relatively small volume began at a time T_0 years ago. If a particular galaxy, that we now find to be at distance r, received for the expansion the velocity v, it requires in order to travel the distance r the time

$$T_0 = r/v = 1/H \tag{30.4}$$

which is the same for all galaxies. This so-called *age of the universe* T_0 is thus the reciprocal of the Hubble constant $1/H$. If as usual we reckon H in km/sec per megaparsec, and T_0 in years, then we have ($1 \text{ Mpc} = 3.084 \times 10^{19}$ km; $1 \text{ year} = 3.156 \times 10^7$ sec)

$$T_0 \text{ [years]} = \frac{978 \times 10^9}{H \text{ km/(sec}\cdot\text{Mpc)}}. \tag{30.5}$$

While Hubble's older value for H led to a value 1.86×10^9 years for

* We restrict ourselves here to $z \ll 1$; otherwise we should have to use relativistic calculations.

the age of the universe that was far too short by comparison with the ages of globular clusters, etc., we obtain from the more recent value of 75 km/(sec·Mpc)

$$T_0 = 13 \times 10^9 \text{ years} \tag{30.6}$$

with an estimated uncertainty of $\pm 5 \times 10^9$ years.

The kinematics of the expanding universe expressed by (30.2) seems at first sight to imply a reversion to heliocentric ideas. However, this is not so. If we write (30.2) as a vector relation

$$\boldsymbol{v} = H \cdot \boldsymbol{r} \tag{30.7}$$

where the origin of the coordinate system lies in our Galaxy, then, seen from another galaxy that has position $\boldsymbol{r}_0$ and velocity $\boldsymbol{v}_0$ relative to ours (so that $\boldsymbol{v}_0 = H \cdot \boldsymbol{r}_0$) we have

$$\boldsymbol{v} - \boldsymbol{v}_0 = H(\boldsymbol{r} - \boldsymbol{r}_0). \tag{30.8}$$

Thus the universe expanding in accordance with (30.7) presents exactly the same appearance to observers in different galaxies. Therefore our *kinematic world-model is homogeneous and isotropic.* We can show that (30.7) represents the *unique* field of motion that satisfies this condition, so long as we require the motion to be irrotational (curl $\boldsymbol{v} = 0$).

E. A. Milne and W. H. McCrea (1934) extended this at first purely kinematic model so as to make it a *newtonian cosmology.* They investigated the motions of a medium (the "gas" of galaxies) that can take place in accordance with newtonian mechanics if one demands throughout *homogeneity, isotropy* and *irrotational* motion *.

Consider at time t a galaxy at distance $R(t)$, then according to Newton's law of gravitation this is attracted by the mass within the sphere of radius R given by $M = \frac{4}{3}\pi R^3 \varrho(t)$, where $\varrho(t)$ is the mass-density at the instant considered. Therefore the equation of motion **

* One shows without difficulty that the requirement of isotropy everywhere implies homogeneity, but not conversely. World-models with curl $\boldsymbol{v} \neq 0$ have been studied, but they must remain outside our present scope.

** Taking explicit account of pressure — here ignored for the sake of simplicity — would not affect the result. The convergence-problem, connected with the slow fall-off Newton's inverse square law of attraction, in the case of infinitely extended systems for long held up the development of newtonian cosmology. Here we have somewhat loosely passed it over. An exact formulation is given, for example, in the Encyclopaedia article by O. Heckmann and E. Schücking, 1959.

of this galaxy is

$$\frac{d^2R}{dt^2} + \frac{GM}{R^2} = 0 \quad \text{where} \quad M = \tfrac{4}{3}\pi R^3 \varrho(t) = \text{const.} \tag{30.9}$$

The solution of this equation leads to world-models which, from a starting-point (singularity) of infinitely great density, either expand monotonically (if the mass M has total energy $h \geqq 0$) or oscillate periodically between $R=0$ and an R_{max} (if $h<0$). Static models are not possible in accordance with (30.9).

World-models which avoid the singularity $R=0$ become possible, if one extends Newton's law of attraction by the *ad hoc* addition of a *repulsion term* that actually does not make itself noticeable in planetary motion but which becomes important over ranges of billions of light-years. In this way, by including a sufficiently small "Λ-term", one can extend the first equation in (30.9) to

$$\frac{d^2R}{dt^2} + \frac{GM}{R^2} - \frac{1}{3}\Lambda R = 0. \tag{30.10}$$

In this formulation world-models also now become possible that expand from an initial radius $R_{t=0}$ slowly at first and then more and more rapidly. Moreover there are models that start from $R=0$ and tend towards a finite $R_{t\to\infty}$. As one sees immediately by taking $d^2R/dt^2=0$ in (30.10), there is then also a static model $R=$ constant.

Generally speaking, newtonian cosmology extends the world-picture of the purely kinematic cosmology, which we discussed first, in that it treats the Hubble constant H as a function of the time t. In a periodic universe, for example, an era with redshifts would be followed by one with blueshifts, and conversely. Rather than discuss the choice between the numerous world-models of newtonian theory, we examine its basic difficulties and their resolution within the framework of relativistic cosmology *.

The theory of special relativity developed by A. Einstein in 1905 starts out from the result of the Michelson experiment in that it requires the propagation of a light-wave in different coordinate systems, which may be in relative translational motion, to present the same appearance, i. e. satisfy the same equation. If the light traverses an element of path-length $dr = (dx^2 + dy^2 + dz^2)^{\frac{1}{2}}$ in a time-element dt

* Our short introduction will not replace a textbook of relativity theory. We wish only to indicate its significance for astronomy and cosmology. The few formulae presented will serve only to give some impression of its theoretical structure.

with vacuum light-speed c, we have

$$dx^2 + dy^2 + dz^2 - c^2 \, dt^2 = 0. \tag{30.11}$$

We may treat the quantity on left as the line-element ds^2 of a four-dimensional space with three spatial coordinates x, y, z and the fourth coordinate $c\,t$ (= light path) or, following H. Poincaré and H. Minkowski, more "intuitively" as $i\,c\,t$ [where $i = \sqrt{(-1)}$]. Then, somewhat more generally, we require that in passing from one cartesian coordinate system to another, which is in translational motion relative to the first, the four-dimensional line-element ds^2 shall remain invariant, where

$$ds^2 = dx^2 + dy^2 + dz^2 - c^2 \, dt^2 \quad \text{or} \quad ds^2 = dx^2 + dy^2 + dz^2 + d\,(i\,c\,t)^2. \tag{30.12}$$

Such a transformation obviously cannot be restricted to the spatial coordinates $x, y, z \to x', y', z'$, but must also transform the time along with these, $x, y, z, t \to x', y', z', t'$. As we see at once from the second form of (30.12), this *Lorentz transformation,* as Einstein named it, is none other than a rotation in the four-dimensional space x, y, z, $i\,c\,t$, in which by definition the line-element ds is invariant. The essential advance made by special relativity as compared with newtonian theory consists in the fact that it takes account of the exceptional status of the (vacuum) velocity of light from the outset. Corresponding to this, its further development leads to the result that no material motion and no signal of any kind can exceed the speed $c = 3 \times 10^{10}$ cm/sec. Thus our newtonian world-models can be trusted only so far as no speed occurs that exceeds the light-speed, i. v. $v < c$ and $z < 1$.

In regard to the whole of physics, special relativity further requires the invariance of all natural laws under Lorentz transformation, i. e. that the laws be physically independent of motions of translation.

Should it not then be possible so to formulate natural laws that they are invariant under *arbitrary* coordinate transformations? In 1916 A. Einstein had the brilliant idea in his theory of *general relativity* of combining this requirement with a theory of *gravitation.* Within the framework of classical theory, the equality of gravitational and inertial mass, independently of the kind of matter (or in modern terms, of the kind of elementary particles) was indeed a feature to be wondered at. Newton himself, then Bessel and later Eötvös had verified it experimentally with increasing accuracy. But what did it mean?

Einstein promoted the experience that in a freely falling reference-system (a lift) the gravitational force (mg) appears to be cancelled by the inertial force ($m\ddot{z}$) to the status of a fundamental postulate. That is to say, gravitational force and inertial force are in the last resort *the same.* We can transform away these forces by a local transformation of the four-dimensional cartesian coordinate system with the euclidean metric * (30.12). Conversely the coefficients g_{ik} of the Riemannian metric $ds^2 = \sum_{i,k} g_{ik}\, dx^i\, dx^k$ with an arbitrary coordinate system determine, as can be shown, at the same time the gravitational field and the inertial field that are in operation there.

In the domain of the planetary system the predictions of general relativity agree closely with those of newtonian mechanics and gravitation-theory. As regards the three classical tests that could discriminate between the two theories, the present observational situation is the following:

1. *Deflection of light.* The light of a star that (as observed during a total solar eclipse) is seen at an apparent distance of R solar radii from the centre of the solar disk should suffer a deflection by the Sun's gravitational field of $1.75/R$ seconds of arc. The extremely difficult measurements give for the value of the constant between $2''.2$ and $1''.75$.

2. *Gravitational redshift.* A photon $h\nu$ that traverses, for example, the potential difference Sun—Earth $G\mathfrak{M}_\odot/R_\odot$, must show a redshift compared with a laboratory light-source expressed by

$$-\Delta(h\nu) = \frac{G\mathfrak{M}_\odot}{R_\odot}\,\frac{h\nu}{c^2}\,. \tag{30.13}$$

This is formally equivalent to a doppler effect $-c\,\Delta\nu/\nu = c\,\Delta\lambda/\lambda = 0.64$ km/sec. The observations provide qualitative confirmation, but it is at present not possible completely to isolate the gravitational redshift from the doppler effect of motions in the solar atmosphere. Also the redshift in the spectra of white dwarf stars (like the companion of Sirius) cannot be measured with sufficient accuracy. An experiment with the exceedingly sharp γ-lines of the Mössbauer effect in the gravitational field of the Earth offers better prospects. Using the recoil-less γ-line of Fe^{57}, R. V. Pound and G. A. Rebka in

* The form (30.12) differs from that of the euclidean metric of ordinary three-dimensional space in that one term is negative. Such a metric ds^2 is called indefinite or pseudo-euclidean.

1960 were able to verify within 10 percent the frequency shift of only $\Delta\nu/\nu = 2.5 \times 10^{-15}$ which is that calculated for an altitude difference of 22.6 metres.

3. *Perihelion-advance of the planets.* The very small advance of the perihelion of Mercury calculated by Einstein agrees well with the old computations of Leverrier. The reduction by G. H. Clemence and R. L. Duncombe ($\sim$1956), using electronic computation, of a vast amount of observational material gave with considerably improved accuracy:

Planet		Mercury	Venus	Earth
Perihelion advance	Observed	$43''.11 \pm 0.45$	$8''.4 \pm 4.8$	$5''.0 \pm 1.2$
per century	Calculated	$43''.03$	$8''.6$	$3''.8$

(30.14)

Since the perihelion-advance is a second-order effect, while the deflection of light-rays and the redshift are only first-order effects, we may claim that all in all there is a remarkable verification of relativity theory.

After this digression we return to our problems of relativistic cosmology.

If from the outset we make the cosmological postulate that the universe must be homogeneous and isotropic throughout, then according to H. P. Robertson and others, we can express the four-dimensional line-element *ds* in the form (after a suitable choice of units of length and time)

$$ds^2 = dt^2 - R^2(t)\,\frac{dx^2+dy^2+dz^2}{\{1+\frac{1}{4}k\,(x^2+y^2+z^2)\}^2}\,. \tag{30.15}$$

The time-dependent function $R(t)$ determines the radius of curvature of the three dimensional space ($t = \text{constant}$), and it is defined quite analogously to the radius of curvature of a two-dimensional surface. The constant k, which can take the values 0 or ± 1, gives the sign of the spatial curvature, which is everywhere the same for any particular value of t. We have in fact:—

a) $k = 0$ well-known euclidean space.

b) $k = +1$ spherical or (otherwise interpreted) elliptical space. This space is closed and has a finite volume.

c) $k = -1$ hyperbolic space. This space is open.

One can best picture these three spaces or geometries through their two-dimensional analogues (Fig. 30.1).

The further development of relativistic cosmology (which we cannot here review in detail) is accomplished in the following way: The Einstein field-equations which relate the g_{ik} of the universe with

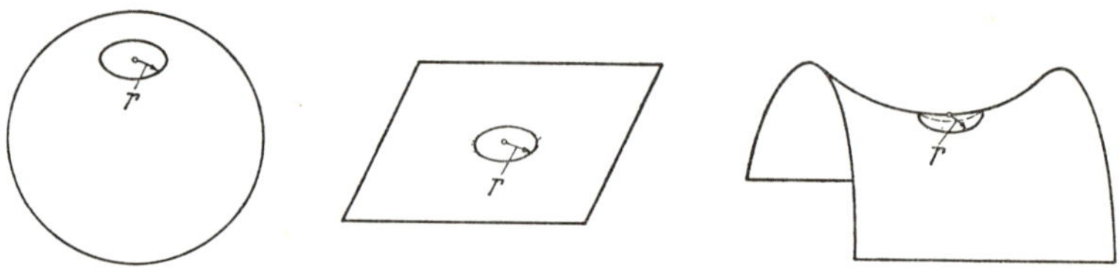

Fig. 30.1. *Surfaces* (i. e. two-dimensional spaces) with *curvatures* $k > 0$, $k = 0$, $k < 0$ as models of curved space.

Curvature k of surface:	$k > 0$	$k = 0$	$k < 0$
Geometry:	spherical or elliptical	euclidian	hyperbolic (Bolyai-Lobatchewski)
Circumference of circle:	$< 2\pi r$	$2\pi r$	$> 2\pi r$
Area of circle:	$< \pi r^2$	πr^2	$> \pi r^2$

its material content and the boundary conditions of the problem (see below), reduce for the Robertson metric (30.15) to a differential equation for $R(t)$. For systems of vanishingly small pressure, this proves to be identical with the differential equation (30.10) of newtonian cosmology, and so we obtain precisely the same choice of models. However, now *a priori* no velocity exceeding that of light can arise. Relativistic cosmology is the first theory to achieve a self-consistent and contradiction-free description of the universe as a whole.

Einstein originally believed that his differential equations of the metric field or gravitational field corresponded to the requirement formulated by E. Mach in 1883 that all intertial and gravitational fields must be traced back explicitly to the action of cosmic masses. That this cannot be immediately correct is shown by de Sitter's model of an empty universe in which nevertheless inertial forces are operative. Clearly the boundary conditions at infinity play a decisive part. H. Hönl and H. Dehnen have recently been able to show that for all finite and closed (i. e. physically reasonable) relativistic model universes the so-called Mach principle is satisfied, i. e. the optical coordinate system of the galaxies here agrees with the inertial system of mechanics, corresponding to astronomical experience in our universe.

What observations (partly actual, partly at anyrate in principle) are at our disposal in order to discover to which model the actual cosmos corresponds? We have:—

1. The *redshift* z in the spectra of galaxies. The measurements on quasars started in 1962 by M. Schmidt have already advanced to redshifts $z > 1$ where quadratic and higher terms in the Hubble relation (30.2) make their appearance. For example, the radio-source 3C9 has $z = 2.012$.

2. As a measure of the *distance* of the galaxies we contemplate a) the apparent magnitude m and b) the angular diameter.

3. The *material contents* of space. These are related to the statistics of apparent magnitudes: How many galaxies per square degree have magnitudes in the range $m \pm \frac{1}{2}$ (after correcting for galactic absorption)? In (23.1), we have already specified the limiting case for a euclidean universe. We know the *masses* of the galaxies as a function of their type, at least as regards order of magnitude.

The problem of *diffuse intergalactic matter* is still only in its earliest stages. F. Zwicky has pointed out that interacting galaxies often appear to be linked together by "filaments". C. Hoffmeister believes he has discovered the first signs of an intergalactic dark cloud in the southern sky. In Chapter 29 we have raised the problem of intergalactic cosmic rays. In brief, we are not yet in a position to say whether in cosmic space, over and above the masses of the stars, more or less unknown masses play a part at the present epoch.

Using the theory of general relativity, the relations between data in categories 1, 2, 3 have been extensively studied by Tolman, Robertson, Sandage, McVittie and others. For example, it is possible in principle (see the end of this Chapter) from the Hubble constant and the mean density of matter to calculate whether our universe corresponds to a closed, i. e. finite ($k = +1$) or an infinite ($k = -1$ or 0) relativistic model. At the moment, however, the accuracy of the available observations is not sufficient for this purpose. The situation is similar in regard to the quadratic term in the redshift relation. However, the unimagined possibilities offered by the investigation of quasars and quasi-stellar galaxies make us expect rapid advances in the near future.

As the most important result hitherto — upon this newtonian and relativistic cosmology are in agreement — we take the finding that the age of the universe $T_0 = H^{-1} \approx 13 \times 10^9$ years derived from Hubble's redshift law is of the same order as the age of the galaxies and particularly of the Milky Way Galaxy got from colour-magnitude diagrams and other indications. The age of the Earth as 4.5×10^9 years is only about three times smaller.

We can therefore conclude — here we obviously have to employ the laws of physics as they are familiar to us — that about T_0 years ago the whole universe (also in thermodynamic respects) was constituted essentially differently from the way it is at present. For even earlier times we all of us lack any records. This is the precise sense of the turn of phrase commonly used in saying that the universe "began" about T_0 years ago.

Our presentation of cosmology, the study of the universe as a whole, has so far proceeded from a heuristic viewpoint. It seems appropriate therefore to interrupt it here with a historical interlude.

H. W. M. Olbers 1826 appears to have been one of the first astronomers to have considered a cosmological problem from an empirical standpoint. Olber's paradox asserts: Were the universe infinite in time and space and (more or less) uniformly filled with stars, then — in the absence of absorption — the whole sky would radiate with a brightness that would match the mean surface brightness of the stars, and thus about that of the surface of the Sun. That this is not the case cannot depend only on interstellar absorption, since the absorbed energy could not actually disappear. However, a finite age of the universe of about 13×10^9 years would already dispose of Olbers's paradox for any otherwise fairly plausible model (W. B. Bonnor 1963). It can be remarked in passing that the paradox assumes a much more interesting aspect with respect to the practically non-absorbable neutrino-radiation.

The modern development of cosmology started on the one hand from the measurements of radial velocities of spiral nebulae and on the other hand from the theory of general relativity.

In connexion with the older radial-velocity measurements by W. M. Slipher ($\sim$1912) already in 1924 C. Wirtz had remarked upon their increase with distance and he had related them to de Sitter's relativistic world-model. In 1917 A. Einstein had shown that his field-equations of general relativity, when extended* to include the

* We have dealt with this extension in (30.10) only in the context of newtonian theory. Actually it was first employed by Einstein in the context of general relativity. For this reason its physical interpretation remained obscure for a long time. At present there exists no physical theory or mea-

Λ-term, have a static cosmological solution (Einstein's spherical universe). Also in the same year W. de Sitter found the solution we have mentioned giving an empty expanding universe.

The fact that we have no knowledge whatever of cosmic structures and happenings more than about a time T_0 in the past can also be interpreted to mean that here we have arrived at a naturally-imposed limit to our knowledge. Such situations are not unknown in the development of physics. The insight that $c = 3 \times 10^{10}$ cm/sec is the greatest possible speed led to *relativity theory*. The further knowledge that $h = 6.62 \times 10^{-27}$ erg sec is the smallest action led to *quantum mechanics*. Analogously, we may hope that the knowledge of the fundamental character of the Hubble constant, or of the "age of the universe $T_0 = H^{-1}$ will lead to a *cosmological physics*. For cosmic space-time such a theory could differ considerably from present-day physics but it would have to include the latter as a limiting case for our own space-time neighbourhood.

The theory of steady-state cosmology developed since 1948 by H. Bondi and T. Gold and by F. Hoyle, and then by others, represents one attempt in this direction. This theory has indeed now been given up by most astronomers, but we wish to describe it briefly and take it as an occasion for certain more general considerations. Steady-state cosmology assumes from the outset that the universe is not only spatially isotropic and homogeneous (as do the usual relativistic models more on mathematical grounds) but also that it is *homogeneous in time* i. e. in a steady state. This is so to say a universe for bureaucrats in which everything at all times shall be regulated in accordance with the same paragraphs. The necessary departure from "ordinary physics" — there are several formulations — is that a mechanism is postulated that makes possible the continual production of hydrogen in the cosmos, since this is needed for replacing the matter that "expands away" and as fuel for stars. Adherents of the steady-state theory could, however, up to now neither connect this process of the creation of matter with nuclear physics nor explain why in all galaxies (where we can test it) the oldest structures have an age of the order of T_0. Even if the theory of the steady-state universe has not yet been worked out in detail, we should nevertheless wish to present a fundamental consideration concerning it. In the sense of

surement of any sort from which the magnitude Λ could be derived. Many people therefore prefer altogether to renounce its use, i. e. to set $\Lambda = 0$.

modern physics, the energy-law is known to be a consequence of the homogeneity of time. However, we can speak of this meaningfully only for a time-interval of the order of T_0. In every theory, which takes the reciprocal of the Hubble constant seriously in the sense we have sketched, we may no longer suppose the energy-law to be valid in connexion with the "creation" of the universe.

A. S. Eddington, P. A. M. Dirac, P. Jordan and others have sought to tackle the problem of "cosmological physics" from a different side. From the elementary constants of physics on the one hand, e, h, c, m (where it remains undecided whether this is to be the mass of the electron, the proton or some other elementary particle) and G, and on the other hand the "constants" of cosmology, the time $T_0 \approx 13 \times 10^9$ years $= 4.1 \times 10^{17}$ sec and the mean density of matter in the universe $\varrho_0 \approx 10^{-30}\,\mathrm{g/cm^3}$, we can form several dimensionless numbers. Factors of the order $2\pi, \ldots$ naturally remain open. In this sense one obtains a set of dimensionless numbers of order of magnitude unity (for example, the Sommerfeld fine-structure constant $\alpha^{-1} = h c/2\pi e^2 = 137$, etc.) and a second set of order 10^{39} to 10^{40}.

[In this connexion it may be remembered that the differences in order of magnitude between the strong (nucleon-nucleon), weak (β-decay) and electromagnetic interactions, — after which gravitation follows — are not yet understood theoretically.]

There thus exist

1. the ratio of electrostatic to gravitational attraction between a proton and an electron

$$\frac{e^2}{G\, m_p\, m_e} = 2.3 \times 10^{39} \tag{30.16}$$

2. the ratio of the length $c\,T_0$ (in a spherical world $\approx$ world-radius) to the classical electron-radius

$$\frac{c\,T_0}{e^2/m\,c^2} = 4.4 \times 10^{40} \tag{30.17}$$

3. the number of nucleons in the universe of the order

$$\varrho_0\, c^3\, T_0^3/m_p = (1.0 \times 10^{39})^2. \tag{30.18}$$

The same facts show, as we see by combining (30.16)—(30.18), that the so-called deceleration parameter (in the case of zero cosmological constant) $q_0 = 4\pi G \varrho_0/3 H_0^2$ of relativistic cosmology is of the order of unity. Since the sign of $2q_0 - 1$ is the sign of the world-curvature $k = \pm 1$ or 0, this means that the actual universe does not differ too much from a euclidean one (which is by no means self-evident).

If we consider the equality as regards order of magnitude of the numbers in (30.16) on the one hand and (30.17) and (30.18) on the other as significant, since the age of the universe T_0 occurs in the latter, we can start speculating upon a cosmological time-dependence of the elementary constant of physics.

In any case we may suppose that the relations (30.16) to (30.18) will play an essential part in any future "cosmological physics". In that connexion also the following consideration naturally arises. Both newtonian cosmology and relativistic cosmology offer us a whole catalogue of possible world-models. However, why is just our own universe realised with certain definite (dimensionless) numerical constants? We can still give no answer to this. As an example, E. A. Milne made an attempt in "Kinematic Relativity" (1948).

31. Origin of the Solar System: Evolution of the Earth and of Life

From the depths of cosmic space we now return to our planetary system with the old question of its origin. So long ago as 1644, in France René Descartes with his vortex theory was able to advance the daring idea that one could come nearer to the answer, not by handing on traditional myths, but by proper investigation. In Germany even in 1755 I. Kant had to let the first edition of his *Allgemeine Naturgeschichte und Theorie des Himmels* appear anonymously because he feared the (protestant) theologians. In the book he treated the formation of the planetary system for the first time "nach *Newton*ischen Grundsätzen". Kant started from a rotating, flattened primeval nebula out of which the planets and later the satellites were then formed. A similar hypothesis was at the basis of the somewhat later (independent) account by S. Laplace 1796 in his popular book "Exposition du Système du Monde". We shall not go into details and differences of these historically important beginnings, but we shall summarize once more the most important facts (see Table 7.1 and Fig. 31.1) which have to be explained:—

1. The orbits of the planets are almost circular and coplanar. Their sense of revolution is the same (direct) and agrees with the sense of rotation of the Sun. The orbital radii (the asteroids are taken

together) form approximately a geometrical progression

$$a_n = a_0 \, k^n \tag{31.1}$$

where $a_0 = 1$ AU, with $n = 0$ for the Earth, and $k \approx 1.85$ (Fig. 31.1).

2. The majority of satellites move in orbits of small eccentricity near to the equatorial planes of their planets. Also the inclinations of these equatorial planes to the ecliptic or, to put it better, to the invariable plane of the planetary system (perpendicular to the resultant angular-momentum vector) are mostly small. The rotation of the planets and the revolution of their satellites proceed for the most part in the direct sense. Exceptions (Pluto and certain satellites) occur mainly near the edge of the system concerned.

3. The terrestrial planets have relatively high densities, the major planets have low densities (Table 7.1 and Fig. 31.1); the former consist (like the Earth) in the main of metals and rocks, the latter of scarcely modified solar material (hydrogen, helium, hydrides). The terrestrial planets have slow rotation and few satellites; the major planets have relatively rapid rotation and numerous satellites.

4. The Sun retains within itself 99.87 percent of the *mass* but only 0.54 percent of the angular momentum ($\Sigma \, m \, r \, v$) of the whole system, while conversely the planets (mainly Jupiter and Saturn) have only 0.135 percent of the mass and 99.46 percent of the angular momentum *.

While the impressive regularity in the structure of the planetary system speaks for an evolution from within itself — in the sense of the theories of Kant, Laplace and later v. Weizsäcker, ter Haar, Kuiper and many others — the paradoxical distribution of angular momentum between the Sun and the planets forms the most important argument for another group (Jeans, Lyttleton and others) who postulate the interaction of the Sun with a passing star or some such occurrence.

The older investigations that were based exclusively upon the basis of celestial mechanics did indeed lead to many important results; we think of E. Roche's criterion for the instability of rotating astronomical bodies or of J. H. Jeans's theory of gravitational instability. As regards the cosmogony of the planetary system,

* As we easily calculate, the angular momentum (page 43) of the planetary orbits (♃, ♄) is 3.15×10^{50} g cm^2 sec^{-1}, and the angular momentum of the Sun (with some uncertainty concerning the increase of angular velocity in the interior) is 1.7×10^{48} g cm^2 sec^{-1}.

however, the discussion of these older theories is unfruitful, since only quite recently two new crucial concepts have gained acceptance:

1. Cosmic bodies of gas are always (partly or wholly) *ionized* and carry magnetic fields with them. Such a plasma obeys the laws of magnetohydrodynamics (H. Alfvén).

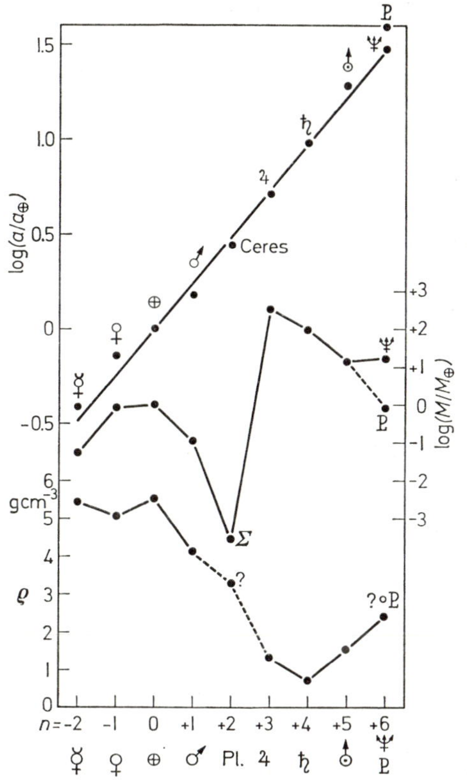

Fig. 31.1. *Orbital semi-major axes a, masses M and mean densities ϱ of the planets.* The orbital semi-major axis and estimated total mass of Ceres have been used to represent the asteroids ($n = 2$). Neptune and Pluto (which was possibly captured later) have both been assigned $n = 6$. The straight line drawn in the upper half corresponds to the relation $\log a = 0.266\, n$ or $a = 1.85^n$

2. The formation of stars from diffuse material and their further evolution is at the present time both observationally (colour-magnitude diagrams, etc. — Chapter 26) and theoretically (internal constitution of the stars, etc. — Chapter 25) to a great extent comprehensible.

We therefore discuss the origin of the planetary system in this context (A. Poveda and others, 1965).

As we saw, all stars originate by the process in which an interstellar gas cloud (which contains also some dust) of some 10^3 solar masses condenses upon itself and then fragments into stars. Also the single systems amongst the stars, such as our Sun, have originally belonged to such multiple systems, associations or clusters. The individual protostar first finds itself in a state of convective equilibrium (Hayashi). When it has contracted to a particular radius (depending on the mass) it attains radiative equilibrium and then it moves towards the left in the HR diagram (Henyey and others) until hydrogen-burning sets in, and the star reaches for the time being a stationary state on the main sequence. Young stars in galactic clusters like NGC 2264 (Fig. 26.7) are recognized first as flare-stars and then as T Tauri or RW Aurigae stars. Strong convection, certainly in combination with appreciable magnetic fields, generates in condensed masses all the typical phenomena of solar activity such as spots and flares and, in particular, large quantities of energetic particles.

As evidenced by their emission lines, young stars have considerable gaseous envelopes. We are perhaps not far off the mark, if we identify these with the primeval nebula of Kant and his successors.

Most young stars i. e. main sequence stars from O to F and, among later spectral types, the stars already mentioned (mostly to the right of the main sequence) in young clusters show rapid rotation with equatorial speeds of the order of 50—500 km/sec.

The older stars, on the other hand, like the F to M main sequence stars of the disk population, rotate only slowly with equatorial speeds <5 km/sec. This feature had to remain inexplicable so long as each star was regarded as an isolated system, whose angular momentum Σmrv (Chapter 6) had to remain constant. Magneto-hydrodynamics first showed a mechanism for the giving up of originally large angular momentum. Stellar magnetic fields which are frozen into the plasma of a star and its envelope — the primeval nebula — and also into the interstellar gas, transfer angular momentum from the star to its (in accordance with Kepler's third law) slowly rotating envelope and ultimately also to the interstellar gas, so that the rotation of the star is slowed down in the course of its evolution (eddy current brake), as observation requires.

What now is the situation as regards angular momentum in the solar system? Were one to transfer the angular momentum of the planets (essentially that of Jupiter) to the Sun, its equatorial speed would be increased from 2 km/sec to about 370 km/sec. This corresponds to about the speed of rotation of young stars. Evidently therefore in the formation of the planetary system angular momentum has been transferred by magneto-hydrodynamic means outwards from the main mass-concentration into the disk (flattened by rotation) i. e. into the region of Jupiter's and Saturn's orbits.

That the formation of such a system is no unusual happening is shown by the discovery of a companion, having only about 1.6 Jupiter-mass, of Barnard's star (M 5 V) (page 132). Also the statistics of mass-ratios in double stars lead us to expect systems with planet-like companions to occur with appreciable frequency.

The rotation and flattening of the initial configuration makes it understandable that the angular momentum vectors of the sub-systems (planets, satellite-systems ...) should be more or less parallel to the vector of the resultant angular momentum.

The significance of the ordering of the orbital radii in a geometric progression (31.1) is not so clear. C. F. v. Weizsäcker (1944) tried to ascribe it to a systematic arrangement of vortices in the initial disk of the planetary system. These should be disposed five in a ring. The planets would have formed in "ball-bearing vortices" where the large vortices bear upon each other. At the present time it may not be profitable to review such details at length. In the same way the hitherto proposed explanations of analogous regularities in satellite-systems still appear very hypothetical.

It is important and fairly certain that the planets have been constructed "in the cold", i. e. that they never were molten throughout. H. C. Urey was probably the first to point out that otherwise one could scarcely account for the internal constitution of the Earth. In the outer parts of the solar system the temperature was so low that water, ammonia and to some extent methane — mainly in the liquid state — were retained, while in the inner part of the system (out to the asteroid belt) the hydrides have largely escaped. In this way we can understand that the major planets should differ basically from the terrestrial planets in their chemical composition and their density.

Within the framework of the theory of stellar evolution, according to Hayashi and others, we can estimate the total time required for the formation of the Sun and of the planetary system to be about 10^7 years. The flattening of the originally nearly spherical primeval nebula can have needed only a small part of this time interval. For free fall (Chapter 6) the requisite time is only of the order of the orbital periods of the major planets.

The meteorites constitute important documents concerning the formation and early history of the planetary system. The earlier view that here we had before us "the" cosmic material with "the" cosmic abundances of the elements has long ago given place to a thorough study of their mineralogical, chemical and isotope structure. At present the only thing certain is that in the large body whose break-up gave the meteorites — maybe there was more than one body — complex processes of magmatic differentiation must have taken place. P. Ramdohr has found in meteorites even microscopic flakes of metallic copper and of gold.

We now trace the evolution of the Earth somewhat further. We have already discussed in Chapter 7 its internal constitution and the sequence of geological layers. Further, we convinced ourselves that our Moon — which already takes an exceptional status in respect of the mass-ratio to its planet — did not come out of the Earth, but must have been formed separately. So we can now turn to questions concerning the history of the Earth's atmosphere that are of special interest with regard to the evolution of life.

The present atmosphere of the Earth cannot date from the time when the Earth and the other planets were formed out of a "T Tauri nebula". For in cosmic matter (Table 19.1) the most abundant among the volatile elements are hydrogen, helium, neon; also the heavier inert gases are still relatively abundant. In the Earth's atmosphere, on the other hand, all the inert elements occur only in minute traces. Also quantitative estimates confirm the view that our atmosphere is a secondary product of volcanic exhalations which supply H_2O, CO_2, N_2, $SO_2 \ldots$, while the most abundant isotope of argon A^{40} was produced by the transformation of K^{40}, and helium by α-decay of known radioactive elements, in the Earth's crust. However, the primitive atmosphere of 4.5×10^9 years ago still contained no free oxygen, since this was fully combined in oxides, silicates, etc., and so was absent in volcanic gases. The formation of O_2 (and with it also ozone O_3) began in the optically thin primitive atmosphere, when water vapour H_2O was separated into $2\,H + O$ by ultraviolet solar radiation (photo-dissociation). But, as H. C. Urey (1959) pointed out, this process could provide only about 10^{-3} of the oxygen now present. A denser layer of oxygen (with its associated ozone) would absorb the short-wave solar radiation so that the amount of gas would not increase further.

The additional oxygen in our atmosphere can have been formed only by photosynthesis in living organisms i. e. in connection with their evolution. An estimate made by E. I. Rabinowitch (1951) from the productive power of the existing plant-world gives the interesting result that the whole of the oxygen in the atmosphere goes through the process of photosynthesis once in only 2000 years.

The history of the atmosphere is thus intimately bound up with the origin of life. As with cosmogony this problem was for long the domain of mythological concepts. After Friedrich Wöhler had in 1828 done away with the boundary between inorganic and organic matter with his synthesis of urea, more recent work in the fields of astrophysics, geology and biochemistry have indeed still not solved the problem of the origin of life, but have nevertheless brought it further into the domain of scientific enquiry.

The complicated molecules, particularly the nucleic acids and proteins, that are characteristic for the structure of living matter, are not stable in the presence of oxygen. Therefore under existing conditions on the Earth, their formation out of inorganic substances is not possible without the cooperation of living organisms. In the "beginning" they could arise only in an oxygen-free atmosphere. Therefore the origin of life seems to be connected with the fact that at first the Earth possessed only a rarefied *reducing* atmosphere composed of H_2O, H_2, CO_2 and very likely, as in the major planets, CH_4, NH_3, etc. Indeed, S. L. Miller and others in 1953 succeeded in producing artificially in such an atmosphere, under the action of electric discharges or suitable ultraviolet irradiation, amino acids and other components of living matter up to porphyrine, etc., in surprising quantities. On account of their very strong sensitivity in the short-wave ultraviolet ($\lambda < 2900$ Å) such molecules could not last long on the surface of the primitive Earth, which was not shielded by absorption due to atmospheric oxygen and ozone; however, they could collect, for instance, on the bottom of shallow lakes about 10 metres deep.

Organic molecules are, nevertheless, still a long way from living organisms. How the first such organisms were created, we cannot determine. The viruses of the present time can propagate themselves only *in* higher organisms, and so we may not regard them as the primitive organisms. Possibly, however, we may look on them as *models* for the early stages of life. Also the simplest living organism form a system in which two essential functions act together:—

a) The power of reproduction and of mutation are built into the genetic material. This contains the information or the steering-mechanism, which — like the store of an electronic computer — takes care that an identical, or only slightly altered, organism is reproduced out of suitable components.

The dying out of unsuitable combinations of molecules gives in itself only a *negative* selection in the *struggle for existence* (Charles Darwin). Only the ability of a "suitable" molecular combination to produce identical or similar offspring (the "know-how") makes it possible for favourable mutations to retain their — if only small — advantages. From the evolutionary viewpoint, they thus profit indirectly by the unfavourable experience of other individuals.

b) With the steering-system there *must* be associated a second mechanism for providing the requisite energy and obviously also the requisite raw material. The building up of a more complicated structure out of simpler ones requires an increase of negative entropy (so-called "negentropy") which according to the second law of thermodynamics is only possible when accompanied by an expenditure of energy.

How — maybe in the lakes we have mentioned — out of a collection of all sorts of organic molecules there originated systems, i. e. living organisms that for once possessed contrivances for a) and b), we do not know. We can quite speculatively think of the regulating cooperation of crystals having large elementary cells and high adsorption-power, like, for example, *zeoliths*.

[For the formation of well-defined macromolecules we must have at all events the existence (largely explicable by quantum theory) of chemical binding forces of very great range. Long-range forces are in general needed in order, so to say, to snap ones fingers at the second law of thermodynamics! For example, the production of superthermal particles of cosmic rays is connected with the internal coherence of larger plasma-clouds by magnetic fields which — in the sense of magneto-hydrodynamics — are frozen into the clouds.]

The earliest living organisms are naturally not preserved for us as fossils. However, in 1965 the biochemist M. Calvin succeeded in detecting the hydrocarbons *pristane* and *phytane* in rocks of ages got by radioactive dating as 2.7×10^9 years. These have to be regarded as (relatively stable and yet sufficiently complex) decomposition products of *organisms*. On the other hand, geological findings of dated rocks with incomplete oxidation reach back as far as the same time. These are important arguments for the view that the origin of atmospheric oxygen is bound up with the beginnings of life on the Earth 2.5 to 3 billion years ago.

At the epoch when our atmosphere had acquired about 1 percent of its present amount of oxygen, important new possibilities for the evolution of life were opened up. The lethal ultraviolet was now shielded off, so that the existence of living organisms became possible in the *oceans,* where such organisms would temporarily be carried to the surface by convection. So the change-over from fermentation to respiration, with its much higher effectiveness, became possible. Thus we may well relate the almost explosive evolution of life in the *Cambrian* age, about 600 million years ago, with the attainment of about this oxygen-level.

An oxygen content of about 10 percent of the present atmosphere would make possible the colonization of the solid ground, which was now in any case protected from dangerous ultraviolet radiation. About 420 million years ago, from the *Silurian* to the *Devonian* (Table 7.2) the first extensive forests came into existence, and in the *Carboniferous* from whose fossil forests we still benefit to-day, the present level of oxygen was undoubtedly reached, if not indeed for a time surpassed.

If we may try to reduce it to its simplest elements, at the basis of the whole evolution of life there is clearly the principle, by storage and retrieval of more and more *information,* to build more complex systems which more and more successfully apply their principle of ordering (production of negentropy) in opposition to the natural tendency of the second law towards the production of statistical-thermodynamic equilibrium, i. e. maximum disorder (increase of entropy). This evolution leads from the fixing of genetic information already by viruses to the construction of instinctive i. e. programmed behaviour-patterns, thence further to memory and the beginnings of intelligent behaviour. The development of man — compared with simpler creatures — rests essentially upon enormously improved techniques for the storage and employment of information, first through speech, then through writing (of every sort) and lastly through the invention of the electronic information-storage device, which may itself translate its "knowledge" into action (from computing to electronically controlled factories).

This "struggle against the entropy law", which becomes ever more gigantic, is inevitably linked with ever greater energy-expenditure. The organisms themselves are, as we say, in a state of flow-equilibrium. That is to say, from the food — and in the case of plants from solar radiation — energy is continually extracted, which is necessary partly for bilding up the system, and partly goes to waste as heat, as in any heat-engine. Man can realize his higher evolution according to the principle described, only if he renders natural sources of energy usable in ever greater measure. Besides the muscle-power of man and animals, the forces of wind and water were employed. The next great step was the use of fossil fuels, coal and oil, in heat-engines. It is certainly no accident that our age of automation has also brought with it the release of nuclear energy.

As regards the question of the *future* of the Sun and with it life on the Earth, the theory of stellar evolution allows us to give a quite definite answer. In a few billion years the Sun will move upwards to the right in the HR diagram and will become a red giant star. Its radius will thereby grow enormously, its bolometric magnitude will increase by several magnitudes and the temperature of the Earth's surface will rise considerably above the boiling point of water, so that the oceans will evaporate. This means without question the end of organic life on the Earth.

On the other hand, one often poses the question as to whether there is life on other astronomical bodies. At present this question is meaningful only if by life we understand the occurrence of organisms whose structure bears some resemblance to terrestrial living beings. We scarcely need to say that their environmental conditions are confined to a fairly narrow range of possibilities.

In our planetary system, one can at all events think of Mars. Its rarefied atmosphere contains at anyrate small amounts of the necessary gases, but it offers no ultra-violet protection; the temperature is a little lower than on the Earth. So the occurrence of extremely primitive organisms seems at least a possibility for discussion, but in our present state of knowledge not very probable. The bands in the infrared spectrum of Mars that were found some years ago and were possibly "organic", have since been assigned to the HDO-molecule (i. e. water with one "heavy" hydrogen atom).

In our Galaxy and in others there are numberless stars indistinguishable from our Sun. There is nothing against the assumption that some of these G-type stars also possess planetary systems, and it seems entirely plausible that here and there in such a system a planet offers conditions on its surface similar to those on the Earth. Why should not living beings have evolved there as well?

From the study of cosmic structures and cosmic evolution we returned to problems of our planetary system, of the origin of life and of our existence. Our historical notes showed how our deeper understanding of the universe is intimately connected with the evolution of the human mind.

What is more admirable and astonishing, all the new facts which have been brought to light or the fact that mankind has developed the ability to understand them? This is a new version of the old problem of "macrocosmos and microcosmos". Whether we penetrate into the depths of the universe or whether we search the mysteries of the human mind, on both sides we view a

New Cosmos.

Stellar Populations in the Galaxy

(see in particular Chapters 23 and 27)

Population:	*Halo population II*		*Disk population*		*Extreme population I*
Transition groups:		Intermediate population II		Older population I	
Typical members:	Subdwarfs. Globular clusters. RR Lyr-variables with $P > 0^d\!.4$.		"Normal" stars of disk and nucleus. Planetary nebulae. Novae. RR Lyr-variables with $P < 0^d\!.4$.		Interstellar matter. Galactic clusters and associations. Bright blue OB stars.
Mean distance from galactic plane:	2000 pc		400 pc		120 pc
Mean velocity $\overline{W}$ perpendicular to galactic plane:	75 km/sec		18 km/sec		8 km/sec
Concentration towards galactic centre:	strong		considerable		weak
Abundance of heavy elements ("metals") relative to Sun	$\frac{1}{200}$ to $\frac{1}{5}$		$\frac{1}{3}$ to 1		~ 1

Physical Constants and Astronomical Quantities *

1. *Astronomical Quantities*

Astronomical unit AU (semi-major axis of Earth's orbit)	$= 1.496 \times 10^{13}$ cm

Parsec pc $= 3.085 \times 10^{18}$ cm $= 206\,265$ AU $= 3.26$ lightyears

A segment of length r astronomical units (or $r/206\,265$ pc) at a distance of r pc subtends an angle 1 second of arc at the observer; a segment of length 0.291 pc at a distance of 1 kpc subtends an angle of 1 minute of arc.

Sidereal year $= 365\overset{d}{.}256$	$= 3.1558 \times 10^7$ s
Earth	
Equatorial radius	$= 6.378 \times 10^8$ cm
Mass	$= 5.977 \times 10^{27}$ g
Sun	
Radius $R_\odot$	$= 6.96 \times 10^{10}$ cm
Mass $\mathfrak{M}_\odot$	$= 1.989 \times 10^{33}$ g
Surface gravity $g_\odot$	$= 2.736 \times 10^4$ cm s^{-2}
Total luminosity $L = 4\pi R^2 \cdot \pi F$	$= 3.90 \times 10^{33}$ erg s^{-1}
Effective temperature $T_{\rm eff}$	$= 5780$ °K

One magnitude corresponds to a luminosity ratio 2.512 (antilog. 0.4). Absolute magnitudes refer to a distance of 10 pc.

2. *Units*

Length 1 mile $= 1.609$ km; 1 foot $= 30.48$ cm; 1 inch $= 2.54$ cm; 1 angstrom $= 1$ Å $= 10^{-8}$ cm.

---

* These are mostly from C. W. Allen *Astrophysical Quantities*, second edition, 1963. In general, values are here not given to the full number of decimals to which they are now confirmed and so we do not quote limits of error.

Power and energy	1 watt (MKS system) = 1 joule $s^{-1} = 10^7$ erg s^{-1}.
	1 electron volt (1 eV) = 1.602×10^{-12} erg corresponding to a wavenumber $\tilde{\nu} = 8067.1$ cm^{-1} or Kayser (ky). or to a wavelength $\lambda = 12\,398$ Å.
	1 unit atomic weight = 1.492×10^{-3} erg = 931×10^6 eV.
Pressure	1 atmosphere (atm) = 760 mm Hg (Torr) = 1.0132×10^6 dyn cm^{-2} or microbar (μb).
Temperature	Absolute temperature T °K = 273.15 + t °C. Thermal energy $kT = 1$ eV corresponds to T = 11 605 °K.

Mathematical constants.

$\pi = 3.1416$; e = 2.7183; $1/M = \ln 10 = 2.3026$.

1 radian = 57°296 und 1° = 0.017 453 radian.

3. *Physical constants*

Light speed	c	$= 2.997\,93 \times 10^{10}$ cm s^{-1}
Gravitation constant	G	$= 6.668 \times 10^{-8}$ dyn cm^2 g^{-2}
Planck's constant	h	$= 2\pi\hbar = 6.625 \times 10^{-27}$ erg sec
Electron charge	e	$= 4.803 \times 10^{-10}$ E.S.U. $= 1.602 \times 10^{-19}$ coulomb (C)
Mass: Unit atomic weight	M	$= 1.660 \times 10^{-24}$ g
Proton	M_p	$= 1.672 \times 10^{-24}$ g
Electron	m	$= 9.108 \times 10^{-28}$ g
Boltzmann constant	k	$= 1.380 \times 10^{-16}$ erg deg^{-1}
Avogadro number	N	$= 6.023 \times 10^{23}$ particles per mole.
Gas constant	$\mathcal{R}$	$= 8.317 \times 10^7$ erg deg^{-1} mole^{-1}
Rydberg constant	R_∞	$= 109\,737.3$ cm^{-1} = 1/911.27 Å
Radiation constants	σ	$= 5.67 \times 10^{-5}$ erg cm^{-2} s^{-1} deg^{-4}
	c_2	$= 1.439$ cm deg
Hydrogen atom		
Atomic weight	μ	$= 1.0080$
Mass	M_H	$= 1.673 \times 10^{-24}$ g
Ionization potential		13.60 eV
Rydberg constant	R_H	$= 109\,677.6$ cm^{-1}
Bohr radius	a_0	$= 0.529 \times 10^{-8}$ cm or 0.529 Å.

Bibliography

The bibliography is limited to important books, journals, etc., that should serve for the further study of particular topics and problems. References to articles on such subjects published in periodicals, observatory publications, etc., can be found in *Astronomischer Jahresbericht.* Historical, national and priority considerations have to be set aside here. An asterisk * is used to indicate works that may prove suitable for the newcomer, but the distinction should not be over-stressed.

[Apart from minor revisions, the bulk of the bibliography is the same as in the German edition. Where an English translation of any work in another language is available, only the translation is listed here. Also a number of other books of more general scope that are not readily accessible to English readers are here omitted but, so far as possible, replaced by roughly corresponding books in English. Translator.]

Introduction to Astronomy in General, Including Astrophysics

Abell, G.: *Exploration of the universe.* New York: Holt, Rinehart, and Winston 1964.

Baker, R. H.: *Astronomy.* 8th edn. New York: Van Nostrand 1964.

* Hoyle, F.: *Frontiers of astronomy.* London: Heinemann 1955, also Mercury Books 1961.

* Hoyle, F.: *Astronomy.* London: Macdonald 1962.

Inglis, S. J.: *Planets, stars, and galaxies* — An introduction to astronomy. 2nd edn. New York: Wiley 1967.

Lynds, B. T.: *Struve's elementary astronomy.* 2nd edn. Oxford Univ. Press 1968.

Smart, W. M.: *The riddle of the universe.* London: Longmans 1968.

An invaluable synopsis of the most important numerical quantities in astronomy and physics (in particular atomic physics and spectroscopy):

Allen, C. W.: *Astrophysical quantities.* 2nd edn. London: Athlone Press 1963.

Chart of the Sky

Norton, A. P.: *A star atlas and reference handbook for students and amateurs.* 15th edn. Edinburgh: Gall and Inglis 1964.

Histories of Astronomy

Abetti, G.: *The history of astronomy.* London: Sidgwick and Jackson 1954.

Berry, A.: *A short history of astronomy.* New York: Dover 1961.

King, H. C.: *The history of the telescope.* London: Griffin 1955.
— *Exploration of the universe.* London: Secker & Warburg 1964.
* Pannekoek, A.: *A history of astronomy.* New York: Signet Science Library 1964 also London: Allen & Unwin 1961.
Struve, O., and V. Zebergs: *Astronomy of the 20th century.* New York-London: Macmillan 1962.

Short Reference Books for the Non-specialist

Baker, R. H., and L. W. Frederick. *An introduction to astronomy.* 7th edn. Princeton, N. J.: Van Nostrand 1968.
* Ernst, B., and T. E. de Vries: *Atlas of the universe.* (English translation) London: Nelson 1961.
* Spitz, A., and F. Gaynor: *Dictionary of astronomy and astronautics.* Paterson, N. J.: Littlefield, Adams 1960.
* Weigert, A., and H. Zimmerman. Trans. J. Home Dickson: *ABC of astronomy.* London: Hilger 1967.

Handbooks: A Selection

Encyclopaedia of physics (ed. S. Flügge). Vols. 50—54 (Astrophysics I—V). Berlin-Göttingen-Heidelberg: Springer 1958—1962. With contributions in German, English and French.
Stars and stellar systems (ed. G. P. Kuiper and B. M. Middlehurst). 9 volumes. Chicago Univ. Press 1960 onwards.
The solar system (ed. G. P. Kuiper and B. M. Middlehurst). 5 volumes. Chicago Univ. Press 1953—1966.
Periodical publications:
Annual review of astronomy and astrophysics (Vol. 1, 1963). Palo Alto: Ann. Reviews.
Advances in astronomy and astrophysics (Vol. 1, 1962). New York-London: Academic Press.
Transactions of the International Astronomical Union. Dordrecht: Reidel.
Space research: Amsterdam: North-Holland 1967.

Annual Review of Literature

Astronomischer Jahresbericht. Berlin: W. de Gruyter.

Journals: A Selection

Astronomy and Astrophysics. Berlin-Heidelberg-New York: Springer.
* *Sky and telescope.* Cambridge, Mass.: Sky Publishing Corp.
The Astrophysical Journal. Chicago Univ. Press.
The Astronomical Journal. New York: American Inst. of Physics.
Publications of the Astronomical Society of the Pacific. San Francisco.
Monthly Notices of the Royal Astronomical Society. Oxford and Edinburgh: Blackwell.
Soviet Astronomy (Translation of Astronomicheskii Zhurnal). New York: Amer. Inst. of Physics.
Annales d'Astrophysique. Paris: C.N.R.S.

Bulletin of the Astronomical Institutes of the Netherlands. Amsterdam: North-Holland.
Zeitschrift für Astrophysik. Berlin-Heidelberg-New York: Springer.
Astronomische Nachrichten. Berlin: Akademie-Verlag.
Astrophysics and Space Science. Dordrecht: Reidel.

Part I: Classical Astronomy

1. Stars and Men: Observing and Thinking. Historical Introduction to Classical Astronomy

With reference to the introductory historical Chapters 1, 10, 22, as well as Chapters 2, 3, see "Histories of astronomy".

2. Celestial Sphere: Astronomical Coordinates: Geographic Latitude and Longitude

3. Motion of the Earth: Seasons and the Zodiac: Day, Year and Calendar

4. Moon: Lunar and Solar Eclipses

See "Introduction to astronomy in general", together with:

Danjon, A.: *Astronomie générale.* J. and. R. Sennac. Paris 1952-1953.
* Kourganoff, V.: *Astronomie fondamentale élémentaire.* Paris: Masson 1961.
Smart, W. M.: *Text-book on spherical astronomy.* 5th edn. Cambridge Univ. Press 1962.
Woolard, E. W., and G. M. Clemence: *Spherical astronomy.* New York-London: Academic Press 1966.
The astronomical ephemeris. London and Washington. (Yearly publication), and
Explanatory supplement to the astronomical ephemeris, (appearing at irregular intervals, most recently in 1961).

5. Planetary System

6. Mechanics and Theory of Gravitation

Brouwer, D., and G. M. Clemence: *Methods of celestial mechanics.* New York-London: Academic Press 1961.
Brown, E. W., and C. A. Shook: *Planetary theory.* Cambridge Univ. Press 1933, also New York: Dover 1964.
— *An introductory treatise on the lunar theory.* New York: Dover reprint 1960.
Poincaré, H.: *Méthodes nouvelles de la mécanique céleste.* 3 volumes. Paris 1892-1899, also New York: Dover.
— *Leçons de mécanique céleste.* 3 volumes. Paris 1905—1910.
Smart, W. M.: *Celestial mechanics.* London: Longmans 1953.

7. Physical Constitution of Planets and Satellites

The solar system (see under "handbooks" above).
Baldwin, R. B.: *The measure of the Moon.* Chicago Univ. Press 1963.
* Brandt, F. C., and P. W. Hodge: *Solar system astrophysics.* New York: McGraw-Hill 1964.
Bruhat, G., and E. Schatzman: *Les planètes.* Paris: Presses Universitaires de France 1952.
Fielder, G.: *Structure of the Moon's surface.* London: Pergamon Press 1961.
Kopal, Z. (ed.): *Physics and astronomy of the Moon.* New York-London: Academic Press 1962.
Kuiper, G. P. (ed.): *The atmospheres of the earth and planets.* 2nd edn. Chicago Univ. Press 1952.
Smith, A. G., and T. D. Carr: *Radio exploration of the planetary system.* Princeton, N. J.: Van Nostrand 1964.
The Moon, IAU Symposium 14. London-New York: Academic Press 1962.
La physique des planètes, Col. Internat. d'Astrophysique Liège 1962.

8. Comets, Meteors and Meteorites, Interplanetary Dust; Structure and Composition

Hawkins, G. S.: *Meteors, comets and meteorites.* New York-San Francisco-Toronto-London: McGraw-Hill 1964.
Lovell, A. C. B.: *Meteor astronomy.* Oxford: Clarendon Press 1954.
Richter, N. B.: *The nature of comets* (English translation). London: Methuen 1963.
La physique des comètes, Col. Internat. d'Astrophysique Liège 1952.

9. Astronomical and Astrophysical Instruments

Stars and stellar systems (see under "handbooks"), Vol. I and II.
Le Galley, D. P. (ed.): *Space science.* New York-London: Wiley 1963.
Selwyn, E. W. H.: *Photography in astronomy.* Rochester, N. Y.: Eastman Kodak, 1950.
The construction of large telescopes, IAU Symposium 27. London-New York: Academic Press 1966.

Part II. Sun and Stars. Astrophysics of Individual Stars

10. Astronomy + Physics = Astrophysics. Historical Introduction

See "Histories of astronomy".

11. Radiation Theory

Chandrasekhar, S.: *Radiative transfer.* Oxford: Clarendon Press 1950, also New York: Dover 1960.
Unsöld, A.: *Physik der Sternatmosphären.* Mit besonderer Berücksichtigung der Sonne. 2. Aufl. Berlin-Göttingen-Heidelberg: Springer 1955.
Woolley, R. v. d. R., and D. W. N. Stibbs. *The outer layers of a star.* Oxford Univ. Press 1953.

12. The Sun

Bray, R. J., and R. E. Loughhead: *Sunspots.* London: Chapman and Hall 1964.

* Ellison, M. A.: *The sun and its influence.* 2nd edn. London: Routledge and Kegan Paul 1959.

Flügge, S. (ed.): Encyclopaedia of Physics, Vol. 52 = Astrophysics III: *The solar system.* Berlin-Göttingen-Heidelberg: Springer 1959.

* Kiepenheuer, K. O.: *Die Sonne.* Berlin-Göttingen-Heidelberg: Springer 1957.

Kuiper, G. P. (ed.): *The Sun. The solar system.* Vol. I. Chicago Univ. Press 1953.

Menzel, D. H.: *Our Sun.* Philadelphia-Toronto: Harvard Books 1949.

Unsöld, A.: *Physik der Sternatmosphären* (see under 11).

Waldmeier, M.: *Ergebnisse und Probleme der Sonnenforschung.* 2. Aufl. Leipzig: Akad. Verlagsgesellschaft 1955.

Zirin, H.: *The solar atmosphere.* Waltham-Toronto-London: Blaisdell 1966.

de Jager, C. (ed.): *The solar spectrum* (Symposium Utrecht 1963). Dordrecht: Reidel 1965.

13. Apparent Magnitudes and Colour Indices of Stars

14. Distances, Absolute Magnitudes and Radii of the Stars

15. Classification of Stellar Spectra: Hertzsprung-Russell Diagram and Colour-Magnitude Diagram

In addition to the works listed under "introduction to astronomy in general, including astrophysics" and "handbooks": —

Morgan, W. W., P. C. Keenan, and E. Kellman: *An atlas of stellar spectra.* With an outline of spectral classification. Chicago Univ. Press 1942.

Russell, H. N., R. S. Dugan, and J. Q. Stewart: *Astronomy.* 2 volumes. New York: Ginn 1926 (revised 1945, 1938).

Catalogue of bright stars. Ed. D. Hoffleit. Yale Univ. Press 1965.

16. Double Stars and the Masses of the Stars

Aitken, R. G.: *The binary stars.* New York: Dover.

Russell, H. N., and Ch. E. Moore: *The masses of the stars.* 2nd edn. Astrophys. Monographs. Chicago Univ. Press 1946.

17. Spectra and Atoms: Thermal Excitation and Ionization

Condon, E. U., and G. H. Shortley: *The theory of atomic spectra.* Cambridge Univ. Press 1963.

Griem, H. R.: *Plasma spectroscopy.* New York: McGraw-Hill 1964.

Herzberg, G.: *Atomic spectra and atomic structure.* New York: Dover.

Moore, Ch. E.: *A multiplet table of astrophysical interest,* 1959; and *Atomic energy levels* (several volumes, 1949 onwards). Washington: Nat. Bureau of Standards.

Spectral classification and multicolour photometry, IAU Symposium 24. London-New York: Academic Press 1966.

18. Stellar Atmospheres: Continuous Spectra of the Stars

19. Theory of Fraunhofer Lines; Chemical Composition of Stellar Atmospheres

Aller, L. H.: Astrophysics. I. *The atmospheres of the Sun and stars.* 2nd edn. II. *Nuclear transformations. Stellar interiors and nebulae.* New York: Ronald 1963 and 1954.

Ambartsumian, V. A., E. R. Mustel et al.: *Theoretical astrophysics.* London: Pergamon Press 1958.

Thackeray, A. D.: *Astronomical spectroscopy.* London: Eyre & Spottiswoode 1961.

Unsöld, A.: *Physik der Sternatmosphären* (see under 11).

Abundance determinations in stellar spectra, IAU Symposium 26. London-New York: Academic Press 1966.

20. Motions and Magnetic Fields in the Solar Atmosphere: Solar Cycle

21. Variable Stars: Motions and Magnetic Fields in Stars

See bibliographies for chapters 11, 12, 18 and 19, together with:—

Campbell, L., and L. Jacchia: *The story of variable stars.* Philadelphia-Toronto: Harvard Books 1945.

Dungey, J. W.: *Cosmic electrodynamics.* Cambridge Univ. Press 1958.

Kundu, M. R.: *Solar radio astronomy.* New York: Wiley 1965.

Payne-Gaposchkin, C.: *The galactic novae.* New York: Macmillan 1963 also New York: Dover 1964.

Aerodynamic phenomena in stellar atmospheres, IAU Symposium 12. Reprinted from *Il Nuovo Cimento* 1961.

The solar corona, IAU Symposium 16. New York: Academic Press 1963.

Stellar and solar magnetic fields, IAU Symposium 22. Amsterdam: North-Holland 1965.

Part III. Stellar Systems
Milky Way and Galaxies: Cosmogony and Cosmology

22. Advance into the Universe. Historical Introduction to Astronomy in the Twentieth Century

See under "Histories of astronomy".

23. Constitution and Dynamics of the Galactic System

Becker, W.: *Sterne und Sternsysteme.* 2nd edn. Dresden and Leipzig: Th. Steinkopff 1950.

Bok, B. J., and P. F. Bok: *The Milky Way.* 3rd edn. Harvard Univ. Press 1957.

Stars and stellar systems (ed. G. P. Kuiper and B. M. Middlehurst), Vol. V. *Galactic structure* (ed. A. Blaauw and M. Schmidt). Chicago Univ. Press 1965.

The Galaxy and the Magellanic Clouds, IAU Symposium 20. Canberra: Australian Academy of Science 1964.

24. Interstellar Matter

Aller, L. H.: *Gaseous nebulae.* London: Chapman and Hall 1956.

Dufay, J.: *Galactic nebulae and interstellar matter.* (English translation). London: Hutchinson 1957.

Middlehurst, B., and L. H. Aller (ed.): *Stars and stellar systems,* Vol. 7: *Nebulae and interstellar matter.* Chicago Univ. Press 1968.

* Pikelner, S.: *Physics of interstellar space.* Moscow: Foreign Languages Publ. House, also New York: Philosophical Lib. 1963.

Van de Hulst, H. C.: *Light scattering by small particles.* New York: Wiley 1957.

Wickramsinghe, N. C.: *Interstellar grains.* The International Astrophysics Series, 9. London: Chapman and Hall 1967.

Interstellar matter in galaxies (ed. L. Woltjer). New York: W. A. Benjamin 1962.

Planetary nebulae, IAU Symposium 34. Dordrecht: Reidel 1968.

25. Internal Constitution and Energy Generation of Stars

Chandrasekhar, S.: *An introduction to the study of stellar structure.* Chicago Univ. Press 1939, also New York: Dover 1957.

Eddington, A. S.: *The internal constitution of the stars.* Cambridge Univ. Press 1926, also New York: Dover 1959.

* Schwarzschild, M.: *Structure and evolution of the stars.* Princeton Univ. Press 1958, also New York: Dover.

Les Processus nucléaires dans les astres. Col. Internat. d'Astrophysique Liège 1953.

26. Colour-Magnitude Diagrams of Galactic and Globular Clusters. Stellar Evolution and Abundances of the Elements

Burbidge, E. M., G. R. Burbidge, W. A. Fowler, and F. Hoyle: *Synthesis of the elements in stars.* Rev. Mod. Phys. **29,** 547 (1957).

Burbidge, G. R.: *Nuclear astrophysics.* Ann. Rev. Nuclear Science **12,** 507 (1963).

—, F. D. Kahn, R. Ebert, S. v. Hoerner, and St. Temesvary: *Die Entstehung von Sternen durch Kondensation diffuser Materie.* Berlin-Göttingen-Heidelberg: Springer 1960.

Fowler, W. A., and F. Hoyle: *Nucleosynthesis in massive stars and supernovae.* Chicago-London: Chicago Univ. Press 1965.

Schwarzschild, M. See Chapter 25.

Struve, O.: *Stellar evolution, an exploration from the observatory.* Princeton Univ. Press 1950.

Stellar populations. (ed. D. J. K. O'Connell.) Conf. Vatican Observ. 1958. Amsterdam: North-Holland.

Modèles d'étoiles et évolution stellaire. Col. Internal. d'Astrophysique Liège 1959.

27. Galaxies

Baade, W.: *Evolution of stars and galaxies* (ed. C. Payne-Gaposchkin). Harvard Univ. Press 1963.

Hodge, P. W.: *Galaxies and cosmology.* New York: McGraw-Hill 1966.
* Hubble, E. P.: *The realm of nebulae.* Yale Univ. Press 1936, also New York: Dover 1958.
* Page, T.: *Stars and galaxies.* Englewood Cliffs, N. J.: Prentice Hall 1962.
Payne-Gaposchkin, C.: *Variable stars and galactic structure.* London: The Athlone Press 1954.
Reddish, V. C.: *Evolution of the galaxies.* Edinburgh: Oliver and Boyd 1967.
* Sandage, A.: *The Hubble atlas of galaxies.* Carnegie Inst. of Washington, Publ. 618, 1961.
De Vaucouleurs, G. & A.: *Reference catalogue of bright galaxies.* Texas Univ. Press 1964.
Problems of extra-galactic research, IAU Symposium 15, 1961. London-New York: Macmillan 1962.

28. Radio Emission from Galaxies: Exploding Galaxies

* Brown, Hanbury R., and A. C. B. Lovell: *The exploration of space by radio.* London: Chapman and Hall 1957.
Burbidge, G., and E. M. Burbidge: *Quasi-stellar objects.* San Francisco: Freeman 1967.
Evans, F. V., and T. Hayfors (ed.): *Radar astronomy.* New York: McGraw-Hill 1968.
Palmer, H. P., R. D. Davies, and M. I. Large (ed.): *Radio astronomy today.* Manchester Univ. Press 1963.
Piddington, J. H.: *Radio astronomy.* London: Hutchinson 1961.
Shklovsky, I. S.: *Cosmic radio waves.* Harvard Univ. Press 1960.
* Smith, F. Graham: Radio astronomy. 3rd edn. London: Pelican books 1966.

29. Cosmic Rays: High Energy Astronomy

Ginzburg, W. L., and S. I. Syrovatskii: *The origin of cosmic rays.* Oxford: Pergamon Press 1964.
Sandström, A. E.: *Cosmic ray physics.* Amsterdam: North-Holland 1965.
* Wolfendale, A. W.: *Cosmic rays.* London: Lewnes 1963.
Progress of elementary particle and cosmic ray physics. (Annual publication since 1952). Amsterdam: North-Holland.

30. Cosmology

Bondi, H.: *Cosmology.* 2nd edn. Cambridge Univ. Press 1960.
Einstein, A.: *Meaning of relativity.* 6th edn. London: Methuen 1960.
* Gamow, G.: *The creation of the universe.* New York: New American Lib. 1957.
Landau, L. D., and E. M. Lifschitz: *Course of theoretical physics.* II. *The classical theory of fields.* 2nd edn. Oxford: Pergamon Press 1962.
Tolman, R. C.: *Relativity, thermodynamics and cosmology.* Oxford: Clarendon Press 1934.
McVittie, G. C.: *Fact and theory in cosmology.* London: Eyre & Spottiswoode 1961.

McVittie, G. C.: *General relativity and cosmology*. Illinois Univ. Press 1965.

The structure and evolution of the universe — 11. Solvay Congress, Brussels 1958.

Witten, L. (ed.): *Gravitation*. New York-London: Wiley 1962.

31. Origin of the Solar System: Evolution of the Earth and of Life

Brancazio, P. J., and A. G. W. Cameron (ed.): *The origin and evolution of atmospheres and oceans*. New York-London-Sydney: Wiley 1963.

Jastrow, R., and A. G. W. Cameron (ed.): *Origin of the solar system*. New York-London: Academic Press 1963.

Kuiper, G. P.: On the origin of the solar system. Article in *Astrophysics*. ed. J. A. Hynek. New York-Toronto-London: McGraw-Hill 1951.

* Lyttleton, R. A.: *Mysteries of the solar system*. Oxford: Clarendon Press 1968.

* Rutten, M. G.: *The geological aspects of the origin of life on Earth*. Amsterdam-New York: Elsevier 1962.

* Urey, H. C.: *The planets, their origin and development*. Yale Univ. Press, also London: Oxford Univ. Press 1952.

von Weizsäcker, C. F.: *The history of nature*. (English translation). London: Routledge & Kegan-Paul 1951.

Figures: Acknowledgments

The numbers are those of the figures appearing in the text.

2.3, 3.1, 3.2, 3.3, 4.1, 4.2, 4.3, 4.4, 5.1, 5.5, 7.1. Seydlitz: Part 5. 7th edn. Allgemeine Erdkunde. Kiel: F. Hirt, and Hannover: H. Schroedel 1961.

6.7. Unsöld, A.: *Physikal. Blätter* **5**, 205 (1964). Mosbach: Physik-Verlag.

7.2. Phot. NASA, Washington.

7.3. Phot. B. Lyot and H. Camichel, Observatoire Pic du Midi.

7.4. Phot. H. Camichel, Observatoire Pic du Midi.

7.5. Phot. Lick Observatory, from *Sky and Telescope* **26**, 342 (1963).

7.6. Phot. NASA, from *Sky and Telescope* **28**, 158 (1964).

7.7. *Soviet Union To-day* **11**, 15 (1966).

8.1. Phot. Mt. Wilson and Palomar Observ.; *Publ. Astron. Soc. of the Pacific* **70**, 202 (Plate III) (1958).

8.2. Swings, P., and L. Haser: *Atlas of Representative Cometary Spectra*, Plate IV. Univ. Liège 1956.

8.3. Gentner, W.: *Die Naturwissenschaften* **50**, 192 (Fig. 1) (1963).

8.4. Hawkins, G. S.: *Annual Rev. of Astronomy and Astrophysics* **2**, 150 (1964).

9.5. Phot. Yerkes Observatory, Williams Bay, Wis.

9.6. Phot. Mt. Wilson and Palomar Observatories.

9.7. *Das Weltall.* Time-Life International, p. 37, 1964.

9.9b. Russell-Dugan-Stewart, *Astronomy* II (Fig. 254). New York: Ginn Co. 1927.

9.10. Eastman Kodak Co., Rochester, N.Y.: "*Kodak Plates and Films*", P. 15 d.

9.11. Dunham Jr., Th., in *Vistas in Astronomy* II (ed. A. Beer), P. 1236. London-New York: Pergamon Press 1956.

9.12. Austral. Nat. Radio Astron. Observ.; 1963.

9.13. After H. M. Schurmeier, R. L. Heacock, and A. E. Wolfe: *Scientific American*, Jan. 1966, P. 57.

11.1, 11.2, 11.3. Unsöld, A.: *Physik der Sternatmosphären*, 2nd edn. Berlin-Göttingen-Heidelberg: Springer 1955.

12.1. Russell-Dugan-Stewart: *Astronomy* I (Fig. 22). New York: Ginn 1926.

12.2. Minnaert-Mulders-Houtgast: *Photometric Atlas of the Solar Spectrum* (section). Amsterdam: Schnabel, Kampfert and Helm 1940.

13.1. Johnson, H. L., and W. W. Morgan: *Astrophys. J.* **114**, 523 (1951).

15.1. Morgan, W. W., P. C. Keenan, and E. Kellman: *An Atlas of Stellar Spectra* (section). Chicago University Press 1942.

15.2. Russell-Dugan-Stewart: *Astronomy* II. New York: Ginn 1927.
15.3. Johnson, H. L., and W. W. Morgan: *Astrophys. J.* **117**, 338 (1953).
16.1. Baker, R. H.: *Astronomy*, 8th edn. Fig. 14.18. New York: Van Nostrand 1964.
16.2. Unsöld, A.: *Physik der Sternatmosphären*, 2nd edn. Berlin-Göttingen-Heidelberg: Springer 1955.
17.3. Merrill, P. W.: *Papers Mt. Wilson Observ. IX*, 118 (1965). Washington: Carnegie Inst.
18.1. Unsöld, A.: *Physik der Sternatmosphären*, 2nd edn. p. 106. Berlin-Göttingen-Heidelberg: Springer 1955.
18.4. Unsöld, A.: *Monthly Not. Roy. Astr. Soc.* **118**, 9 (1958).
18.5. Unsöld, A.: *Physik der Sternatmosphären*, 2nd edn. Berlin-Göttingen-Heidelberg: Springer 1955.
19.1, 19.2, 19.3. Unsöld, A.: *Angewandte Chemie* **76**, 281—290 (1964).
20.1. Danielson, R. E.: *Astrophys. J.* **134**, 280 (1961).
20.2. Unsöld, A.: *Physik der Sternatmosphären*, 2nd edn. Berlin-Göttingen-Heidelberg: Springer 1955.
20.3. Houtgast, J.: *Rech. Astron.* Utrecht XIII, 3; Utrecht 1957.
20.4. Biesbrock, G. van, in *The Sun* (ed. G. P. Kuiper). **I**, 604 (1953). Univ. Chicago Press.
20.5. *Sky and Telescope* **20**, 254 (1960).
20.6. Royds, T.: *Monthly Not. Roy. Atsron.* Soc. **89**, 255 (1929).
20.7. Jager, C. de: *Encyclopaedia of Physics* **52**, 136. Berlin-Göttingen-Heidelberg: Springer 1959.
20.8. Unsöld, A.: *Physik der Sternatmosphären*, 2nd edn. Berlin-Göttingen-Heidelberg: Springer 1955.
20.9. Cape Observatory. *Proc. Roy. Inst.* **38**, No. 175, Pl. I, 1961.
20.10. Palmer-Davies-Large: *Radio Astronomy Today*, p. 19. Manchester Univ. Press 1963.
21.1. Becker, W.: *Sterne u. Sternsysteme*, p. 108. Darmstadt: Steinkopff 1950.
21.2. Minkowski, R.: *Ann. Rev. of Astronomy and Astrophysics* **2**, 248 (1964).
23.4. Phot. Mt. Wilson and Palomar Observ. in O. Struve: *Astronomie.* Berlin: W. de Gruyter 1962.
23.5. Duncan, J. C.: *Astronomy*, p. 408. New York: Harper 1950.
23.6. Oort, J. H., in *Stars and Stellar Systems* **5**, 484 (1965). Univ. Chicago Press.
23.7. Becker, W.: *Z. Astrophys.* **58**, 205 (1964).
23.10. Eggen, O. J.: *Roy. Observ. Bull.* **84**, 114 (1964).
24.2. Oort, J. H., in *Interstellar Matter in Galaxies* (ed. Woltjer). New York: Benjamin 1962.
24.3. Westerhout, G.: *Bull. Astron. Inst. Netherlands* **14**, 254 (1958).
24.4. Phot. Mt. Wilson and Palomar Observatories, in P. W. Merrill, *Space Chemistry*, p. 122, Univ. Michigan Press 1963.
24.5. Goldberg, L., and L. H. Aller: *Atoms, Stars and Nebulae*, p. 182, Philadelphia: Blackiston 1946.

24.6. Phot. Harvard Observ., in R. H. Baker: *Astronomy*, 8th edn., Fig. 17.2. New York: Van Nostrand 1964.
25.1. Fowler, W. A., in *Liège Astrophys. Sympos.* 1959, p. 216.
26.1. Sandage, A. R.: *Publ. Astron. Soc. Pacific* **68,** 499 (1956).
26.2. Johnson, H. L.: *Astrophys. J.* **116,** 646 (1952).
26.3. Murray, C. A., P. M. Corben, and M. C. Allchorn: *Roy. Observ. Bull.* No. 91, p. E 357, London 1965.
26.4. Baade, W.: *Stars and Galaxies,* p. 91. Harvard Univ. Press 1963.
26.5. and 26.6. Kippenhahn, R., H. C. Thomas, and A. Weigert: *Z. Astrophys.* **61,** 246 (1965).
26.7. Walker, M.: *Astrophys. J. Suppl.* **2,** 376 (1956).
26.9. Walker, M.: *Astrophys. J.* **125,** 651 (1957).
27.1. Phot. Mt. Wilson and Palomar Observ., from The Hubble Atlas of Galaxies, p. 18. Carnegie Inst. of Washington 1961.
27.2. Hubble, E.: *Astrophys. J.* **69,** 120 (1929).
27.3. After E. Hubble: *The Realm of Nebulae,* p. 45. Yale Univ. Press 1936.
27.4. Reference as 27.1, p. 38.
27.5. Babcock, H. W., in *Lick Observ. Bull.* No. 498, Pl. III, 1939.
27.6. Lallemand, A., M. Duchesne, and M. Walker: *Publ. Astron. Soc. Pacific* **72,** 79 (1960).
27.7. Morgan, W. W., and N. U. Mayall: *Publ. Astron. Soc. Pacific* **69,** 295 (1957).
27.8. Zwicky, F.: *Astrophys. J.* **140,** 1627 (1964).
27.9. Baade, W., and H. Swope: *Astron. J.* **66,** 326 (1961).
28.2. Hill, E. R., O. B. Slee, and B. Y. Mills: *Austral. J. Physics* **11,** 542 (1958).
28.3. Hogg, D. E.: *Astrophys. J.* **140,** 992 (figure 2) (1964).
28.4. Cooper, B. F. C., R. M. Price, and D. J. Cole: *Austral. J. Phys.* **18,** 602 (1965). — P. Maltby et al.: *Astrophys. J.* **140,** 44 (1964).
28.5. Sandage, A. R.: *Scientific American,* Nov. 1964, p. 39.
29.1. Morrison, P.: *Encyclopaedia of Physics* **46,** 1, p. 7. Berlin-Göttingen-Heidelberg: Springer 1961.

Index

Herstellung: Konrad Triltsch, Graphischer Betrieb, Würzburg